Aprender

Eureka Math®
4.º grado
Módulo 4

Publicado por Great Minds®.

Copyright © 2019 Great Minds®.

Impreso en los EE. UU.
Este libro puede comprarse en la editorial en eureka-math.org.
1 2 3 4 5 6 7 8 9 10   BAB   25 24 23 22 21

ISBN 978-1-64054-992-0

G4-SPA-M4-L-05.2019

Aprender • Practicar • Triunfar

Los materiales del estudiante de *Eureka Math*® para *Una historia de unidades*™ (K–5) están disponibles en la trilogía *Aprender, Practicar, Triunfar*. Esta serie apoya la diferenciación y la recuperación y, al mismo tiempo, permite la accesibilidad y la organización de los materiales del estudiante. Los educadores descubrirán que la trilogía *Aprender, Practicar y Triunfar* también ofrece recursos consistentes con la Respuesta a la intervención (RTI, por sus siglas en inglés), las prácticas complementarias y el aprendizaje durante el verano que, por ende, son de mayor efectividad.

Aprender

Aprender de *Eureka Math* constituye un material complementario en clase para el estudiante, a través del cual pueden mostrar su razonamiento, compartir lo que saben y observar cómo adquieren conocimientos día a día. *Aprender* reúne el trabajo en clase—la Puesta en práctica, los Boletos de salida, los Grupos de problemas, las plantillas—en un volumen de fácil consulta y al alcance del usuario.

Practicar

Cada lección de *Eureka Math* comienza con una serie de actividades de fluidez que promueven la energía y el entusiasmo, incluyendo aquellas que se encuentran en *Practicar* de *Eureka Math*. Los estudiantes con fluidez en las operaciones matemáticas pueden dominar más material, con mayor profundidad. En *Practicar*, los estudiantes adquieren competencia en las nuevas capacidades adquiridas y refuerzan el conocimiento previo a modo de preparación para la próxima lección.

En conjunto, *Aprender* y *Practicar* ofrecen todo el material impreso que los estudiantes utilizarán para su formación básica en matemáticas.

Triunfar

Triunfar de *Eureka Math* permite a los estudiantes trabajar individualmente para adquirir el dominio. Estos grupos de problemas complementarios están alineados con la enseñanza en clase, lección por lección, lo que hace que sean una herramienta ideal como tarea o práctica suplementaria. Con cada grupo de problemas se ofrece una Ayuda para la tarea, que consiste en un conjunto de problemas resueltos que muestran, a modo de ejemplo, cómo resolver problemas similares.

Los maestros y los tutores pueden recurrir a los libros de *Triunfar* de grados anteriores como instrumentos acordes con el currículo para solventar las deficiencias en el conocimiento básico. Los estudiantes avanzarán y progresarán con mayor rapidez gracias a la conexión que permiten hacer los modelos ya conocidos con el contenido del grado escolar actual del estudiante.

Estudiantes, familias y educadores:

Gracias por formar parte de la comunidad de *Eureka Math*®, donde celebramos la dicha, el asombro y la emoción que producen las matemáticas.

En las clases de *Eureka Math* se activan nuevos conocimientos a través del diálogo y de experiencias enriquecedoras. A través del libro *Aprender* los estudiantes cuentan con las indicaciones y la sucesión de problemas que necesitan para expresar y consolidar lo que aprendieron en clase.

¿Qué hay dentro del libro Aprender?

Puesta en práctica: la resolución de problemas en situaciones del mundo real es un aspecto cotidiano de *Eureka Math*. Los estudiantes adquieren confianza y perseverancia mientras aplican sus conocimientos en situaciones nuevas y diversas. El currículo promueve el uso del proceso LDE por parte de los estudiantes: Leer el problema, Dibujar para entender el problema y Escribir una ecuación y una solución. Los maestros son facilitadores mientras los estudiantes comparten su trabajo y explican sus estrategias de resolución a sus compañeros/as.

Grupos de problemas: una minuciosa secuencia de los Grupos de problemas ofrece la oportunidad de trabajar en clase en forma independiente, con diversos puntos de acceso para abordar la diferenciación. Los maestros pueden usar el proceso de preparación y personalización para seleccionar los problemas que son «obligatorios» para cada estudiante. Algunos estudiantes resuelven más problemas que otros; lo importante es que todos los estudiantes tengan un período de 10 minutos para practicar inmediatamente lo que han aprendido, con mínimo apoyo de la maestra.

Los estudiantes llevan el Grupo de problemas con ellos al punto culminante de cada lección: la Reflexión. Aquí, los estudiantes reflexionan con sus compañeros/as y el maestro, a través de la articulación y consolidación de lo que observaron, aprendieron y se preguntaron ese día.

Boletos de salida: a través del trabajo en el Boleto de salida diario, los estudiantes le muestran a su maestra lo que saben. Esta manera de verificar lo que entendieron los estudiantes ofrece al maestro, en tiempo real, valiosas pruebas de la eficacia de la enseñanza de ese día, lo cual permite identificar dónde es necesario enfocarse a continuación.

Plantillas: de vez en cuando, la Puesta en práctica, el Grupo de problemas u otra actividad en clase requieren que los estudiantes tengan su propia copia de una imagen, de un modelo reutilizable o de un grupo de datos. Se incluye cada una de estas plantillas en la primera lección que la requiere.

¿Dónde puedo obtener más información sobre los recursos de Eureka Math?

El equipo de Great Minds® ha asumido el compromiso de apoyar a estudiantes, familias y educadores a través de una biblioteca de recursos, en constante expansión, que se encuentra disponible en eureka-math.org. El sitio web también contiene historias exitosas e inspiradoras de la comunidad de *Eureka Math*. Comparte tus ideas y logros con otros usuarios y conviértete en un Campeón de *Eureka Math*.

¡Les deseo un año colmado de momentos "¡ajá!"!

Jill Diniz

Jill Diniz
Directora de matemáticas
Great Minds®

El proceso de Leer-Dibujar-Escribir

El programa de *Eureka Math* apoya a los estudiantes en la resolución de problemas a través de un proceso simple y repetible que presenta la maestra. El proceso Leer-Dibujar-Escribir (LDE) requiere que los estudiantes

1. Lean el problema.

2. Dibujen y rotulen.

3. Escriban una ecuación.

4. Escriban un enunciado (afirmación).

Se procura que los educadores utilicen el andamiaje en el proceso, a través de la incorporación de preguntas tales como

- ¿Qué observas?

- ¿Puedes dibujar algo?

- ¿Qué conclusiones puedes sacar a partir del dibujo?

Cuánto más razonen los estudiantes a través de problemas con este enfoque sistemático y abierto, más interiorizarán el proceso de razonamiento y lo aplicarán instintivamente en el futuro.

Contenido

Módulo 4: Medición de ángulos y figuras en el plano

Nombre _____ Fecha _____

1. Usa las siguientes instrucciones para dibujar una figura en el recuadro de la derecha.

   a. Dibuja dos puntos: A y B.

   b. Usa una regla para dibujar $\overrightarrow{AB}$.

   c. Dibuja un punto nuevo que no esté sobre $\overrightarrow{AB}$. Márcalo como C.

   d. Dibuja $\overline{AC}$.

   e. Dibuja un punto que no esté sobre $\overrightarrow{AB}$ ni $\overline{AC}$. Llámalo D.

   f. Dibuja $\overleftrightarrow{CD}$.

   g. Usa los puntos que acabas de marcar para nombrar un ángulo. _____

2. Usa las siguientes instrucciones para dibujar una figura en el recuadro de la derecha.

   a. Dibuja dos puntos: A y B.

   b. Usa una regla para dibujar $\overline{AB}$.

   c. Dibuja un punto nuevo que no esté sobre $\overline{AB}$. Márcalo como C.

   d. Dibujare $\overrightarrow{BC}$.

   e. Dibuja un punto nuevo que no esté sobre $\overline{AB}$ ni $\overrightarrow{BC}$. Márcalo como D.

   f. Dibuja $\overleftrightarrow{AD}$.

   g. Identifica $\angle DAB$ dibujando un arco para indicar la posición del ángulo.

   h. Identifica otro ángulo relacionando los puntos que acabas de dibujar. _____

Lección 1:   Identificar y dibujar puntos, rectas, segmentos de rectas, semirrectas y ángulos. Reconocerlos en varios contextos y figuras conocidas.

1

© 2019 Great Minds®. eureka-math.org

3.   a.   Observa las figuras conocidas de abajo. Marca algunos puntos en cada figura.

b.   Usa esos puntos para marcar y nombrar las representaciones de cada uno de los siguientes elementos en la tabla de abajo: semirrecta, recta, segmento de recta y ángulo. Extiende los segmentos para mostrar rectas y semirrectas.

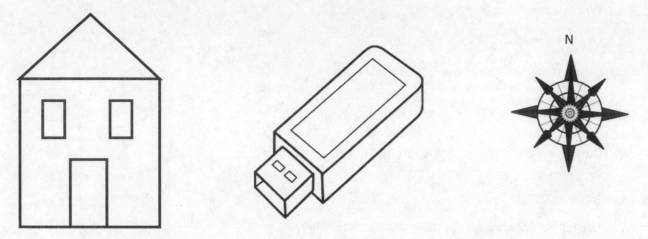

	Casa	Memoria USB	Brújula
Semirrecta			
Recta			
Segmento de recta			
Ángulo			

Extensión: Dibuja una figura conocida. Márcala con puntos y luego identifica las semirrectas, rectass, segmentos de recta y ángulos, según sea pertinente.

Nombre _____   Fecha _____

1. Dibuja un segmento de recta para conectar la palabra con su imagen.

Semirrecta

Recta

Segmento de recta

Punto

Ángulo

2. ¿En qué es diferente una recta de un segmento de recta?

Lección 1:   Identificar y dibujar puntos, rectas, segmentos de rectas, semirrectas y
               ángulos. Reconocerlos en varios contextos y figuras conocidas.

© 2019 Great Minds®. eureka-math.org

3

1. La figura 1 tiene tres puntos. Conecta los puntos A, B y C con tantos segmentos de recta como sea posible.

2. La figura 2 tiene cuatro puntos. Conecta los puntos D, E, F y G con tantos segmentos de recta como sea posible.

Figura 1   Figura 2

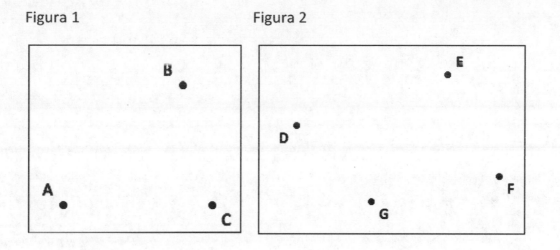

Lee   Dibuja   Escribe

Lección 2:   Usar los ángulos rectos para determinar si los ángulos son mayores, menores
                 o iguales al ángulo recto. Dibujar ángulos rectos, obtusos y agudos.

© 2019 Great Minds®. eureka-math.org

5

Nombre _____   Fecha _____

1. Usa la plantilla de ángulo recto que hiciste en clase para determinar si cada uno de los siguientes ángulos es mayor que, menor que o igual a un ángulo recto. Marca cada uno como *mayor que, menor que* o *igual a,* luego conecta cada ángulo con el nombre correcto: agudo, recto u obtuso. El primer ejercicio ya está resuelto.

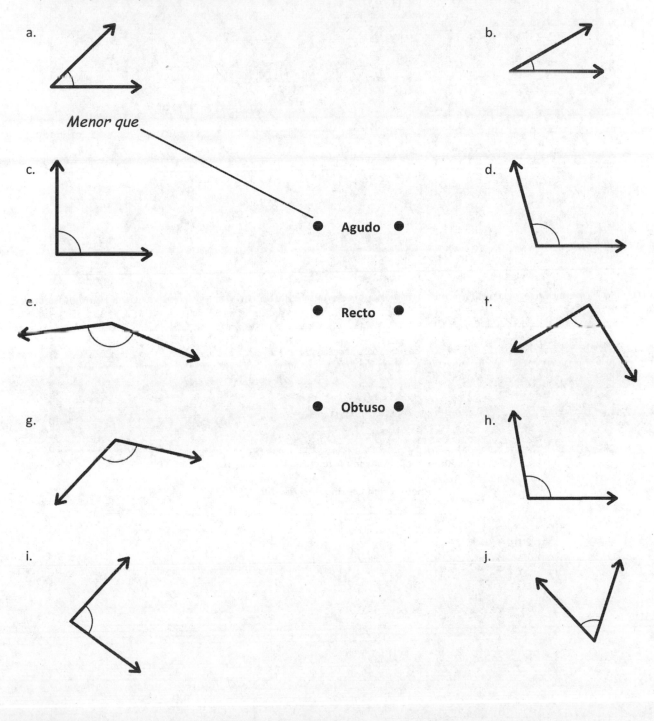

a.

Menor que

b.

c.

● **Agudo** ●

d.

e.

● **Recto** ●

f.

● **Obtuso** ●

g.

h.

i.

j.

Lección 2:   Usar los ángulos rectos para determinar si los ángulos son mayores, menores o iguales al ángulo recto. Dibujar ángulos rectos, obtusos y agudos.

7

© 2019 Great Minds®. eureka-math.org

2. Usa tu plantilla de ángulo recto para identificar ángulos agudos, rectos y obtusos dentro de la pintura de Picasso, *Fábrico, Horto de Ebbo.* Traza al menos dos de cada uno, marca con puntos y, luego, nómbralos en la tabla debajo de la pintura.

© 2013 Estate of Pablo Picasso / Artists Rights Society (ARS), New York

Photo: Erich Lessing / Art Resource, NY.

Ángulo agudo		
Ángulo obtuso		
Ángulo recto		

Lección 2: Usar los ángulos rectos para determinar si los ángulos son mayores, menores o iguales al ángulo recto. Dibujar ángulos rectos, obtusos y agudos.

3. Dibuja cada uno de los siguientes ángulos usando una regla y la plantilla de ángulo recto que creaste. Explica las características de cada uno comparando el ángulo con un ángulo recto. Usa las palabras *mayor que, menor que* o *igual a* en tus explicaciones.

    a. Ángulo agudo

    b. Ángulo recto

    c. Ángulo obtuso

Lección 2:   Usar los ángulos rectos para determinar si los ángulos son mayores, menores
o iguales al ángulo recto. Dibujar ángulos rectos, obtusos y agudos.

9

© 2019 Great Minds®. eureka-math.org

Nombre _____ Fecha _____

1. Llena los espacios en blanco para hacer verdaderas las afirmaciones usando una de las siguientes palabras: *agudo, obtuso, recto, llano.*

   a. En clase, hicimos un ángulo _____ doblando dos veces un papel.

   b. Un ángulo _____ es menor que un ángulo recto.

   c. Un ángulo _____ es mayor que un ángulo recto, pero menor que un ángulo llano.

2. Usa una plantilla de ángulo recto para identificar los siguientes ángulos.

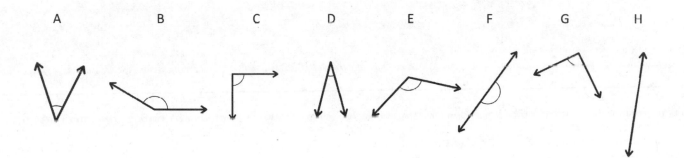

   A   B   C   D   E   F   G   H

   a. ¿Cuáles son ángulos rectos? _____

   b. ¿Cuáles son ángulos obtusos? _____

   c. ¿Cuáles son ángulos agudos? _____

   d. ¿Cuáles son ángulos llanos? _____

Lección 2:   Usar los ángulos rectos para determinar si los ángulos son mayores, menores o iguales al ángulo recto. Dibujar ángulos rectos, obtusos y agudos.

11

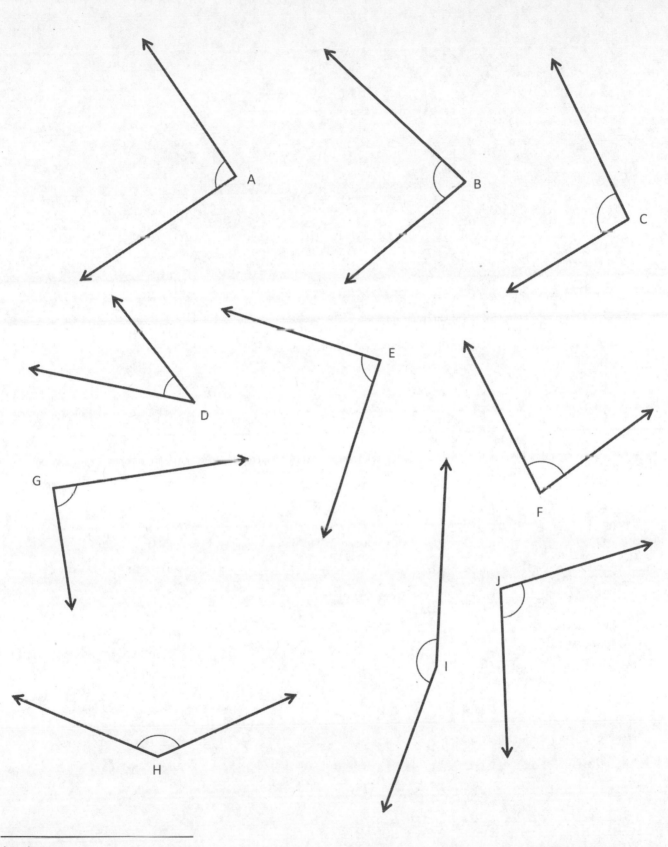

ángulos

Lección 2:   Usar los ángulos rectos para determinar si los ángulos son mayores, menores
o iguales al ángulo recto. Dibujar ángulos rectos, obtusos y agudos.

13

© 2019 Great Minds®. eureka-math.org

a. Calcula aproximadamente para dibujar el punto X lo más cerca de la mitad de $\overline{AB}$ que puedas.

b. Calcula aproximadamente el punto Y a la mitad de $\overline{CD}$.

c. Dibuja el segmento de recta horizontal XY. ¿Qué palabra crean los segmentos?

d. Borra el segmento XY. Dibuja el segmento CF. ¿Qué palabra crean los segmentos?

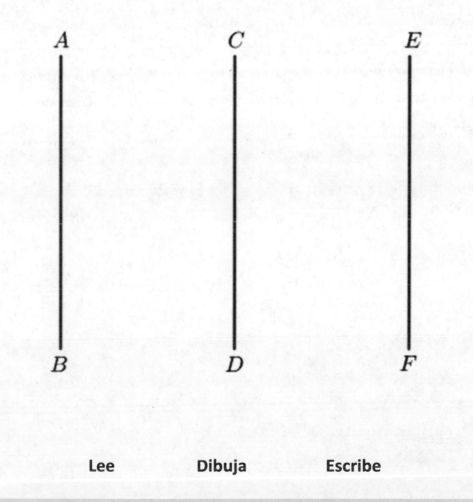

Lee   Dibuja   Escribe

Nombre _____ Fecha _____

1. En cada objeto, traza al menos un par de rectas que aparenten ser perpendiculares.

2. ¿Cómo sabes si dos rectas son perpendiculares?

3. En las matrices cuadradas y triangulares de abajo, usa los segmentos proporcionados en cada matriz para dibujar un segmento que sea perpendicular. Usa una regla de borde recto.

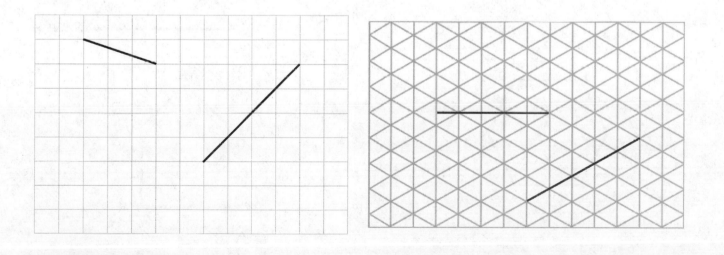

4. Usa la plantilla de ángulo recto que creaste en clase para determinar cuáles de las siguientes figuras tienen un ángulo recto. Marca cada ángulo recto con un cuadrado pequeño. En cada ángulo recto que encuentres, nombra el par de lados perpendiculares correspondientes. (El Problema 4(a) ya está empezado).

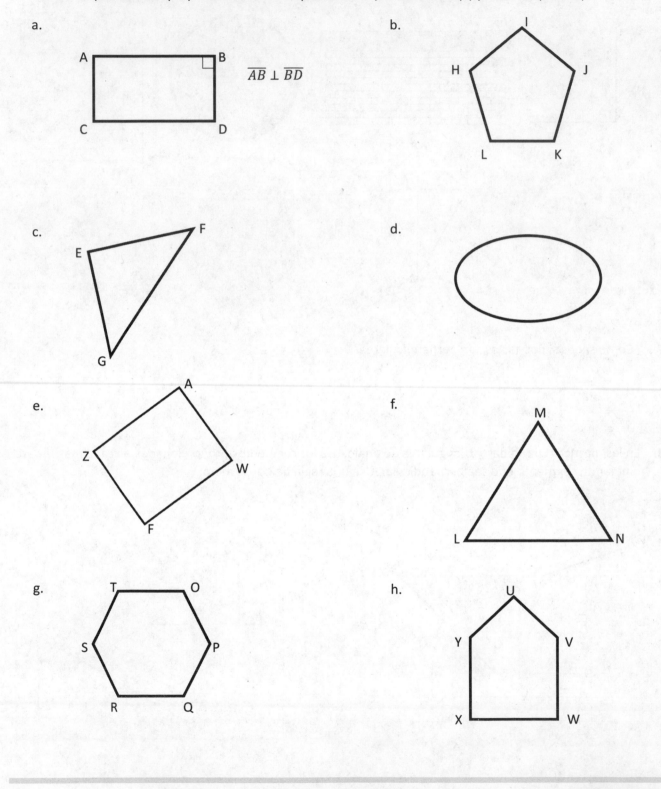

a.

$\overline{AB} \perp \overline{BD}$

b.

c.

d.

e.

f.

g.

h.

Lección 3:   Identificar, definir y dibujar rectas perpendiculares.

5. Marca cada ángulo recto de la siguiente figura con un cuadrado pequeño. (Nota: El ángulo recto no tiene que estar dentro de la figura). ¿Cuántos pares de lados perpendiculares tiene la figura?

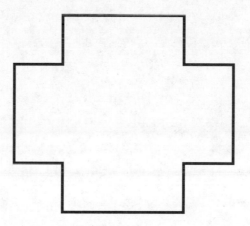

6. ¿Verdadero o falso? Las figuras que tienen al menos un ángulo recto también tienen al menos un par de lados perpendiculares. Explica tu razonamiento.

Nombre _____   Fecha _____

Usa una plantilla de ángulo recto para medir los ángulos de las siguientes figuras. Marca cada ángulo recto con un cuadrado pequeño. Luego, nombra todos los pares de lados perpendiculares.

1.

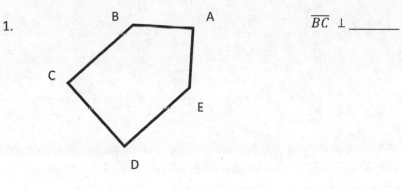

$\overline{BC} \perp$ _____

2.

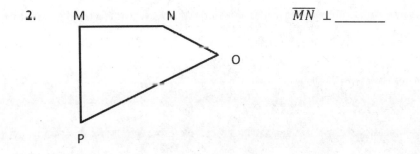

$\overline{MN} \perp$ _____

Observa las letras *R*, *E*, *A* y *L*.

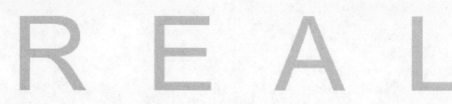

a. ¿Cuántas rectas son perpendiculares? Descríbelas.

b. ¿Cuántos ángulos agudos hay? Descríbelos.

c. ¿Cuántos ángulos obtusos hay? Descríbelos.

Lee   **Dibuja**   **Escribe**

Nombre _____   Fecha _____

1. En cada objeto, traza al menos un par de rectas que aparenten ser paralelas.

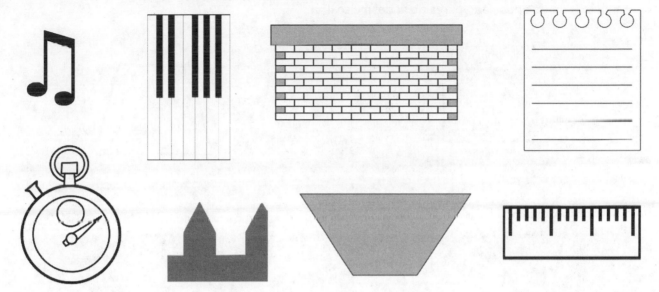

2. ¿Cómo sabes si dos rectas son paralelas?

3. En la rejilla cuadrada y la triangular de abajo, usa los segmentos dados en cada rejilla para dibujar un segmento que sea paralelo. Usa una regla de borde recto.

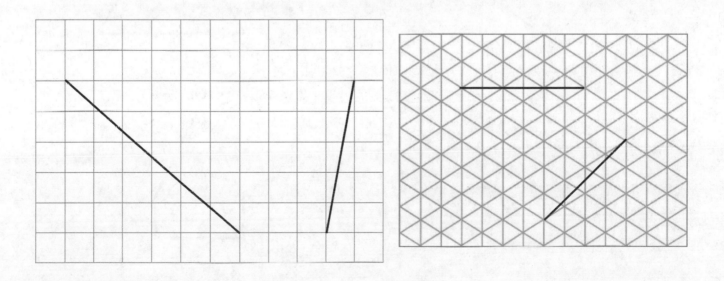

4. Determina cuál de las siguientes figuras tiene lados que son paralelos usando una regla y la plantilla de ángulo recto que creaste. Encierra en un círculo la letra de las figuras que tiene al menos un par de lados paralelos. Marca cada par de lados paralelos con puntas de flecha y, luego, identifica los lados paralelos con un enunciado parecido al del inciso 4(a).

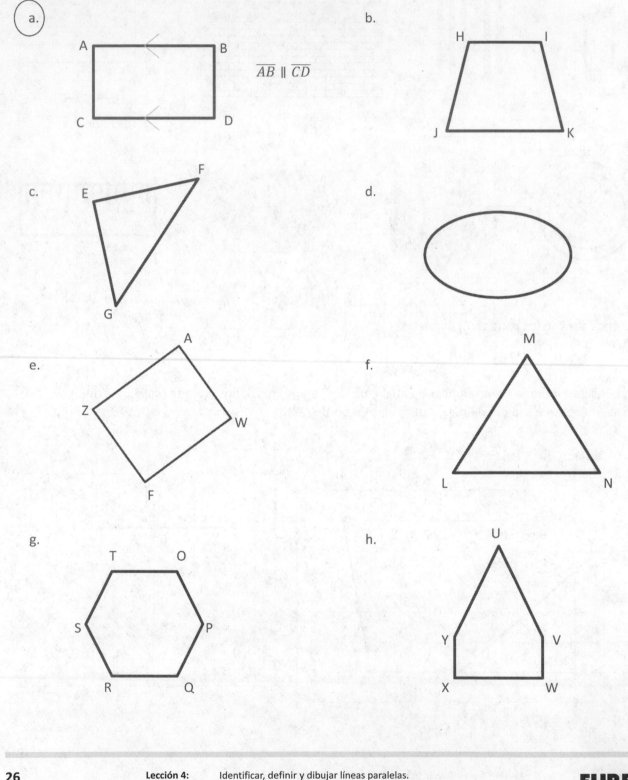

$$\overline{AB} \parallel \overline{CD}$$

5. ¿Verdadero o falso? Un triángulo no puede tener lados que sean paralelos. Explica tu razonamiento.

6. Explica por qué $\overline{AB}$ y $\overline{CD}$ son paralelos, pero $\overline{EF}$ y $\overline{GH}$ no lo son.

7. Dibuja una recta usando tu regla. Ahora, usa tu plantilla de ángulo recto y la regla para dibujar una recta paralela a la primera recta que dibujaste.

Nombre _____   Fecha _____

Observa los siguientes pares de rectas. Identifica si son paralelas, perpendiculares o que se intersecan.

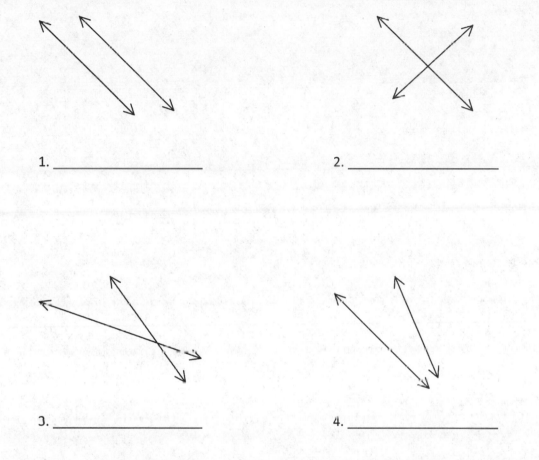

1. _____

2. _____

3. _____

4. _____

Coloca la plantilla de ángulo recto encima del círculo para determinar cuántos ángulos rectos se pueden acomodar alrededor del punto central del círculo. (No se permite que se sobrepongan). ¿Cuántos ángulos rectos se pueden acomodar?

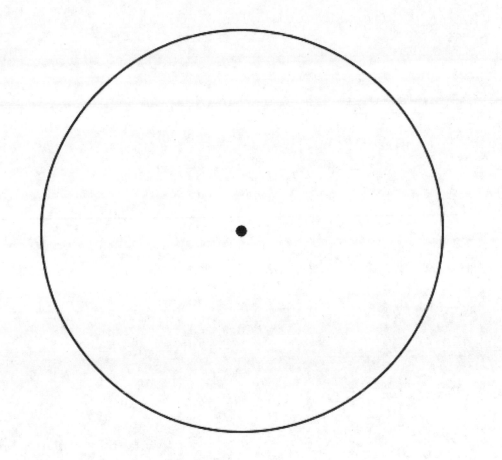

_____

_____

_____

Lee   **Dibuja**   **Escribe**

Lección 5:   Usar un transportador circular para comprender al ángulo de 1 grado como un $\frac{1}{360}$ de giro. Explorar los ángulos de referencia usando el transportador.

31

Nombre _____   Fecha _____

1. Haz una lista de las medidas de los ángulos de referencia que dibujaste, empezando con el Conjunto A. Redondea la medida de cada ángulo a los 5° más cercanos. Los dos conjuntos ya están empezados.

   a. Conjunto A: 45°, 90°,

   b. Conjunto B: 30°, 60°,

2. Encierra en un círculo cualquier medida angular que aparezca en las dos listas. ¿Qué notas acerca de las listas?

3. Haz una lista con las medidas angulares del Problema 1 que sean agudas. Traza cada ángulo con tu dedo mientras dices su medida.

4. Haz una lista con las medidas angulares del Problema 1 que sean obtusas. Traza cada ángulo con tu dedo mientras dices su medida.

Lección 5:   Usar un transportador circular para comprender al ángulo de 1 grado como un $\frac{1}{360}$ de giro. Explorar los ángulos de referencia usando el transportador.

33

© 2019 Great Minds®. eureka-math.org

5. Hoy descubrimos que 1° es $\frac{1}{360}$ de un giro completo. Es 1 de 360°. Eso significa que un ángulo de 2° es igual a $\frac{2}{360}$ de un giro completo. ¿Qué fracción de un giro completo es cada uno de los ángulos de referencia que enumeraste en el Problema 1?

6. ¿Cuántos ángulos de 45° se necesitan para hacer un giro completo?

7. ¿Cuántos ángulos de 30° se necesitan para hacer un giro completo?

8. Si no tuvieras un transportador, ¿cómo construirías un cuarto del mismo desde 0° hasta 90°?

Lección 5: Usar un transportador circular para comprender al ángulo de 1 grado como un $\frac{1}{360}$ de giro. Explorar los ángulos de referencia usando el transportador.

© 2019 Great Minds®. eureka-math.org

Nombre _____   Fecha _____

1.   ¿Cuántos ángulos rectos hacen un giro completo?

2.   ¿Cuál es la medida de un ángulo recto?

3.   ¿Qué fracción de un giro completo es 1°?

4.   Nombra al menos cuatro medidas angulares de referencia.

Lección 5:   Usar un transportador circular para comprender al ángulo de 1 grado como
un $\frac{1}{360}$ de giro. Explorar los ángulos de referencia usando el transportador.

35

© 2019 Great Minds®. eureka-math.org

Corta los círculos de la plantilla que están en la página siguiente. Dobla el Círculo A y el Círculo B como si fueras a formar una plantilla de ángulo recto. Traza las rectas perpendiculares de los dobleces. ¿Cuántos ángulos rectos ves en el centro de cada círculo? ¿Importa el tamaño del círculo?

_____

_____

_____

_____

Lee   **Dibuja**   **Escribe**

EUREKA
MATH®

Lección 6:   Usar diferentes transportadores para distinguir la medida angular de la medida de longitud.

37

© 2019 Great Minds®. eureka-math.org

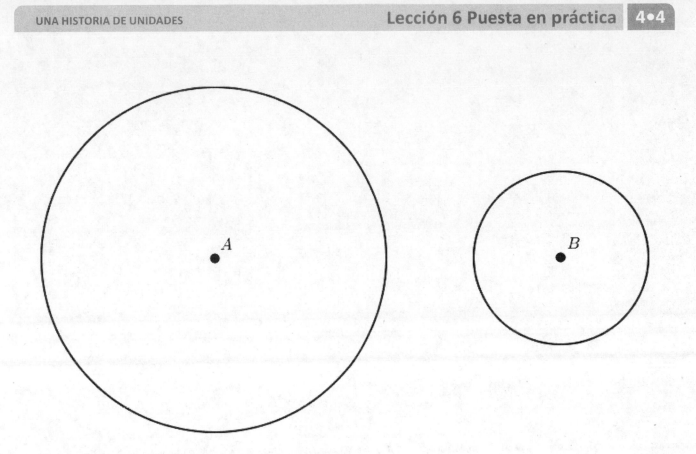

Lee   Dibuja   Escribe

Lección 6:   Usar diferentes transportadores para distinguir la medida angular de la
                medida de longitud.

© 2019 Great Minds®. curcka-math.org

39

Nombre _____ Fecha _____

C

D

E

EUREKA MATH

Lección 6: Usar diferentes transportadores para distinguir la medida angular de la medida de longitud.

41

© 2019 Great Minds®. eureka-math.org

Nombre _____   Fecha _____

1. Usa un transportador para medir los ángulos y, luego, registra las medidas en grados.

    a.

    b.

    c.

    d.

Lección 6:   Usar diferentes transportadores para distinguir la medida angular de la
              medida de longitud.

© 2019 Great Minds®. eureka-math.org

43

e.   f.

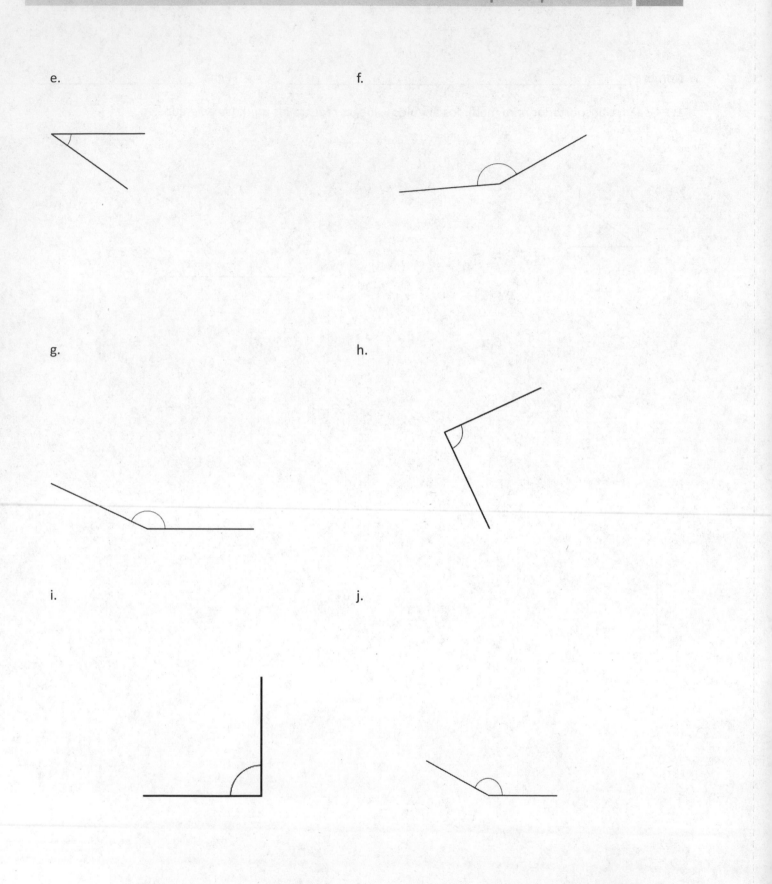

g.   h.

i.   j.

Lección 6:   Usar diferentes transportadores para distinguir la medida angular de la medida de longitud.

2. a. Usa tres transportadores de diferentes tamños para medir el ángulo. Si es necesario, extiende las líneas usando regla.

Transportdor #1: _____°

Transportdor #2: _____°

Transportdor #3: _____°

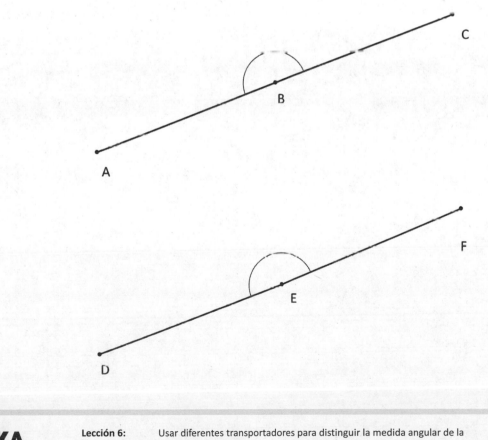

b. ¿Qué notas acerca de la medida del ángulo de arriba usando cada uno de los transportadores?

3. Usa un transportador para medir cada ángulo. Extiende la longitud de los segmentos si es necesario. Cuando extiendes los segmentos, ¿la medida angular es la misma? Explica cómo lo sabes.

a.

C

B

A

b.

F

E

D

Lección 6:   Usar diferentes transportadores para distinguir la medida angular de la medida de longitud.

45

© 2019 Great Minds®. eureka-math.org

Nombre _____   Fecha _____

Usa cualquier transportador para medir los ángulos y, luego, registra las medidas en grados.

1.

2.

3.
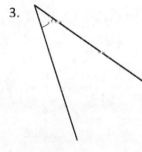

4,

EUREKA
MATH®

Lección 6:   Usar diferentes transportadores para distinguir la medida angular de la
medida de longitud.

47

© 2019 Great Minds®. eureka-math.org

Predice la medida de $\angle XYZ$ usando la plantilla de ángulo recto. Luego, encuentra la medida real de $\angle XYZ$ usando un transportador circular y un transportador de 180°.

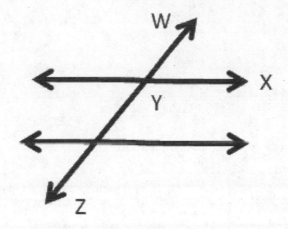

_____

_____

_____

Lee   **Dibuja**   **Escribe**

Lección 7:   Medir y dibujar ángulos. Dibujar las medidas angulares dadas y verifícalas con un transportador.

49

EUREKA MATH

© 2019 Great Minds®. eureka-math.org

Nombre _____   Fecha _____

Figura 1

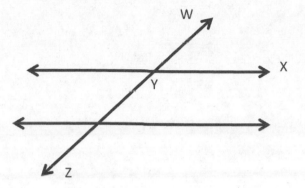

Figura 2

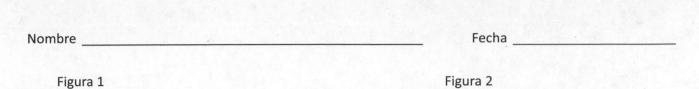
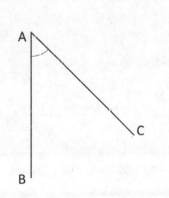

Figura 3

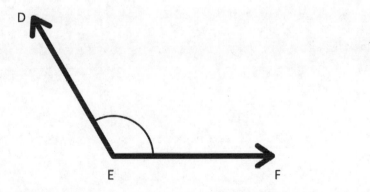

Figura 4

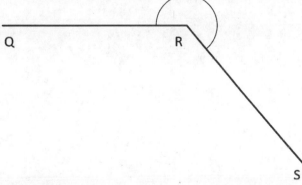

Lección 7:   Medir y dibujar ángulos. Dibujar las medidas angulares dadas y
              verifícalas con un transportador.

51

© 2019 Great Minds®. eureka-math.org

Nombre _____   Fecha _____

Construye ángulos que midan la cantidad de grados proporcionada. Para los Problemas 1–4, usa la semirrecta mostrada como una de las semirrectas del ángulo con su extremo como el vértice del ángulo. Dibuja un arco para indicar el ángulo que se midió.

1.   30°

2.   65°

3.   115°

4.   135°

Lección 7:   Medir y dibujar ángulos. Dibujar las medidas angulares dadas y
verifícalas con un transportador.

53

© 2019 Great Minds®. eureka-math.org

5. 5°

6. 175°

7. 27°

8. 117°

9. 48°

10. 132°

Lección 7:   Medir y dibujar ángulos. Dibujar las medidas angulares dadas y
verifícalas con un transportador.

Nombre _____   Fecha _____

Construye ángulos que midan la cantidad de grados proporcionada. Dibuja un arco para indicar el ángulo que se midió.

1. 75°

2. 105°

3. 81°

4. 99°

Lección 7:   Medir y dibujar ángulos. Dibujar las medidas angulares dadas y
              verifícalas con un transportador.

© 2019 Great Minds®. eureka-math.org

55

Dibuja una serie de relojes que muestren las 12:00, 3:00, 6:00 y 9:00. Usa un arco para identificar un ángulo y calcula aproximadamente el ángulo que se creó entre las dos manecillas del reloj.

_____

_____

_____

Lee   Dibuja   Escribe

EUREKA MATH®

Lección 8:   Identificar y medir ángulos como giros y reconocerlos en varios contextos.

57

© 2019 Great Minds®. eureka-math.org

Nombre _____   Fecha _____

1. Joe, Steve y Bob se pararon en medio del patio viendo hacia la casa. Joe giró 90° hacia la derecha. Steve giró 180° hacia la derecha. Bob giró 270° hacia la derecha. Nombra el objeto que tiene enfrente cada niño.

Joe _____

Steve _____

Bob _____

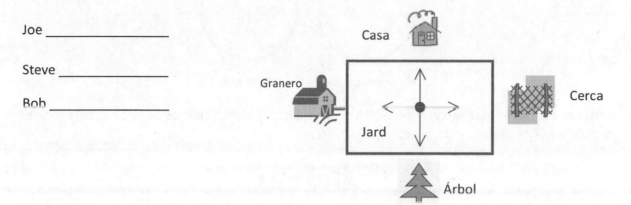

2. Monica vio el reloj al inicio de la clase y al final de la clase. ¿Cuántos grados giró el minutero desde el inicio hasta el final de la clase?

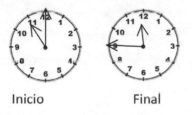

Inicio   Final

3. El patinador saltó en el aire e hizo un 360. ¿Qué quiere decir eso?

4. El Sr. Martin salió de su casa manejando y olvidó su cartera. Hizo un 180. ¿Hacia dónde va ahora?

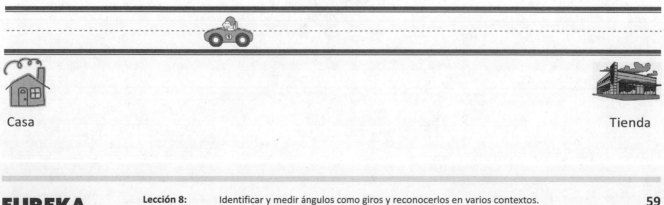

Casa   Tienda

5. Juan giró la llave de la regadera 270° a la derecha. Haz un dibujo mostrando la posición de la perilla después de que la giró.

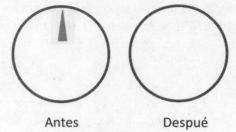

Antes   Despué

6. Barb usó sus tijeras para recortar un cupón del periódico. ¿Cuántos cuartos de giro necesita rotar el papel para mantenerse en las líneas?

7. ¿Cuántos cuartos de giro hay que rotar la figura para que esté derecha?

8. Meredith estaba viendo hacia el norte. Giró 90° a la derecha y luego 180° más. ¿En qué dirección está viendo ahora?

Nombre _____   Fecha _____

1. Marty está parado de manos. Describe cuántos grados girará su cuerpo para estar de pie otra vez.

2. Jeffrey empezó a andar en su bici en la 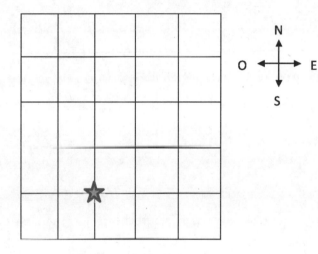. Viajó 3 cuadras hacia el norte, luego giró 90° a la derecha y viajó otras 2 cuadras. ¿En qué dirección iba? Dibuja su ruta en la rejilla de abajo. Cada unidad cuadrada representa 1 cuadra.

Lección 8:   Identificar y medir ángulos como giros y reconocerlos en varios contextos.

61

© 2019 Great Minds®. eureka-math.org

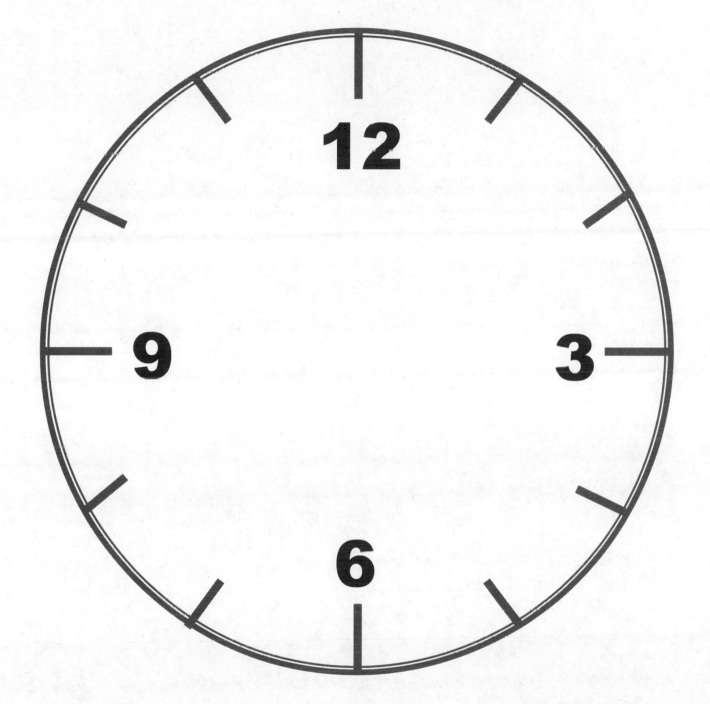

reloj

Lección 8:   Identificar y medir ángulos como giros y reconocerlos en varios contextos.

63

EUREKA MATH®

Haz una lista con las horas del reloj en las que la manecilla de las horas y la de los minutos forman un ángulo de 90°. Verifica tu trabajo usando un transportador.

Estate atento a este posible error: ¿por qué las manecillas a las 3:30 no forman un ángulo de 90° como era de esperarse?

_____

_____

_____

_____

Lee   **Dibuja**   **Escribe**

EUREKA MATH®

Nombre _____   Fecha _____

1. Completa la tabla.

Bloque de patrón	Total de bloques que se ajustan alrededor de 1 vértice	Un ángulo interior mide...	La suma de los ángulos alrededor del vértice
a.		360° ÷ ____ = ____	____ + ____ + ____ + ____ = 360°
b.			
c.			____ + ____ + ____ + = 360°
d. (Ángulo agudo)			
e. (Ángulo obtuso)			
f. (Ángulo agudo)			

Lección 9:   Descomponer ángulos usando bloques de patrón.   67

© 2019 Great Minds®. eureka-math.org

2. Encuentra la medida de los ángulos indicados por los arcos.

Bloques de patrón	Medida angular	Enunciado de suma
a. 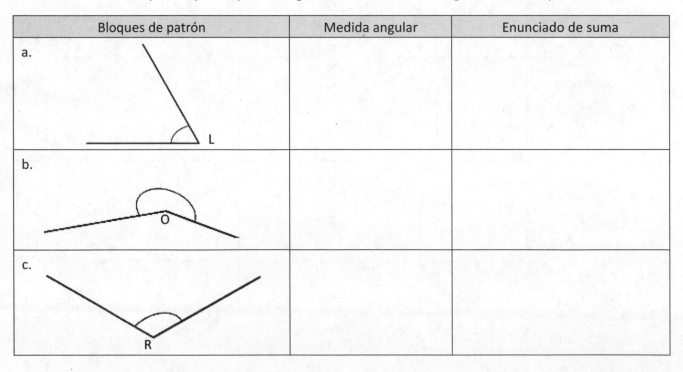		
b.		
c.		

3. Usa dos o más bloques de patrón para averiguar las medidas de los ángulos indicados por los arcos.

Bloques de patrón	Medida angular	Enunciado de suma
a.		
b.		
c.		

Lección 9:   Descomponer ángulos usando bloques de patrón.

© 2019 Great Minds®. eureka-math.org

Nombre _____   Fecha _____

1. Describe y dibuja dos combinaciones de bloques de patrón con rombos azules de manera que creen un ángulo llano.

2. Describe y dibuja dos combinaciones de bloques de patrón con el triángulo verde y el hexágono amarillo de manera que creen un ángulo llano.

Usando bloques de patrón con la misma figura o con figuras diferentes, crea un ángulo llano. ¿Qué figuras usaste? ¿Qué bloque de patrón puedes agregar a tu figura actual para crear un ángulo de 270°? ¿Cómo lo sabes?

_____

_____

_____

Lee   Dibuja   Escribe

EUREKA MATH®

Lección 10:   Usar la suma de las medidas angulares adyacentes para resolver
              problemas utilizando un símbolo para la medida angular desconocida.

71

© 2019 Great Minds®. eureka-math.org

Nombre _____   Fecha _____

Escribe una ecuación y resuélvela para la medida de $\angle x$. Verifica la medida usando un transportador.

1. $\angle CBA$ es un ángulo recto.

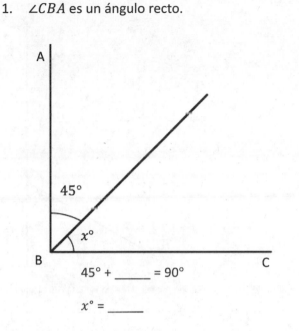

$45° + $ _____ $= 90°$

$x° = $ _____

2. $\angle GFE$ es un ángulo recto.

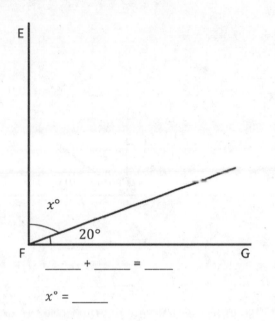

_____ $+$ _____ $=$ _____

$x° = $ _____

3. $\angle IJK$ es un ángulo llano.

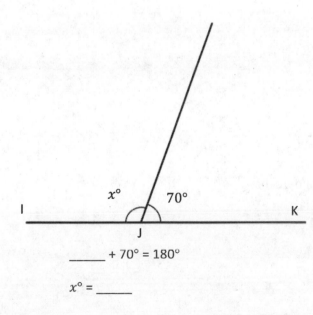

_____ $+ 70° = 180°$

$x° = $ _____

4. $\angle MNO$ es un ángulo llano.

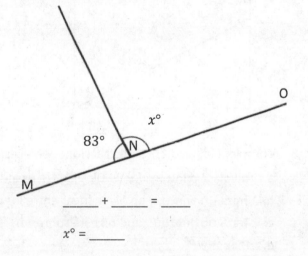

_____ $+$ _____ $=$ _____

$x° = $ _____

Lección 10:   Usar la suma de las medidas angulares adyacentes para resolver
              problemas utilizando un símbolo para la medida angular desconocida.

73

© 2019 Great Minds®. eureka-math.org

Encuentra las medidas angulares desconocidas. Escribe una ecuación para resolver el problema.

5. Resuelve para la medida de ∠TRU.
    ∠QRS es un ángulo llano.

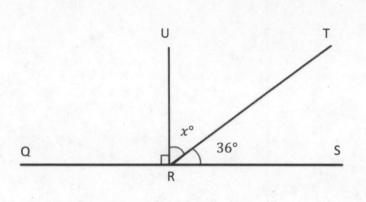

6. Resuelve para la medida de ∠ZYV.
    ∆XYZ es un ángulo llano.

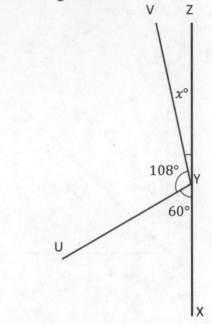

7. En la siguiente figura, ACDE es un rectángulo. Sin usar transportador, determina la medida de ∠DEB.
    Escribe una ecuación que pueda usarse para resolver el problema.

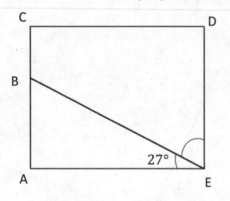

8. Completa las siguientes instrucciones en el espacio de la derecha.

    a. Dibuja 2 puntos: M y N. Dibuja $\overleftrightarrow{MN}$ usando una regla.

    b. Traza un punto O en algún lugar entre los puntos M y N.

    c. Traza un punto P, que no esté sobre $\overleftrightarrow{MN}$.

    d. Dibuja $\overline{OP}$.

    e. Encuentra la medida de ∠MOP y ∠NOP.

    f. Escribe una ecuación para mostrar que los ángulos suman la medida de un ángulo llano.

Lección 10:   Usar la suma de las medidas angulares adyacentes para resolver problemas utilizando un símbolo para la medida angular desconocida.

© 2019 Great Minds®. eureka-math.org

Nombre _____ Fecha _____

Escribe una ecuación y resuelve para x. $\angle TUV$ es un ángulo llano.

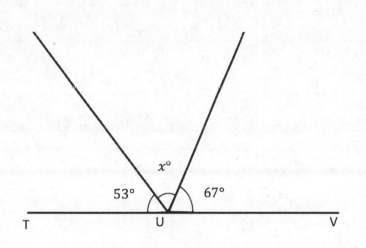

Ecuación: _____

x° = _____

Lección 10:   Usar la suma de las medidas angulares adyacentes para resolver
              problemas utilizando un símbolo para la medida angular desconocida.

© 2019 Great Minds®. eureka-math.org

75

Usa bloques de patrón de varios tipos para crear un diseño en el que puedas ver una descomposición de 360°. ¿Qué figuras usaste? Escribe una ecuación para mostrar cómo descompusiste 360°.

_____

_____

_____

_____

Lee   Dibuja   Escribe

EUREKA MATH®

Lección 11:   Usar la suma de las medidas angulares adyacentes para resolver
problemas utilizando un símbolo para la medida angular desconocida.

77

© 2019 Great Minds®. eureka-math.org

Nombre _____   Fecha _____

Escribe una ecuación y resuélvela numéricamente para las medidas angulares desconocidas.

1.

20° $d°$

_____° + 20° = 360°

$d° =$ _____°

2.

$c°$

_____° + _____° = 360°

$c° =$ _____°

3.

74°

$e°$

_____° + _____° + _____° = _____°

$e° =$ _____°

4.

$f°$

60°

_____° + _____° + _____° = _____°

$f° =$ _____°

Lección 11: Usar la suma de las medidas angulares adyacentes para resolver
problemas utilizando un símbolo para la medida angular desconocida.

79

© 2019 Great Minds® eureka-math.org

Escribe una ecuación y resuélvela numéricamente para las medidas angulares desconocidas.

5. O es la intersección de $\overline{AB}$ y $\overline{CD}$. $x°=$ _____ $y°=$ _____
    $\angle DOA$ es de 160°, y $\angle AOC$ es de 20°.

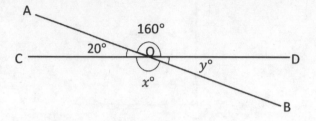

6. O es la intersección de $\overline{RS}$ y $\overline{TV}$. $g°=$ _____ $h°=$ _____ $i°=$ _____
    $\angle TOS$ es de 125°.

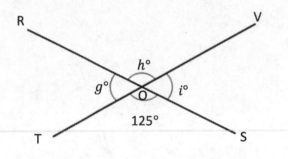

7. O es la intersección de $\overline{WX}$, $\overline{YZ}$, y $\overline{UO}$. $k°=$ _____ $m°=$ _____ $n°=$ _____
    $\angle XOZ$ es de 36°.

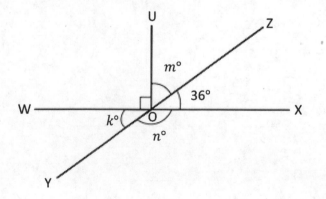

Lección 11:   Usar la suma de las medidas angulares adyacentes para resolver
              problemas utilizando un símbolo para la medida angular desconocida.

Nombre _____   Fecha _____

Escribe ecuaciones usando variables para representar las medidas angulares desconocidas. Encuentra las medidas angulares desconocidas numéricamente.

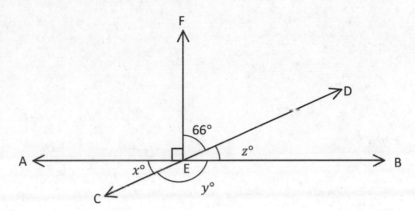

1. $x° =$

2. $y° =$

3. $z° =$

Lección 11:   Usar la suma de las medidas angulares adyacentes para resolver
problemas utilizando un símbolo para la medida angular desconocida.

81

© 2019 Great Minds®. eureka-math.org

Corta a lo largo de la línea punteada y desdobla la figura. Nota cómo cada lado dividido por la línea doblada coincide con el otro. Dobla de otra manera y ve si los lados coinciden. Observa los atributos de la figura y escribe un resumen de tus observaciones.

_____

_____

_____

_____

_____

_____

_____

_____

_____

_____

_____

Lee   **Dibuja**   **Escribe**

EUREKA MATH®

Lección 12: Reconocer las líneas de simetría de las figuras bidimensionales proporcionadas. Identificar figuras con línea de simetría y dibujar las líneas de simetría.

© 2019 Great Minds®. eureka-math.org

83

pentágono

Lección 12:   Reconocer las líneas de simetría de las figuras bidimensionales
proporcionadas. Identificar figuras con línea de simetría y dibujar las
líneas de simetría.

© 2019 Great Minds®. eureka-math.org

85

Nombre _____   Fecha _____

1. Encierra en un círculo las figuras que tienen dibujadas correctamente las líneas de simetría.

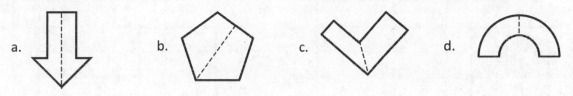

a.   b.   c.   d.

2. Encuentra y dibuja todas las líneas de simetría de las siguientes figuras. Escribe la cantidad de líneas de simetría que encontraste en el espacio en blanco debajo de la figura.

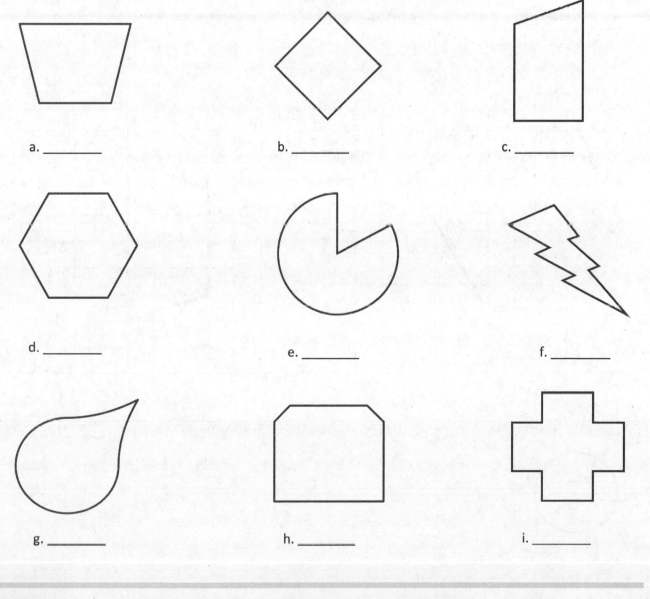

a. _____   b. _____   c. _____

d. _____   e. _____   f. _____

g. _____   h. _____   i. _____

Lección 12:   Reconocer las líneas de simetría de las figuras bidimensionales
proporcionadas. Identificar figuras con línea de simetría y dibujar las
líneas de simetría.

© 2019 Great Minds®. eureka-math.org

87

3. Abajo está dibujada la mitad de cada figura. Usa la línea de simetría, representada por la línea punteada, para completar cada figura.

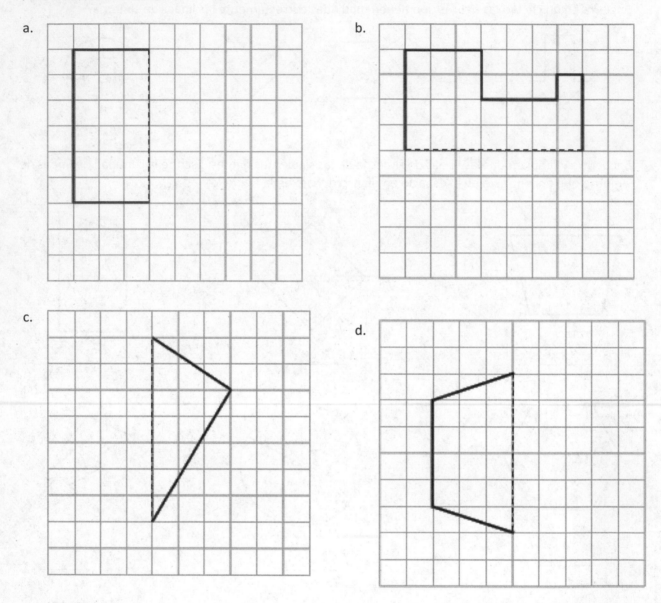

a.

b.

c.

d.

4. La figura de abajo es un círculo. ¿Cuántas líneas de simetría tiene la figura? Explica.

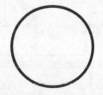

Lección 12: Reconocer las líneas de simetría de las figuras bidimensionales proporcionadas. Identificar figuras con línea de simetría y dibujar las líneas de simetría.

© 2019 Great Minds®. eureka-math.org

Nombre _____   Fecha _____

1. ¿La línea dibujada es una línea de simetría? Encierra en un círculo tu elección.

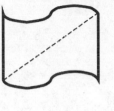

     Sí   No   Sí   No   Sí   No

2. Dibuja tantas líneas de simetría como puedas encontrar en la figura de abajo.

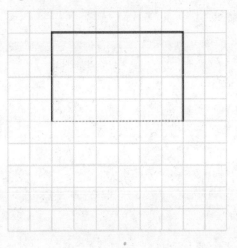

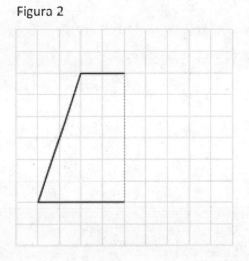

Figura 1

Figura 2

líneas de simetría

Corta a lo largo de la línea punteada en la plantilla que está en la hoja siguiente. Dobla los triángulos *A*, *B* y *C* para mostrar sus líneas de simetría. Usa una regla para trazar cada pliegue. Observa las relaciones de las figuras simétricas con los ángulos y las longitudes laterales. Escribe un resumen de tus observaciones.

_____

_____

_____

_____

_____

_____

_____

_____

_____

_____

Lee   **Dibuja**   **Escribe**

Lección 13:   Analizar y clasificar triángulos con base en las longitudes laterales, las medidas angulares o ambas.

© 2019 Great Minds®. eureka-math.org

93

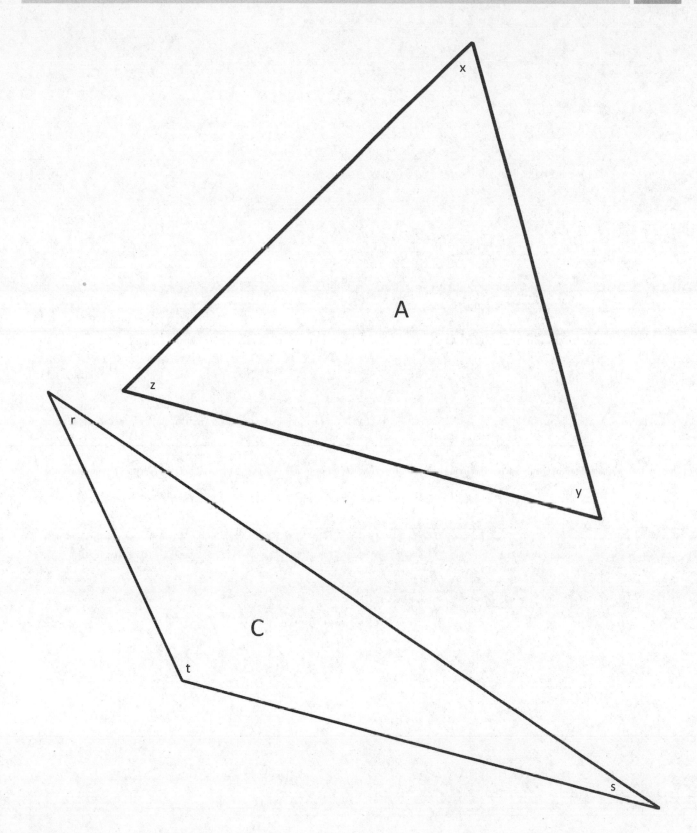

triángulos

Lección 13:   Analizar y clasificar triángulos con base en las longitudes laterales, las
medidas angulares o ambas.

95

© 2019 Great Minds®. eureka-math.org

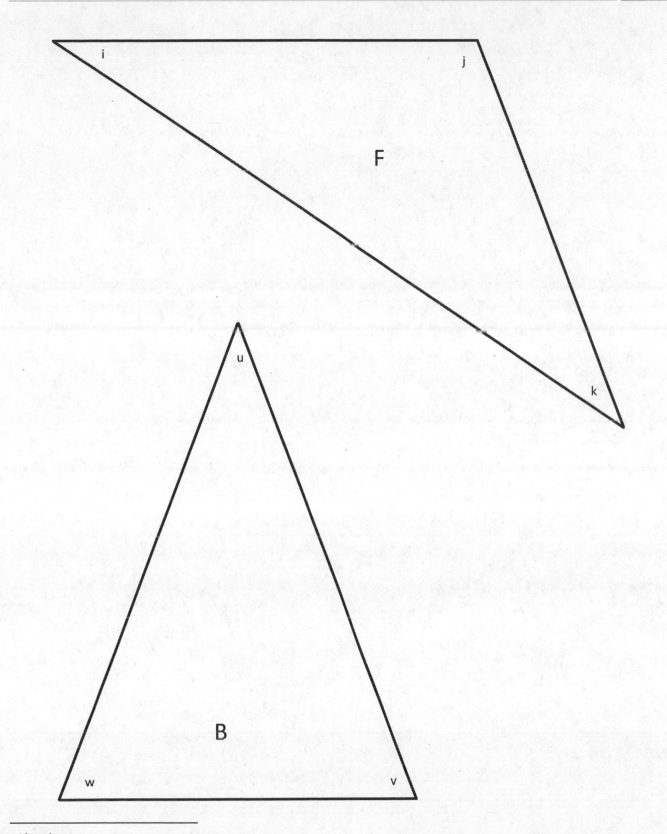

triángulos

Lección 13:   Analizar y clasificar triángulos con base en las longitudes laterales, las
medidas angulares o ambas.

97

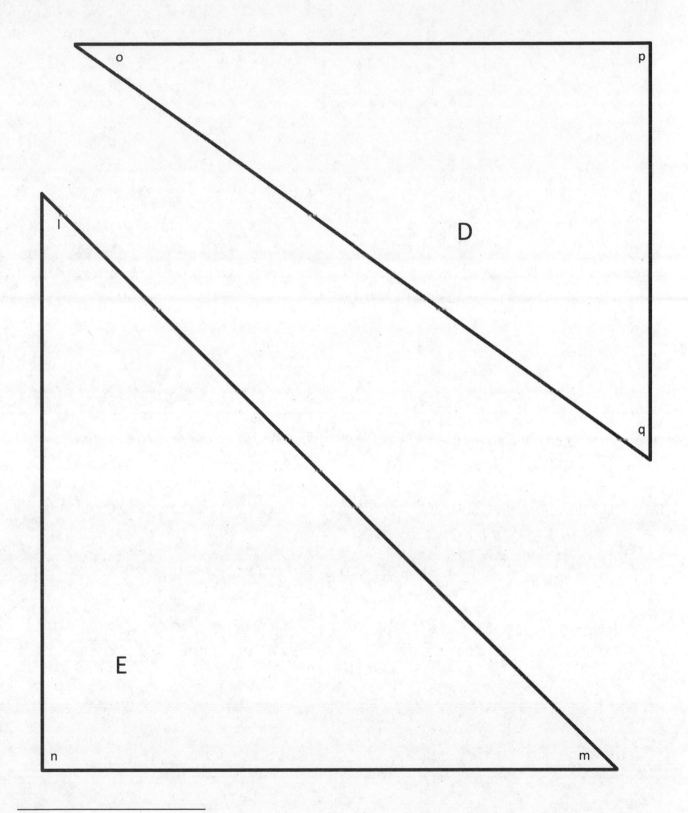

triángulos

EUREKA MATH

Lección 13:   Analizar y clasificar triángulos con base en las longitudes laterales, las medidas angulares o ambas.

99

© 2019 Great Minds®. eureka-math.org

Nombre _____   Fecha _____

Dibujo de un triángulo	Atributos (Incluir longitudes laterales y medidas angulares).	Clasificación	
A			
B			
C			
D			
E			
F			

Nombre _____ Fecha _____

1. Clasifica cada triángulo por sus longitudes laterales y sus medidas angulares. Encierra en un círculo los nombres correctos.

	Clasificación por las longitudes laterales			Clasificación por las medidas angulares		
a.	Equilátero	Isósceles	Escaleno	Agudo	Recto	Obtuso
b.	Equilátero	Isósceles	Escaleno	Agudo	Recto	Obtuso
c.	Equilátero	Isósceles	Escaleno	Agudo	Recto	Obtuso
d.	Equilátero	Isósceles	Escaleno	Agudo	Recto	Obtuso

2. $\triangle ABC$ tiene una línea de simetría como se muestra. ¿Qué te dice esto acerca de las medidas de $\angle A$ y $\angle C$?

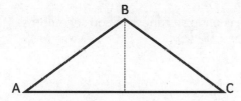

3. $\triangle DEF$ tiene tres líneas de simetría como se muestra.

   a. ¿Cómo te pueden ayudar las líneas de simetría para averiguar qué ángulos son iguales?

   b. $\triangle DEF$ tiene un perímetro de 30 cm. Marca las longitudes laterales.

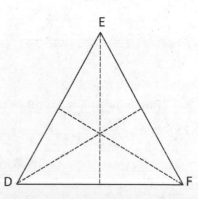

Lección 13: Analizar y clasificar triángulos con base en las longitudes laterales, las medidas angulares o ambas.

103

© 2019 Great Minds®. eureka-math.org

4. Usa una regla para conectar puntos y formar otros dos triángulos. Usa cada punto una sola vez. Los triángulos no deben sobreponerse. Uno o dos puntos no se usarán. Nombra y clasifica los tres triángulos. El primer ejercicio ya está resuelto.

Nombre de los triángulos usando los vértices	Clasificación por la longitud lateral	Clasificación por la medida angular
Δ FJK	Escaleno	Obtuso

5. a. Haz una lista con tres puntos de la gráfica de arriba que, cuando se conecten con segmentos, no resulten en un triángulo.

    b. ¿Por qué esos tres puntos que enumeraste no resultan en un triángulo si se conectan con segmentos?

6. ¿Puede un triángulo tener dos ángulos rectos? Explica.

Lección 13:   Analizar y clasificar triángulos con base en las longitudes laterales, las medidas angulares o ambas.

© 2019 Great Minds®. eureka-math.org

Nombre _____   Fecha _____

Usa las herramientas adecuadas para resolver los siguientes problemas.

1. Los triángulos de abajo se han clasificado según sus atributos compartidos (longitud lateral o tipo de ángulo). Usa las palabras *agudo, recto, obtuso, escaleno, isósceles* o *equilátero* para rotular los encabezados e Identificar la forma en que se ordenaron los triángulos.

2. Dibuja líneas para identificar cada triángulo según el tipo de ángulo y su longitud lateral.

a.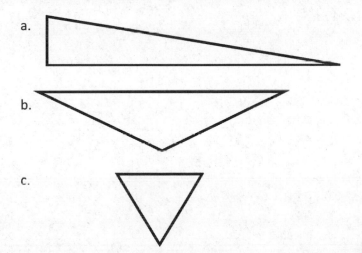

b.

c.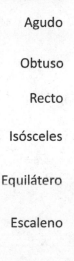

Agudo

Obtuso

Recto

Isósceles

Equilátero

Escaleno

3. Identifica y dibuja cualquier línea de simetría en los triángulos del Problema 2.

Lección 13: Analizar y clasificar triángulos con base en las longitudes laterales, las medidas angulares o ambas.

105

© 2019 Great Minds®. eureka-math.org

Dibuja tres puntos en el papel cuadriculado para que, cuando se conecten, formen un triángulo. Usa tu regla para conectar los puntos y formar un triángulo. Determina cómo se puede clasificar el triángulo que construiste: rectángulo, agudo, obtuso, equilátero, isósceles o escaleno.

a. ¿Cómo puedes clasificar tu triángulo?

Lee   **Dibuja**   **Escribe**

Lección 14:   Definir y construir triángulos con los criterios proporcionados. Explorar
la simetría en triángulos.

© 2019 Great Minds®. eureka-math.org

107

b. ¿Qué atributos consideraste para clasificar el triángulo?

c. ¿Qué herramientas usaste para poder dibujar y clasificar el triángulo?

Lee   **Dibuja**   **Escribe**

Lección 14:   Definir y construir triángulos con los criterios proporcionados. Explorar
                    la simetría en triángulos.

Nombre _____ Fecha _____

1. Dibuja triángulos que se ajusten a las siguientes clasificaciones. Usa una regla y un transportador. Marca las longitudes laterales y los ángulos.

    a. Rectángulo e isósceles

    b. Obtuso y escaleno

    c. Agudo y escaleno

    d. Agudo e isósceles

2. Dibuja todas las líneas de simetría posibles en los triángulos de arriba. Explica por qué algunos triángulos no tienen líneas de simetría.

Lección 14:   Definir y construir triángulos con los criterios proporcionados. Explorar
la simetría en triángulos.

© 2019 Great Minds®. eureka-math.org

109

¿Son verdaderas o falsas las siguientes afirmaciones? Explica usando imágenes o palabras.

3. Si Δ *ABC* es un triángulo equilátero, $\overline{BC}$ debe medir 2 cm. ¿Verdadero o falso?

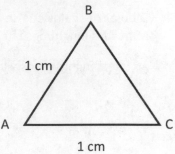

4. Un triángulo no puede tener un ángulo obtuso y uno recto. ¿Verdadero o falso?

5. Δ *EFG* se puede describir como un triángulo rectángulo y como un triángulo isósceles. ¿Verdadero o falso?

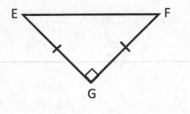

6. Un triángulo equilátero es isósceles. ¿Verdadero o falso?

Extensión: En Δ *HIJ*, a = b. ¿Verdadero o falso?

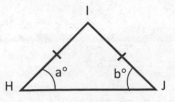

Lección 14:   Definir y construir triángulos con los criterios proporcionados. Explorar la simetría en triángulos.

Nombre _____ Fecha _____

1. Dibuja un triángulo isósceles obtuso y, luego, dibuja cualquier línea de simetría que exista.

2. Dibuja un triángulo rectángulo escaleno y, luego, dibuja cualquier línea de simetría que exista.

3. Cada triángulo tiene al menos _____ ángulos agudos.

Lección 14:   Definir y construir triángulos con los criterios proporcionados. Explorar
la simetría en triángulos.

111

© 2019 Great Minds®. eureka-math.org

a. En el papel cuadriculado, dibuja dos segmentos de recta perpendiculares, cada uno de 4 unidades, que se extiendan desde un punto V. Identifica los segmentos como $\overline{SV}$ y $\overline{UV}$. Dibuja $\overline{SU}$. ¿Qué figura construiste? Clasifícala.

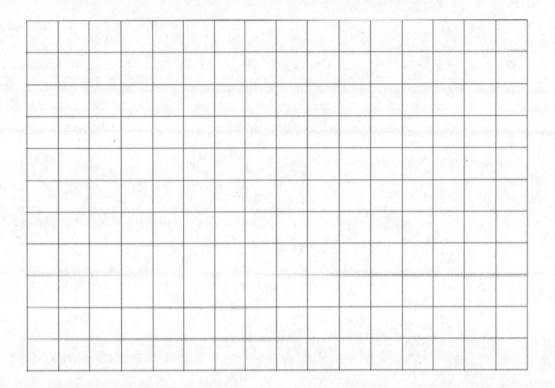

b. Imagina que $\overline{SU}$ es una línea de simetría. Construye la otra mitad de la figura. ¿Qué figura construiste? ¿Cómo lo sabes?

Lee   Dibuja   Escribe

Lección 15: Clasificar cuadriláteros con base en rectas paralelas y perpendiculares y la presencia o ausencia de ángulos de un tamaño especificado.

113

Nombre _____ Fecha _____

Construye las figuras con los atributos proporcionados. Nombra la figura que creaste. Sé tan específico como puedas. Usa una hoja extra en blanco si es necesario.

1. Construye cuadriláteros con al menos un conjunto de lados paralelos.

2. Construye un cuadrilátero con dos conjuntos de lados paralelos.

3. Construye un paralelogramo con cuatro ángulos rectos.

4. Construye un rectángulo con todos los lados de la misma longitud.

Lección 15:   Clasificar cuadriláteros con base en rectas paralelas y perpendiculares y la presencia o ausencia de ángulos de un tamaño especificado.

© 2019 Great Minds®. eureka-math.org

115

5. Usa el banco de palabras para nombrar cada figura, sé tan específico como sea posible.

| Paralelogramo | Trapezoide | Rectángulo | Cuadrado |

a.

_____

b.

_____

c.

_____

d.

_____

6. Explica el atributo que hace que un cuadrado sea un rectángulo especial.

7. Explica el atributo que hace que un rectángulo sea un paralelogramo especial.

8. Explica el atributo que hace que un paralelogramo sea un trapezoide especial.

Lección 15: Clasificar cuadriláteros con base en rectas paralelas y perpendiculares y la presencia o ausencia de ángulos de un tamaño especificado.

© 2019 Great Minds®. eureka-math.org

EUREKA
MATH

Nombre _____   Fecha _____

1. Dibuja un paralelogramo en el espacio de abajo.

2. Explica por qué un rectángulo es un paralelogramo especial.

Lección 15:   Clasificar cuadriláteros con base en rectas paralelas y perpendiculares y
la presencia o ausencia de ángulos de un tamaño especificado.

© 2019 Great Minds®. eureka-math.org

117

En las estrellas, encuentra al menos dos ejemplos diferentes de los siguientes polígonos. Explica qué atributos usaste para identificar a cada uno.

- Triángulos equiláteros

- Trapecios

- Paralelogramos

- Rombos

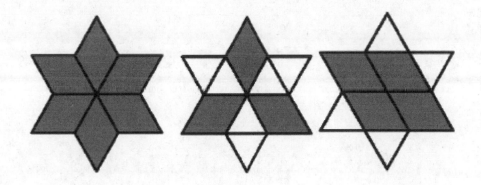

_____

_____

_____

_____

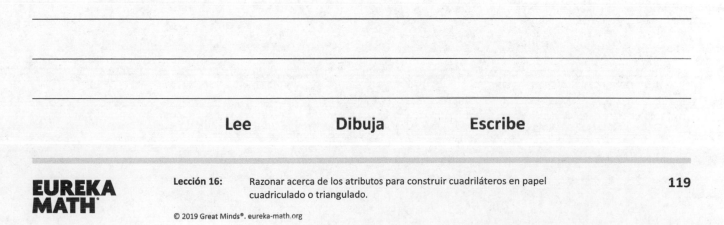

Lee   **Dibuja**   **Escribe**

Lección 16: Razonar acerca de los atributos para construir cuadriláteros en papel cuadriculado o triangulado.

119

Nombre _____   Fecha _____

1. En el papel cuadriculado, dibuja al menos un cuadrilátero que se ajuste a la descripción. Usa el segmento proporcionado como uno de los segmentos del cuadrilátero. Nombra la figura que dibujes usando uno de los siguientes términos.

Paralelogramo	Trapezoide	Rectángulo
Cuadrado		Rombo

a. Un cuadrilátero que tenga al menos un par de lados paralelos.

b. Un cuadrilátero que tenga cuatro ángulos rectos.

c. Un cuadrilátero que tenga dos pares de lados paralelos.

d. Un cuadrilátero que tenga al menos un par de lados perpendiculares y al menos un par de lados paralelos.

Lección 16:   Razonar acerca de los atributos para construir cuadriláteros en papel cuadriculado o triangulado.

121

© 2019 Great Minds®. eureka-math.org

2. En el papel cuadriculado, dibuja al menos un cuadrilátero que se ajuste a la descripción. Usa el segmento proporcionado como uno de los segmentos del cuadrilátero. Nombra la figura que dibujes usando uno de los siguientes términos.

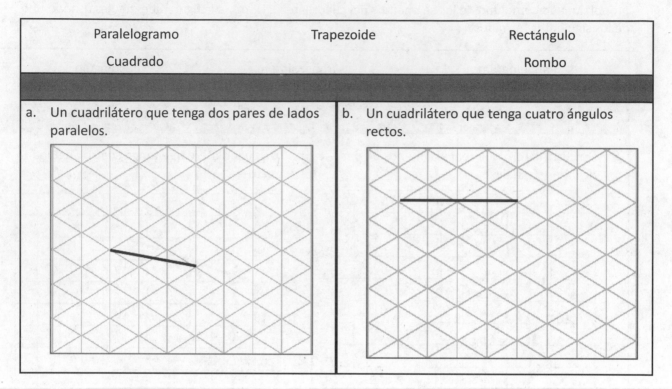

Paralelogramo   Trapezoide   Rectángulo

Cuadrado   Rombo

a. Un cuadrilátero que tenga dos pares de lados paralelos.

b. Un cuadrilátero que tenga cuatro ángulos rectos.

3. Explica los atributos que hacen que un rombo sea diferente de un rectángulo.

4. Explica el atributo que hace que un cuadrado sea diferente de un rombo.

Nombre _____   Fecha _____

1. Construye un paralelogramo que no tenga ningún ángulo recto en una rejilla rectangular.

2. Construye un rectángulo en una rejilla triangulada.

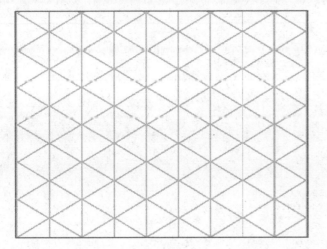

Lección 16:   Razonar acerca de los atributos para construir cuadriláteros en papel
cuadriculado o triangulado.

© 2019 Great Minds®. eureka-math.org

123

Créditos

Great Minds® ha hecho todos los esfuerzos para obtener permisos para la reimpresión de todo el material protegido por derechos de autor. Si algún propietario de material sujeto a derechos de autor no ha sido mencionado, favor ponerse en contacto con Great Minds para su debida mención en todas las ediciones y reimpresiones futuras.

Wilderness
First Responder

Wilderness
First Responder

*How to Recognize, Treat, and Prevent
Emergencies in the Backcountry*

SECOND EDITION

Buck Tilton, M.S., W.E.M.T.

FALCONGUIDE®

GUILFORD, CONNECTICUT
HELENA, MONTANA
AN IMPRINT OF THE GLOBE PEQUOT PRESS

Text and page design by Casey Shain

Photo credits: Pages 15 (both), 78, and 116, by Gates Richards Jr., courtesy of Wilderness Medicine Institute; pp. 37, 56, 79, 118, and 249 by Melissa Gray, courtesy of Wilderness Medicine Institute.

Illustration credits: Page 17, by Janet Colandrea; pp. 43 (right), 50, 81 (top), 84 (left), 89, 97, 119, 203, and 208, by Diane Blasius; chapter 36, Wilderness Transportation of the Sick or Injured, by Marc Bohne; all others by Bethany P. Crittendon.

ISSN: 1546-9654
ISBN-13: 978-0-7627-2801-5

Manufactured in the United States of America
Second Edition/Fourth Printing

To buy books in quantity for corporate use or incentives, call **(800) 962–0973** or e-mail **premiums@GlobePequot.com.**

The author and publisher have made every effort to ensure the accuracy of the information in this book at press time. However, they cannot accept any responsibility for any loss, injury, or inconvenience resulting from the use of information contained in this guide. Readers are encouraged to seek medical help whenever possible. This book is no substitute for a doctor's advice.

To everyone who contributed their expertise to this book,
and to all those who have helped make wilderness medicine so much more
than a couple of words, this book is gratefully dedicated.

Contents

List of Figures

Acknowledgments

The first edition of this book, published in 1998, would not have been possible without the tremendous efforts of Shana Tarter, who read and reread; Bethany Crittendon, who drew and redrew; and Tom Burke, MD, who generously gave his time to add a final and critical touch of medical science and art. Deepest thanks to you three. This second edition reflects more effort from others, especially Tod Schimelpfenig—who nagged and picked at fine points until I gave in and used his earnest and wise suggestions—and, once again, Shana, who never fails.

To others who contributed generously go great thanks:

- Joel Buettner, WEMI
- Joe Costello, MS, CSCS, CMT
- Mark Crawford, WEMT-P
- Daniel DeKay, RN, WEMT
- Kate Dernocoeur, EMT-P
- William Forgey, MD
- Melissa Gray, WEMT
- Charles Gregg, Esq.
- Colin Grissom, MD
- Peter Hackett, MD
- Murray Hamlet, DVM
- Linda Lindsey, RN
- Gates Richards, WEMT
- Richard Sugden, MD
- Ken Thompson, WEMT

Section One

Introduction

CHAPTER ONE

Wilderness Emergency Medical Care

You Should Be Able To:

1. Describe the need for well-trained providers of wilderness medicine.
2. Outline a brief history of wilderness medicine.
3. Describe the difference between wilderness medicine and urban medicine.

It Could Happen to You

After two days of late summer hiking under heavy backpacks into the Bighorn Crags of Idaho, you and three friends near the point on the map where an unnamed lake supposedly abounds with fine fishing and pleasant campsites tucked into the shadows of a dense forest. Clouds that collected over the afternoon start to spill a thin shower, and you stop to put on rain gear. With only a short series of switchbacks separating you from your destination, your group arrives at the scene of an accident. A lone hiker sits against a tree, pack by his side, face wearing a grimace of pain. He complains of lower right leg pain and the inability to bear weight on the injury. The hiker says he slipped on a wet rock while descending the trail. He is wearing a cotton T-shirt and shorts, and you note his lower right leg appears bloody and bruised. Occasional shivers disrupt his ability to speak.

Welcome to the world of wilderness medicine! It's an extraordinary world, a world filled with mountains and deserts, lakes and rivers, broad expanses of tundra and thin passages through serpentine canyons, fields of ice and fields of flowers, deep oceans with distant shorelines, and undeveloped lands where English is a foreign language. It's a world of cold and heat, wet and dry, high and low, dark and light, rushing noise and immense quiet, and, sometimes, utter solitude. It takes an hour to hike to this place, or a day of paddling, or a week of climbing, or a month of sailing. What common thread weaves through the world of wilderness medicine? The *Wilderness Medical Society Practice Guidelines for Wilderness Emergency Care* states you are "more than one hour from definitive medical care." Who stands beyond the one hour? Outdoor leaders and educators do, wilderness guides and enthusiasts, military personnel, remote researchers, field journalists, and many more. And they share the fact that hospitals and, usually, physicians are far enough away that the closest thing to anything definitively medical could be *you.*

First Response and Responsibility

Wilderness medicine is often difficult and demanding. The wilderness can turn little emergencies into big emergencies. There is a smaller margin for error than in an urban environment. The Wilderness First Responder (WFR) must be able to recognize, treat, and, whenever possible, prevent problems created by and within a wilderness environment. Anticipating and preventing problems and managing the risks that are inherent in wilderness travel are at least as important as recognizing and treating problems. Indeed, taking action to prevent emergencies, especially emergencies related to the wilderness environment such as heat, cold, altitude, hygiene, and blisters, is a particular responsibility of the WFR. The best guideline for the WFR might well be stated as this: *Plan ahead and prepare.*

To travel beyond the trailhead or put-in is to accept responsibility for your health and well-being, as well as the health and well-being of those you lead. The ability to be as self-reliant as possible is critical. The WFR must know how to travel in wilderness, how to dress and eat and drink, how to choose and care for gear—in short, how to live properly and safely in wilderness.

Whatever the terms you choose to define wilderness,

broad or narrow, physical or spiritual, all tracts of the world's wild lands share two common truths: They are decreasing in size and increasing in value. To journey into wilderness is to accept the responsibility to leave what you find as untouched by your passing as possible, to leave *zero impact* in the backcountry.

A Brief History of Wilderness Medicine

Before there was any medicine, there was "wilderness medicine," if we use that term to describe care given people far from a hospital. Evidence in prehistoric skeletal remains indicates that broken bones were set and adequate healing occurred. Almost every culture that has left record of its existence has left signs that some form of medicine was practiced, often well outside of a medicine man or woman's "office."

In the frigid winter of 1811–12, Napoleon's surgeon general, Baron Larrey, trained soldiers to care for their wounded comrades at the battlefront. Here was the first known orchestrated effort to keep participants in action by providing immediate attention in the field. Hypothermia and frostbite drove Larrey and the tiny remainder of Napoleon's troops out of Russia without the defenders having to fire too many shots, but the precedent had been set, and pre-hospital emergency medicine took its initial organized leap forward.

During the early years of the United States, almost all medicine was wilderness medicine, care provided in remote environments. Doctors were few, self-reliance was necessary for survival, and people learned to provide treatment for themselves and others when it was required. To early American pioneers, wilderness was a constant presence and medicine was a regular activity. When this nation went to war, management of injuries in the field continued to be an immediate necessity; our battlegrounds were another type of wilderness, and much of the modern growth in emergency medicine originated with the U. S. military.

As populations congregated in cities, medical needs were met more and more by physicians and hospitals, but accidents and sudden illnesses remained a major health problem. The American Red Cross, founded in 1905, began teaching first-aid classes that continue today, classes that have affected hundreds of thousands of people. First aid, then and now, has been designed to provide care over a relatively short time span, until a physician can be reached.

The next official leaps in urban pre-hospital care were taken in 1966, a year that was important for two significant developments. First, the United States government passed the National Highways Safety Act, giving the U.S. Department of Transportation the responsibility for developing an emergency medical services (EMS) system. From this act came the first emergency medical technician (EMT) course. The EMT program became extremely popular but was oriented (and remains today) toward the "golden hour"—the goal of getting the patient to the hospital within 60 minutes. Second, The American Heart Association (AHA) began to teach cardiopulmonary resuscitation (CPR) courses to the public.

After World War II, with the appearance of more and more leisure time, people began to return to remote areas for recreation. Their medical problems followed them. Prior to 1966 a significant unofficial realization had taken place, an awareness of the inadequacy of Red Cross first-aid courses to prepare outdoor enthusiasts for extended wilderness ventures. In the 1950s training programs were initiated that adapted the growing knowledge of medicine to wilderness settings. These early "mountaineering first-aid" programs were written by physicians and managed by outdoor organizations such as The Mountaineers in Seattle. A grand addition to the almost nonexistent literature of wilderness medicine appeared in 1967: the first edition of *Medicine for Mountaineering,* edited by James Wilkerson, MD. (It is now available in its fifth edition.)

In 1976 Stan Bush, a wilderness search-and-rescue director in Colorado, proposed the first Wilderness EMT course. In 1977 the Appalachian Search and Rescue Conference (ASRC) began offering wilderness-oriented EMT classes at the University of Virginia, and the National Outdoor Leadership School (NOLS) began to offer advanced first-aid courses designed especially to meet the needs of their instructors. In February 1977, Stonehearth Open Learning Opportunities (SOLO), a training center in Conway, New Hampshire, began offering wilderness first-aid courses, specializing in the needs of outdoor leaders. The first edition of *Wilderness Medicine* by William Forgey, M.D., the "Father of Wilderness Medicine," was published in 1979. Founded in 1983, the Wilderness Medical Society (WMS), a physician-oriented group, began to offer wilderness medical training through conventions and scientific meetings. The first edition of *Management of Wilderness and Environmental Emergencies,* edited by Paul Auerbach and Edward Geehr, both M.D.s, the definitive piece of wilderness medicine literature to date, also appeared in 1983. (It is now available in its fourth edition under the title *Wilderness Medicine.*) In 1984 SOLO developed the first Wilderness First Responder curriculum and, in January 1985, taught the first WFR course to Outward Bound instructors in Florida. In 1990 the first edition of *Medicine for the Backcountry* was published, a book written by Buck Tilton and Frank Hubbell, cofounder of SOLO, to provide the first practical guide for WFR students. (It is now available in its third edition.)

The 1980s saw the birth of other wilderness pre-hospital emergency medicine training organizations, such as Wilderness Medical Associates (WMA), and, in 1990, the Wilderness

Medicine Institute (WMI) was founded. In 1999 WMI became a part of NOLS. And the list keeps growing.

Research and development in wilderness medicine continues to be a dynamic area of the medical world. Providers of wilderness medical treatment and prevention are reaching out further and further into the field to save lives and limit suffering. To be the best possible WFR will require you to learn well now and to keep up with the steady advancement of wilderness medicine.

Wilderness Medicine vs. Urban Medicine

Wilderness medicine involves standard medical principles provided in a context that requires attention to extended contact time with the patient, environmental extremes, treatment with limited and nonspecialized equipment that may require improvisation, and the possible lack of communication.

Extended Contact Time

Patient needs change over time. Problems may become worse and the patient's life or limb may be threatened by the changes. Open wounds, for instance, require little additional attention from the urban first responder because they are handled by the hospital, typically within an hour. In the wilderness an open wound may lead to life-threatening infection before an evacuation can be accomplished.

Over hours, through nights, and sometimes for days, attention to the patient's general well-being must be considered. Such factors as urination, defecation, hydration, thermoregulation, and physical and psychological comfort must be monitored.

A patient's injury in the wilderness may merit a different approach than in an urban environment to improve the long-term outcome. A dislocation, for instance, may necessitate an attempt at reduction.

Environmental Extremes

Cold, heat, wind, rain, snow, ice, rough terrain, high altitude, darkness, and other environmental extremes often increase the stress and risks to the patient and to the rescuers. In addition to the physical risk, harsh conditions may complicate even the simplest care.

Limited Equipment/Improvisation

In the wilderness there is often little or no medical equipment available. The principles of treatment do not change, but care may have to be provided with improvised gear.

Communication

Urban rescuers rarely make transport decisions. The patient either gets transported or refuses transport. In the wilderness rescuers must make independent decisions regarding not only patient treatment but also whether or not (and how) to evacuate the patient, often without any communication with the rest of the world.

One of the greatest challenges of wilderness medicine, however, is the variety of situations the rescuer may find, unique circumstances that often defy a "cookbook" approach to medicine. To choose the treatment "recipe," the WFR needs training and common sense as a foundation for making decisions.

Another unique aspect of wilderness medicine is this: The decisions made and treatment provided by a WFR often enable the patient to remain in the field to enjoy a wilderness experience.

Training

What a WFR needs to know is well established. Training should include the underlying general anatomy and physiology and the foundational skill of a thorough patient assessment. The recognition, treatment, and, where applicable, prevention of all the most likely traumatic, medical, and environmental problems need to be addressed. Students need to be trained in the management of these emergencies over a long period of time, and such training should include, to name a few priority considerations: the cleaning and closing of wounds; the reduction of dislocations and angulated fractures; the effects on care of cold, heat, and other environment-related problems; a focused assessment of the spine to determine whether or not to take long-term spinal precautions; the management of anaphylaxis; and the cessation of CPR. Instruction should include the handling of minor as well as major complications. In addition, students need to learn principles of general patient care over hours to days, which includes keeping a patient warm, clean, and comfortable; paying attention to nutrition and hydration needs; and monitoring body functions. Training should include basic rescue considerations from a wilderness environment because all patients will require answers to these questions: Can they stay or go, and should it be fast or slow?

The first-aid kit that saves lives, prevents disability, and eases suffering is carried, for the most part, in the human brain. The safest plan for anyone who works or plays far from definitive medical intervention is to be well trained in wilderness medicine.

Conclusion

After a quick initial assessment, you determine that your patient in the Bighorn Crags has no immediate threats to life. Within a few minutes you've helped ease him off the wet ground onto his sleeping pad. You've dug dry clothing and rain gear from his pack, and, protected from the environment, his shivering begins to subside.

A focused assessment reveals no concerns other than the lower right leg and the potential for moderate to severe hypothermia. Your assessment of the leg reveals the possibility of a fracture. With the wound thoroughly cleaned and bandaged, you and one of your friends immobilize the leg using extra clothing for padding and a Crazy Creek chair for rigid support.

While you're treating the patient, your two other companions have set camp as close as comfortably possible. It's no problem for the four of you to carry the patient into the tent. You help him into his sleeping bag and begin dinner preparations.

Over dinner you and your group decide the best plan of action includes going out for help. You write, complete with details concerning the status of the patient, a request for aid. Two of your friends will start out with the message tomorrow morning.

All this, and much more, is the stuff of wilderness medicine. The instruction offered by this book will help you learn to provide care in urban and wilderness settings, but the emphasis in every case remains on response to the sick or injured when definitive treatment is far away, and, whenever possible, on steps to take to prevent emergencies.

CHAPTER TWO

Legal Issues in Wilderness Medicine

You Should Be Able To:

1. Describe what the law requires of the Wilderness First Responder.
2. Define "negligence" and describe the four elements relative to proof of negligence.
3. Describe important legal considerations relative to the Wilderness First Responder, including consent, assault and battery, abandonment, and confidentiality.
4. Describe legal protection for the Wilderness First Responder.

It Could Happen to You

On a trek in Nepal, one of your clients, Mr. Brown, slips on a treacherous stretch of trail, tumbles a couple of dozen feet, and ends up looking as if he was trampled by a herd of yaks. Most of the damage is superficial—except the dislocation of his left shoulder. According to the protocols written by your medical adviser, supported by your training and current certification, you reduce the dislocation and evacuate Mr. Brown under his own power. During the walk out, the patient complains of numbness in the affected arm. Back in the United States, Mr. Brown's physician diagnoses lingering nerve damage, mild but bothersome, a result of your reduction.

Mr. Brown had signed a pretrip assumption of risk and release from liability and indemnity agreement that stated (1) he understood injury was one of the dangers of the trek, and he assumed that risk, and (2) he released you from all liability, claims, and causes of action connected with his participation in the trek. Despite the release Mr. Brown sues you, claiming negligence in your failure to properly care for his dislocated shoulder. You move to have the claim dismissed, arguing Mr. Brown understood the risk and signed a valid release.

"Reasonable and prudent actions" should describe the medical care you provide in wilderness settings. The conscientious and responsible Wilderness First Responder will concentrate on the opportunities to be of service and not let a concern for liability affect his or her performance. The law requires only that you provide the care that patients require, and it protects those care providers who do their jobs well.

You could find yourself, however, defending a lawsuit that claims that you should have done more, or less, for a patient. Failure to defend such a claim successfully can hurt you professionally and financially. You should, therefore, understand the legal issues involved in this new and important field of medicine of which you have chosen to be a part.

This chapter will deal with general legal concepts, not the laws of particular states. You should seek legal advice regarding the applicable laws of your state and consult with your medical adviser, if you have one, regarding the medical aspects of the discussion that follows. A *medical adviser* is a licensed physician who advises an unlicensed medical practitioner.

The areas of the law that are most important to the care provider, whether in the city or the wilderness, are contract and tort law. These are branches of civil law, as opposed to criminal law.

Civil Law and the WFR

Contract Law

Contracts are promises that are expressed or implied, written or oral. A person can be sued to enforce these promises or to pay money if they are broken. All parts of a contract should be clearly expressed and understood: who is to do what for whom, when, and for what consideration or payment, and the remedy if a person does not perform as promised. At some time in your career, you have had and/or will have a contract with someone, perhaps your employer or a person in your care. Some states may consider that you have entered into an implied oral contract as soon as you or someone on your behalf causes another to believe you will give medical help if needed. Other contracts with which you may be involved are releases, in which a person forgives you in advance for a wrong you might later commit, and contracts of insurance, which allow you to acquire protection from claims of persons who may be injured by you.

Tort Law

The other area of the law, the one in which you will probably be involved if you are ever named in a suit, is tort law. Tort law deals with wrongs to people and property not usually involving contracts. The word *tort* comes from a French word meaning "wrong" or "harm." While the most familiar of these torts are intentional bodily injuries and fraud, the focus of our discussion will be the tort of *negligence*—the careless, unintentional act that harms another person to whom you owe a duty of care.

Negligence

The good news is that, generally, you will be protected from legal liability for negligence if you do your job well and in accordance with the standards of your profession. Typically, and in your favor, people who participate in outdoor ventures are more likely than others to accept responsibility for a risky activity and, therefore, are less likely to sue. Nevertheless this area of the law is of considerable interest to the wilderness care provider, whose scope of responsibility and authority may vary from state to state and whose role in a particular situation may not always be well defined by law.

The elements of a claim of negligence are (1) a duty to act; (2) a breach of duty, or failure to perform that duty; and (3) a physical or psychological injury that was caused or contributed to by the breach or failure.

Duty to Act

In most states you have a duty to act if you have a prior relationship with the injured or ill person. If the person is in your direct care or is a participant in an activity (a summer camp or outdoor program activity, for example) for which you have been hired to provide medical care services as all or part of your job, you clearly have a duty to that person. Even if you work as a volunteer, you bear the same duty to act as if you held a paid position. If you know a person is relying on you for assistance, you, once again, have a duty to that person. You have a special relationship with that person, who is no longer a "stranger."

The *standard of care* in pre-hospital medicine—the care you are asked to provide for those to whom you have a duty—is largely determined by the specific training you have been given, the training that has provided you with the skills and knowledge of how to do what, and when. If your patient assessment, for instance, reveals the possibility of a fractured leg, the standard of care, generally, is to appropriately splint the leg in question and monitor the patient.

Because wilderness medicine is a newer profession, the standards may be less clear than for those operating in ambu-

lances, emergency rooms, and other urban medicine situations. *What* you are allowed to do will depend on the laws of the state where you work and/or the medical protocols written for you by your medical adviser, if you have one. Be sure you are operating within those laws and protocols as you consider reducing a dislocation, for example. *How well* you perform will be measured by standards that are much broader. Never do more, or less, than you are trained to do.

Breach of Duty

The second element of negligence is a violation, or breach, of the duty of care. A breach can be an act (commission) or the failure to act (omission). In most cases that have gone to court, wilderness medicine providers have been sued for failure to act, for omitting care that should have been provided.

Generally speaking, the law will consider you at fault and liable for payment of damages to the injured person if you have not performed as a reasonable person would with your background and training in the same or similar circumstances. Examples might be the misreading of obvious vital signs or the failure to splint a fracture. *Gross* negligence might be attempting to provide care when you are under the influence of drugs or alcohol. Gross negligence is a crime.

Injury Caused by the Breach

The third element of a negligence claim is a loss or injury for which the breach of duty or wrongful act is either the cause or a contributing cause. A loss can be the result of an accident or illness or the result of the care you provide. The loss can include fright and other emotional/psychological trauma and certainly includes loss of property, personal injury, and death. You will not be liable if another person or event is shown to have caused the injury—for example, if a qualified person to whom you transfer the patient acts negligently. Also, the loss must have been a reasonably foreseeable result of the breach of duty. You should not be liable if a person, because of some preexisting condition of which you could not have been aware, reacts badly to a regular procedure applied by you. In this event, a person with your training could not have foreseen the result and *should* not be liable.

Legal Considerations

Consent

Before care is given, the *informed* consent of an adult, or the parent or guardian of a minor, is required by law and should be in writing or at least witnessed by a third party whenever possible. Informed consent means the patient is advised of the problem, the proposed treatment, the risk of treatment, and what to expect if no treatment is given, and the patient gives consent, actual or implied. Failure to acquire informed consent could possibly result in a suit against you for assault and/or battery. Fortunately, the law recognizes *implied* consent in emergency situations when it can be reasonably assumed that the patient, if conscious and reliable (or a parent, if the patient is a minor), would have agreed to the assistance offered. If you work with minors, which means, in most cases, anyone under 18 years of age, you are well advised to carry a document signed by the parent or guardian allowing you to provide medical care in an emergency. If you find yourself in an emergency involving a minor and without prearranged consent, go ahead and treat to the best of your ability.

Assault/Battery

Assault is defined as any act of commission or omission that places a patient in fear of bodily harm. *Battery* is defined as providing medical care or otherwise touching a patient without consent.

Abandonment

You may be liable to a patient, or to a patient's family, for abandonment if you terminate care prematurely and the patient suffers harm later, or if you transfer care to a less qualified or unqualified person and the patient suffers harm later.

Example 1: You, the WFR, stop on a steep trail to aid an out-of-shape hiker "dead on his feet" from extreme fatigue and nausea; the hiker begins to depend on your assistance. You know this to be a busy trail, and you leave the hiker unattended because you want to make it over the next pass before dark. Arguably, you have abandoned a patient who may suffer harm.

Example 2: You, the WFR, stop to aid the hiker in Example One and leave him in the care of the next passerby, who happens to have no medical training. Arguably, you have abandoned a patient who may suffer harm.

Example 3: You, the WFR, stop to aid the hiker in Example One, and you decide the hiker will not be able to continue on his own. Your decision is to leave him unattended to hike out for more help. You leave him well supplied with food, water, and extra clothing. Arguably, you have not abandoned a patient because you have acted in a reasonable and prudent manner.

Confidentiality

Defaming a person's character or reputation verbally is called *slander*. Doing so in writing is called *libel*. Communication between you and your patient is considered confidential. Inappropriate disclosure of information regarding assessment, history, or treatment of your patient—in verbal or written form—can be considered a breach of confidentiality that makes you vulnerable to a claim of slander or libel. Do not, in other words, share information about your patient inappropriately or with inappropriate parties. Be cautious of what you say to other members of an expedition and via radio communications. Report

and/or document only the facts and not your assumptions. For instance, report that the patient states, "I've had a few drinks," but do *not* say, "The patient is drunk."

Legal Protection

If you are sued for negligence, you have defenses. These defenses include the absence of one or more of the necessary elements of a claim of negligence: duty to act, breach of duty, and causation of loss or injury. The negligence of others, including the person injured, also can reduce or eliminate your liability. In many states the judge or jury is allowed to compare, on a percentage basis, the fault of all who may have contributed to the injury. This is usually referred to as *comparative negligence* or *comparative fault*, and ensures that you are obligated only for that part of the loss that you caused.

The Good Samaritan

To encourage trained people to offer care, most states have so-called Good Samaritan laws that provide that a person who voluntarily gives emergency assistance will not be liable for simple carelessness, such as negligence. There is no such protection for gross negligence, which is carelessness that is so extreme as to show a complete disregard for the person injured. Note that Good Samaritan care must be voluntary and performed in an emergency. In a wilderness setting such a statute might control your voluntary care of a stranger found injured on the trail, but if you have a duty to act the Good Samaritan laws do not protect you.

Documentation

You may have heard the saying, "If it isn't written down, it didn't happen." This means that, in a lawsuit, if an important event is not recorded by the care provider, the judge and jury will probably assume that it did not occur. It is important, therefore, to avoid guessing about what happened in the field, that you make a written record at the time of the event or shortly thereafter. At minimum the record should include dates and times, patient history, a description of the scene, your physical assessment and treatment, and changes in the patient while in your care (see Chapter 3: Patient Assessment, "The SOAP Note"). It is also important to document, with a witness if possible, a refusal of treatment by an informed patient.

The Care Provider as an Employee

The fact that you are an employee is no protection from liability, except to the extent the employer's insurance may take care of your legal liability and expenses. Insurance is a matter you should carefully consider, whether you are acting independently or for an employer. It is your responsibility to know whether or not you are insured and how well the insurance protects you.

Summary

What is the practical legal effect of all this? The care provider assesses the emergency; removes the patient from harm; stabilizes the patient; provides other limited, essential care; and prepares proper reports and records. As a wilderness care provider you probably will have more responsibilities (for expedition medicines, for example) and provide more treatments that might not be indicated if hospital care were more available.

Most of the issues facing the care provider on wilderness expeditions relate to small wounds, athletic injuries, environmental emergencies, and hygiene-related problems, but there is always the possibility of severe trauma or illness, a difficult-to-assess stomach cramp, a diabetic reaction, or a severe laceration. Such occurrences, an hour or longer from the attention of a licensed physician, are much more serious than if encountered in an urban setting. If you accept the responsibility of care in the wilderness, you must be prepared with appropriate training, equipment, and medical protocols for such emergencies.

You are well advised, then, if you work for an outdoor program or search-and-rescue team, to seek out and work under the authority of a medical adviser. The medical adviser, or base-station physician, will share responsibility with you for the adequacy (or inadequacy) of your performance if she or he authorizes you to administer drugs, reduce dislocations, close wounds, or otherwise engage in procedures that may exceed the customary role of the urban first responder. If it is done well, there will be no complaint. If done improperly, questions of training, technique, authority, consent, and alternative remedies will be carefully examined by investigators, lawyers, experts, and a judge and jury.

In a wilderness or outdoor program setting, additional issues important to you as a staffperson will include (1) the screening and supervision of participants; (2) the adequacy of equipment and supplies; and (3) the presence of a carefully designed plan or set of protocols for medication administration, evacuation, and other emergencies.

Of Critical Importance

As a Wilderness First Responder, you are certified to provide care, not licensed to practice medicine. To administer drugs and perform medical procedures and techniques that go beyond the scope of urban practice, you need to work under the authority of a medical adviser/director.

Protection for Wilderness Trip Leaders and Organizations

1. Use release/indemnification forms that take into account specific state laws.
2. Medically screen participants thoroughly prior to a trip.
3. Supply staff with well-stocked first-aid kits.
4. Supply staff with incident/accident report forms.
5. Provide written protocols to aid in evacuation decisions and plans.
6. Give care under the authority of and with written protocols from a medical adviser.
7. Encourage or require staff to obtain and maintain sufficient levels of medical training.

Conclusion

The court determines it is reasonable to assume that you, as a leader of a trek to Nepal, would have knowledge of reducing dislocations in the field. An expert witness explains to the court that the protocols written by your medical adviser were precise and that your documentation indicates you followed them precisely. Another expert witness explains that the future of Mr. Brown's shoulder looks far better than it would if it had remained dislocated. The pretrip form Mr. Brown signed, says the court, indicates he understood an injury of this nature could occur and that he assumed that risk. You go back to work in the wilderness.

Charles R. Gregg, Esq., contributed his expertise to this chapter.

Section Two

Patient Assessment
and
Initial Concerns

CHAPTER THREE

Patient Assessment

You Should Be Able To:

1. Describe how to immediately establish control of the scene.
2. Describe the importance of and how to establish a safe scene, including the use of body substance isolation.
3. Describe what it means to "size up the scene," including the mechanism of injury, the number of patients, and a general impression.
4. Describe the importance of and methods of establishing an effective relationship with the patient.
5. Describe the importance of and demonstrate how to do an initial assessment.
6. Describe the importance of and demonstrate how to do a focused assessment, including a physical examination, measuring vital signs, and obtaining a patient history.
7. Describe the importance of and demonstrate how to document patient assessment and treatment in writing and report on patient assessment and treatment verbally.

It Could Happen to You

Only the rocks know what happened, and maybe a couple of trees, but they aren't talking. Spring warms the air, the cottonwoods have sprouted new leaves, and this quiet section of the lower Green River you and a friend are canoeing, the river that now passes beneath a high sandstone cliff, has carried you within sight of a young man sprawled on the ground near the water's edge. In the stern of the canoe, you back-paddle while your friend in the bow draws. The canoe eases to shore.

The young man lies face up, unmoving, a smear of blood from mouth to ear. No other people are evident. Stillness reigns. What happened? What do you do? There are clues everywhere, but where is Sherlock Holmes when you need him?

Your ability to adequately manage an emergency is rooted in your ability to properly assess the scene and the patient. The patient assessment is made through the gathering and processing of information in a series of logical steps that end with a complete "picture" of the patient. The information, for purposes of gathering and documenting, is divided into a series of steps, and the steps are referred to, collectively, as the *patient assessment system*. The steps include (1) a scene size up, (2) an initial assessment of the patient, (3) a focused exam and history of the patient, (4) documenting and reporting the event, and (5) monitoring the patient's condition. You should follow the system, step by step, until you have mastered it and are able to perform it at a high level of competence. Then any decisions to vary from the system can be made from an educated foundation, and not because something was forgotten. Following the patient assessment system allows you to gather all the "clues." Imagine the ineffable Mr. Holmes, pipe clamped between his teeth, eyes missing nothing, mind shifting into high gear. But this is not Baker Street, Dr. Watson. This is the wild outdoors.

Assessment is the foundation of all medicine, but assessment is far more than a medical skill. When you watch someone who does an excellent job of assessing, you watch someone who does an excellent job of relating to people. Your assessment will come easiest and best for all concerned if you establish an effective relationship with the patient and the other people on the scene and maintain a healthy working relationship with coworkers and others for whom you have responsibility.

On the scene, a rush of adrenaline may threaten your ability to think clearly and act responsibly. To help you, consider stopping momentarily in your assessment to take a deep calming breath. *Stop! Breathe. Go.*

STOP! SIZE UP THE SCENE

Establish Control

Emergencies are often charged with emotion and confusion. Even minor chaos increases the risk of injury to rescuers and bystanders and the risk of inadequate care for the patient. Although many different styles of leadership and models for decision-making exist, emergencies often call for someone to be directive, at least until the scene is safe and the patient is stabilized. This is best accomplished if you have discussed leadership styles in case of an emergency with coworkers and others for whom you have responsibility prior to a critical situation.

Two qualities describe the best Wilderness First Responder in an emergency. The first is *competence:* You know your stuff, and you are capable and ready to act. The second is *confidence:* You appear able to deal with the situation. You don't have to feel confident, but you should appear confident and sound confident. Avoid shouting. Speak with quiet authority. Then listen. Your goals should be to (1) provide the greatest good for the greatest number in the shortest time and (2) do no harm.

Survey the Scene for Hazards

Every assessment should start before you even reach the patient's side: an assessment of the scene for hazards. Is there immediate danger from the environment—rockfall, thin ice over water, carbon monoxide filling the tent? Have you taken precautions to prevent being a giver or taker of communicable diseases? (See "Isolate Body Substance," this page).

Your desire to rush in and help as soon as possible must be tempered with your need to maintain your own personal safety, the safety of other members of your party, and the safety of the patient, to prevent further injury. A second patient is always a tragedy, not to mention a dramatic increase in the difficulty of the situation. In the wilderness a second patient is not only tragic but also represents a loss of resources. Every person in a wilderness situation is an irreplaceable resource, someone who can help carry and care for the patient. Protect your resources. Hazards must be eliminated, or at least minimized, before approaching the patient.

Never create a second patient!

Determine the Mechanism of Injury

Scene assessment includes assessing the *mechanism of injury* (MOI) the patient has undergone. If the climber fell, what did he land on? Rocks? Compacted soil? Grass? In what position did you find him?

Much of the information concerning MOI may have to be gathered through questioning the patient and/or eyewitnesses. How far did the climber fall? What body part hit the ground first? Did a rope slow his descent? Was he wearing a helmet? Sometimes substantial forces can produce injuries that the patient will not be aware of early in your assessment. Knowledge of the MOI will help in your initial assessment and make you aware of and able to anticipate possible changes in your patient later.

Isolate Body Substances

Communicable, or infectious, diseases must be considered a potential risk to rescuers. Therefore, you should *assume* that every patient could spread germs that cause illness. You should also protect every patient from germs that *you* could be carrying (see Chapter 30: Communicable Diseases and Camp Hygiene).

Use barrier devices to ensure **body substance isolation (BSI)** to prevent your skin and mucous membranes from contacting the blood or other body fluids of a patient. The BSI methodology includes the following recommendations:

1. *Keep protective disposable gloves available at all times, and wear them when there is the slightest chance you may contact a patient's blood, other body fluids, mucous membranes, or broken skin.* Wear gloves when you handle bandages, clothing, or other items contaminated with blood or other body fluids. When you no longer need them, pull the gloves off carefully from your wrist, leaving them inside out. Dispose of the gloves as soon as possible after use by sealing them in a plastic bag with other contaminated items and marking the bag as contaminated material. Wash your hands as soon as possible after removing the gloves.

   Note: Without gloves, you may be able to improvise protection by slipping plastic bags over your hands.

2. *Wear protective glasses if the scene could involve the spraying of contaminated droplets* such as might erupt from a coughing patient, an aggressively bleeding patient, and/or a vomiting patient. Wear your sunglasses if nothing else is available.

3. *Wear a protective mask over your nose and mouth if the scene could involve the spraying of contaminated droplets—* a bandanna will do when nothing else is available.

4. *Use a rescue mask with a one-way valve if the patient requires you to breathe for him or her.*

5. *Consider putting on your rainwear if the scene is especially messy* with blood, vomit, or other body substances.

Count the Number of Patients

An assessment of the scene should include scanning the area for other patients. One patient demanding attention, someone obviously afraid and/or in pain, can cause you to immediately focus on that patient. Meanwhile, sitting or lying quietly nearby, a second patient slowly dies from blood loss or loss of an airway. Make sure, as soon as possible, that you know how many patients are involved in the scene.

What's Your General Impression?

What is your general impression of the patient? Hurt, or not hurt? Sick, or not sick?

If your impression is, "This is a seriously hurt or seriously ill person," you should be preparing yourself for a rapid assessment, rapid treatment, and a rapid decision concerning transport of the patient from the wilderness. Keep in mind, however, that "rapid" transport from the wilderness is more often a wishful thought than a reality.

Experience plays a role in helping you get a general impression, but examples of "serious" problems include (1) a patient unable to breathe or unable to breathe adequately; (2) a patient without a pulse; (3) a patient with pale, blue, or gray skin; (4) a patient with an altered level of consciousness; (5) blood spurting into the air or pooling rapidly under the patient; (6) extremities with obvious deformities; and (7) extremities that are missing.

If your impression is, "This person is not seriously hurt or seriously ill," then you can prepare yourself for a more relaxed assessment and treatment and for a more relaxed decision concerning the need for transport.

Figure 3-1: *Establish a relationship.*

Establish a Relationship

If the patient is conscious, there is an excellent chance she or he will be frightened, anxious, and in pain. If there are no obvious immediate threats to life (see "Perform an Initial Assessment," this page), take a minute to establish an open, clear, honest, effective relationship. Introduce yourself, if needed, and state your qualifications to provide care. It is polite and respectful, and it allows you to get permission to treat. Position yourself as close to eye level as possible with the patient, and make eye contact.

If you're wearing sunglasses, take them off if there is no danger of contamination from body substances. Speak reassuringly, quietly, but loud enough to be heard easily. If you don't know the patient, ask for his or her name . . . and use it. If the patient is older, it is usually best to use the last name: "Ms. Smith" or "Mr. Jones." Maintain an open dialogue throughout your assessment and treatment by listening to the patient, answering questions, and asking permission before you do something.

Touch is a universal way to show concern and provide comfort. Your hand on the patient's arm or shoulder, if you are comfortable doing it, can give valuable reassurance, but do not assume the patient wants you to touch him or her during your upcoming examination. Whenever possible, before touching in order to assess, be sure to ask permission.

If the patient is unconscious and/or there are obvious immediate threats to life, relationship building may have to wait. Remember, however, that the patient, even if unconscious, should still be treated with respect and compassion.

Compassion? You don't *have* to be compassionate. You can be effective by acting with competence and without showing an awareness of human distress and a desire to alleviate it. Arguably, however, compassionate care providers are better care providers. Few, if any, people would choose to be treated without compassion.

STOP! ASSESS THE PATIENT FOR IMMEDIATE THREATS TO LIFE

Perform an Initial Assessment

The goal of the initial assessment is to find and treat any immediate threats to the patient's life. This is a "stop-and-fix" phase. If you find something that needs treatment during the initial assessment, you should immediately stop and fix it. Although

Figure 3-2: *Patient assessment.*

the initial assessment is presented here in a systematic manner—**ABCDE**—you may find what you actually do varies somewhat from alphabetical order. As you approach the patient, you should quickly scan for obvious threats. If your visual scan reveals blood squirting from an open artery, you might find yourself treating "C" (Circulation) before "A" (Airway). Still, the concept of ABCDE provides a sound basis from which to work.

Establish Responsiveness/ Control the Cervical Spine

The first step in the initial assessment is to determine the patient's ability to respond to stimuli. This is the first step because a patient with a mental status altered from normal may need airway management or other early life-saving intervention.

Once you have introduced yourself and gained consent to treat, it is highly recommended that you place one of your hands gently on the patient's head, a reminder to stay still until a further assessment is made of the MOI.

This may be an appropriate time to ask what happened. If you learn that serious forces have been involved, your level of concern over the spine will rise—appropriately.

If the MOI and/or the patient's mental status indicates the possibility of spine or head trauma, take immediate and complete manual control of the patient's head and neck by placing both of your hands on the patient's head or by having an assistant rescuer take manual control of the patient's head and neck.

A Is for Airway

Is the patient's airway open? If the patient speaks, the airway is at least temporarily open. If the patient is unconscious *and* if the patient's breathing sounds indicate difficulty moving air in and out, open the airway immediately with a **head-tilt/chin-lift** if cervical spine damage is *not* suspected, or a **jaw thrust** if cervical spine damage is suspected (see Chapter 4: Airway and Breathing). Check the airway for blockages or potential blockages, such as blood or foreign bodies. Anything blocking the airway must be removed immediately. Ask conscious patients to open their mouths while you take a quick look inside. Ask them to spit out anything in the mouth, such as gum, tobacco, or anything that could later become an obstruction. With the unconscious patient, you will have to manually remove obstructions and potential obstructions.

B Is for Breathing

With a conscious patient, ask her or him to take a deep breath. Is breathing painful? Is it difficult? Do you need to expose the chest now to check for life-threatening chest injuries?

If the patient appears unconscious and/or does not speak, while maintaining manual stabilization of the patient's head and with the airway opened, place your ear against the patient's mouth in order to **look** across the chest and upper abdomen for movement, **listen** for the sound of breathing, and **feel** for the brush of moving air against your ear. If the patient is wearing bulky clothing, your second hand placed over the lower chest/upper abdomen will allow you to feel for chest movement. Assess breathing for about 10 seconds. Is the patient moving air in and out of the chest? If not, you will have to begin breathing for the patient (see Chapter 4: Airway and Breathing).

You may find your patient in a position that makes it impossible to assess airway and breathing or in a position that makes breathing for the patient impossible (such as face buried in snow). In these cases you will have to reposition the patient. Even though the possibility of spinal cord injury may exist, assurance of breathing takes precedence. Kneeling beside the patient, control the head and neck as best you can by cradling the head with one hand. Straighten out the legs. Roll the patient gently onto one side, and then into a *supine* (face up) position. This is easier and safer for the patient if two or more people are available to perform the roll. With two or more rescuers, the one controlling the head should be in charge of the movement (see Chapter 8: Spine Injuries).

C Is for Circulation (and Bleeding)

With a conscious patient, check for a **radial pulse,** an indication of adequate circulation. The radial pulse is found on the inside of the wrist, on the thumb side of the arm, just **proximal** (the side toward the heart) of the wrist bones. With an unconscious patient, check for a **carotid pulse** on the side of the neck you are on, in the valley beside the trachea (windpipe). Do not reach across the trachea (windpipe) to check the other carotid pulse,

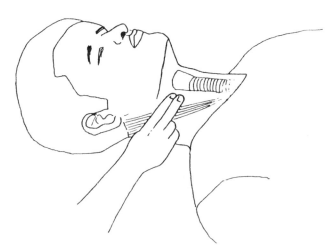

Figure 3-3: *Checking the carotid pulse.*

and do not try to check both carotid pulses at once. These maneuvers may partially block the trachea or reduce blood flow to the brain. If you find no carotid pulse, you will have to start CPR (see Chapter 5: Cardiopulmonary Resuscitation). You are not stopping to count the pulse at this point—you are making sure it is there.

If the patient's heart is beating, check for severe bleeding with more than a quick visual scan. Run your free hand, preferably a gloved hand, under the patient and inside of bulky clothing. Check your gloved hand for blood. If you see blood, check the wound. Bleeding can look deceptively serious and still be minor. Generally, only blood loss that is spurting or flowing heavily should be attended to in the initial assessment (see Chapter 6: Bleeding).

D Is for Disability

Since damage to the central nervous system—the brain and the spinal cord that runs down the neck and back—can cause permanent disability or death, the patient should be kept immobile, preventing further damage, as long as there is any suspicion of spinal involvement. "D" reminds you to assume there is spinal cord injury until the pace of the emergency has slowed and sufficient information gathered to allow you to make a decision to treat or not treat the spine.

If you haven't ascertained the MOI already, you can do so at this point by asking the patient what happened. "D" then may also stand for Decision, the decision to maintain manual control of the head and neck or not. If you decide the MOI could have possibly caused spine damage, keep the patient immobile until you have investigated further in the focused assessment.

Remember: You stop and fix threats to life, but any problems you discover that do not threaten life should be treated after the assessment is complete.

E Is for Exposure/Environment

Clothing will often hide injuries you need to be managing. If, at this point, your initial assessment has revealed a possibly serious injury, such as a bleeding site, remove and/or cut the patient's clothing away to expose the injury for a better look. Keeping in mind that in a wilderness situation the patient may need those clothes later, remove no more clothing than is necessary to expose the injury. With a conscious patient, tell him or her exactly what you are doing and why. Protect the patient's modesty as much as possible.

In extremes of cold, too much exposure of the patient can itself become a problem. Rely on your sense of touch through the clothing as much as possible. Look for convenient access points through winter clothing, such as zippers and Velcro closures.

Cold, heat, wind, rain, and other environmental conditions are almost always factors that require your consideration in the wilderness. Consider them early and attend to them early, if they are a concern.

STOP! COMPLETE A FOCUSED EXAM AND PATIENT HISTORY

When you have completed a thorough initial assessment, you have to make a decision about the further care of the patient. If the patient has any life-threatening problems that have not been managed in the field, or if the mechanism of injury suggests serious underlying problems, or if your intuition tells you more is going on than you are ready to handle, the patient should be on the way to the nearest hospital—in an ideal world. The focused assessment could then be performed during transport. Rapid evacuation, however, is rarely possible in the wilderness, which means two things: (1) Critical patients have a high mortality rate in the wilderness, and (2) you will usually have time to do a complete focused assessment before moving the patient.

The focused assessment has three parts: (1) a physical examination, (2) checking of vital signs, and (3) determining relevant medical history. The order may vary. If the patient is ill, for

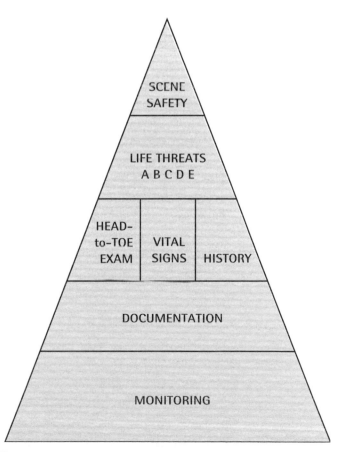

Figure 3-4: *The pyramid of patient assessment.*

instance, it is often more important to take the history first. Treatment for problems discovered in the initial assessment is immediate. Treatment for problems discovered in the focused assessment, unless they are ABCDE problems, can wait until the assessment is complete.

Perform a Physical Examination

If you haven't asked yet, the physical examination should begin by asking the patient for a *chief complaint*. This is done with a simple question, such as, "Can you tell me what hurts?" or, "Can you tell me why you need help?" This is important because the patient may have an obvious and severely angulated fracture of the lower leg but may respond to the question with, "I [gasp] can't [gurgle] breathe!" If the patient responds with two complaints, ask which is the more severe. If the patient is unable to answer, "altered level of consciousness" may be recorded as the chief complaint.

Knowing the chief complaint will give you another clue when you make your final decisions, as well as tell you what areas of the patient's body to be especially careful with during the exam.

The physical exam is a head-to-toe investigation during which you will **look** for, **ask** about, and **feel** for (LAF), as well as **listen** for and, sometimes, **smell** for clues. In medical terminology, you will *inspect* (look), *inquire* (ask), *palpate* (feel), and *auscultate* (listen). The patient *is* most of the questions, and the patient *has* most of the answers.

Look for bruises and other discolorations, bleeding, swelling and other deformities, or anything else out of the ordinary. Look for grimaces in response to pain from a patient with an altered level of consciousness. If a potential problem presents itself, take a look beneath clothing. Most medical care is performed at skin level.

Ask about pain (it hurts all the time), tenderness (it hurts only in response to being touched), or any unusual sensations.

Feel for unusual softness or hardness, rigidity, heat or cold, or anything out of the ordinary.

Listen for labored or unusual breathing sounds, grunts, or groans from the patient when you touch a specific spot.

Smell for unusual body odors, breath odors, or odors from clothing or the environment.

The Head-to-Toe Examination

Position yourself at the patient's shoulder. If you have someone who can work with you, have that person manually stabilize the patient's head during the exam if such immobilization is required. If you are alone, ask the conscious patient to refrain from moving his or her head. If you are alone with an unconscious patient, proceed with the examination, moving the patient as little as possible. Maintain a calm dialogue with the

patient, explaining what you are doing and asking for information. Even if the patient is unconscious, a calm, reassuring voice will often be of benefit. If the patient is seated, the same guidelines apply. Pressure from your hand and/or fingers during the exam should be firm but not overly aggressive.

Head and neck: Carefully remove a hat, cap, or helmet. Run your fingers gently through the patient's hair. If hair is matted to the head, leave it in place but look closely for damage. If you discover a depression in the skull, be very careful not to move the bone fragments (see Chapter 9: Head Injuries). Feel gently along the muscles and bones of the neck. Check for proper alignment of the trachea. Look for bruising behind and below the ears. Look for a medical identification tag at the neck, a tag that might relate important information such as allergies and preexisting conditions.

Face: If the patient is wearing sunglasses, carefully remove them to assess the eyes for damage. The pupils of the eyes should be approximately equal in size and responsive to light. Replace sunglasses if the patient is more comfortable with the glasses on. Apply gentle pressure to the bony structures of the face and jaw. You may ask the patient to clench the teeth to check for integrity of the jaw. Without moving the head, check the eyes, ears, nose, and mouth for damage or unusual fluids. Note any unusual breath odors, such as alcohol.

Chest and shoulders: Spread your hands widely over the shoulders, one at a time, including the clavicle and shoulder blade, sort of like a shoulder sandwich, and squeeze. Spread your hands over the rib cage and check for instability of the ribs during respirations and asymmetry of respirations. Ask the patient to take a deep breath. A normal chest should rise and fall easily and equally on both sides without pain. Check high and low on the chest wall. Press gently but firmly on the sternum (breastbone) with the side of your hand (see Chapter 10: Chest Injuries).

Abdomen: Press gently with one hand with the flat of your fingers, not your fingertips, on the four quadrants of the patient's abdomen. The four quadrants are the patient's upper right, upper left, lower right, and lower left, with the navel as the central point. Do not press on obvious injuries. Check for pain, guarding (the patient protecting the abdomen from your palpations), rigidity in abdominal muscles, and distention of the abdomen. These are usual indicators of internal damage (see Chapter 11: Abdominal Injuries). You may also, while checking the abdomen, run your hands under the patient's lower back as far as they will reach without moving the patient.

Pelvis: With your hands cupped over the iliac crests (pelvic crests), press gently downward, then inward.

Genitals: Unless damage is indicated, there is no need to check the genitals. You may wish to inquire, "Is there any reason I should check your genital area?" If checking the genitals is required, state your reason for checking and ask permission first, if possible. Try to have a rescuer of the same gender and/or witnesses when checking genitals, especially with minors.

Lower extremities: Take a look at both legs for signs of injury, such as angulations, protruding bones, legs shortened and/or rotated abnormally at the hip. Check each leg, one at a time, from hip to foot. You will need to remove the shoes or boots and socks to inspect the feet of an unreliable patient, but you do not typically need to visually inspect the feet of a reliable patient who denies altered sensations in the feet. In a cold environment, however, visually check the feet for frostbite. And at least slide your hand into the shoes to check for warmth on all patients. Be prepared to protect the foot as soon as possible from cold. Check to see if the patient has sensation in the feet and the ability to wiggle the toes. Impairment of sensation and/or mobility may indicate spinal damage. In all patients compare the strength in each leg by asking the patient to push and pull his or her feet against the resistance of your hands.

Upper extremities: Check each arm separately. Check to see if the patient has sensation in the hands and the ability to wiggle the fingers. Impairment of sensation and/or mobility may indicate spinal damage. In all patients compare the strength in each arm by asking the patient to squeeze your hands. Assess the hands for warmth.

Back and buttocks: When the patient is sitting or lying in a side position, assess the back prior to moving the patient. To make an accurate assessment of the back of a patient lying supine, you will need to roll the patient. This is typically an uncomplicated process with a patient who has not revealed a reason to suspect spine damage. You should, however, take care to not add further harm to injuries that have been revealed. Rolling a patient with a possible spinal injury requires much greater care (see Chapter 8: Spine Injuries), but it is necessary before you can complete the head-to-toe examination. When you have access to the back, walk your fingers carefully but fully down the length of the spine, and check the flanks for damage.

Note: Consider moving all patients onto an insulating pad at the end of the patient exam. At this point, having assessed for injuries from head to toe, you know what specific care to take during the move to the pad.

Check Vital Signs

Vital signs relate how well the patient's basic life support systems—nervous system, circulatory system, and respiratory system—are doing their jobs. An accurate measurement of a body's vital functions, a "set of vitals," do not tell you what is wrong with your patient, but they do relate how well your patient is doing. If your first set of vitals is grossly abnormal, whatever the injury or illness, your level of concern will skyrocket. In the wilderness, however, the second set of vital signs is often more important than the first set, and the third more important than the second. Whatever the injury or illness, a stable patient's vital signs stay the same; an improving patient or a deteriorating patient will be indicated by changes in vital signs.

When vital signs are measured and the *speed at which* vitals change are often critical factors in patient management. It is important, especially in the long-term care that a wilderness situation often demands, that you note and record the exact time that you take a set of vital signs. The time at which you take a set of vitals may indeed be considered a vital sign. Circumstances will dictate how often the signs are taken, but the change over time is the key to using vital signs in long-term patient care. One set of vitals is like a photograph while several sets are like a movie.

Constant monitoring of patients should be maintained until they are no longer in your care. To better ensure adequate monitoring of vital signs, it is best to have the same rescuer check each time. Only the same rescuer will be able to ascertain, for instance, changes in quality. And remember: Do not isolate on one vital sign, but take them as a part of the whole picture.

The vital signs are (1) level of consciousness (LOC); (2) heart rate (HR), or pulse; (3) respiration rate (RR), or breathing; (4) skin color, temperature, and moisture (SCTM); (5) blood pressure (BP); (6) pupils (P); and (7) body core temperature (T). During a patient's progress through an injury or illness, the first four vital signs (LOC, HR, RR, SCTM) will change early and sometimes, if the patient is seriously in trouble, very fast. The last three vital signs (BP, P, T) will change late and may indicate a more serious condition when and if they do change.

Vital signs, including time, should be recorded as soon as possible after they are measured.

Early-Changing Vital Signs

■ LEVEL OF CONSCIOUSNESS

The level of consciousness is a measure of a brain's ability to relate to the outside world. A normal level of consciousness allows a patient to answer in a way that tells you if he or she is oriented in person, place, time, and event (who, where, when, and what happened). LOC is often the first vital sign to noticeably change—and it is consistently the most difficult sign to assess accurately. It can be recorded by using the **AVPU scale,** a descending scale that assigns the patient a status ranging from "alert" to "unresponsive." But the AVPU scale alone will not provide all the information you want to note about the LOC (see "Communicating the Assessment of LOC," this chapter).

A for Alert: The patient seems normal and answers, with reasonable intelligence, questions about person, place, time, and event: Who are you? Where are you? Approximately what time is it? What happened to you?

In determining LOC, note that there are degrees of alertness. A patient who responds appropriately to all four questions is fully alert and should be classified with an alert-and-oriented-times-four status, or A+O x 4. A patient, however, may be awake but not fully alert.

LOC=A+O x 4: Patient answers all four questions correctly (person, place, time, and event).

LOC=A+O x 3: Patient answers three questions correctly (person, place, and time).

LOC=A+O x 2: Patient answers two questions correctly (person and place).

LOC=A+O x 1: Patient answers one question correctly (knows who he or she is).

It is also important to note that a patient can alter in level of consciousness without dropping in status. For instance, a patient may be described as A+O x 4 and calm, or as A+O x 4 and irritable or anxious. So it is important to describe any alterations from "normal" in the patient as well as his or her status on the AVPU scale.

V for Verbal: Any patient with a level of consciousness below "A" is considered an unconscious patient. The verbal patient (LOC=V) is not alert, but he or she does respond in some way when spoken to—a yell from the rescuer may stimulate a grimace, grunt, or rolling away from the noise. The patient may even follow simple commands.

You also need to note and describe the specific response of the patient after a verbal stimulus. A patient who turns toward you when you speak softly is at a different level of consciousness than a patient who grimaces slightly when you scream in his or her ear.

P for Pain: If the patient does not react to verbal stimuli, but he or she does react to being pinched or rubbed in sensitive areas, such as a pinch on the back of the arm or a knuckle rub on the sternum, the status of the patient's LOC is "P" (responds to painful stimuli). Pulling away from a pinch or sternal rub or pushing away the hand that pinches or rubs are appropriate responses to pain, as is a groan or moan. An inappropriate response to pain, such as curling toward a fetal position, indicates an even deeper level of unconsciousness.

Once again, it is important to note the specific response to a specific painful stimulus. For example, did your patient moan a little when you applied your pinch—or did the patient try to swap your hand away?

U for Unresponsive: A patient who does not respond to any stimuli is unresponsive (LOC = U).

Communicating the Assessment of Level of Consciousness

EXAMPLE 1

You approach a scene and the patient is sitting up, eyes open, looking around.
You say, "Hello, I'm Shana, a Wilderness First Responder. Can I help you?"
The patient responds, "Yes, I guess so."
"OK. What's your name?"
"Todd."
"Do you know where you are, Todd?"
"I'm in Lander."
"Do you know what day and time it is?"
"Sure, it's Saturday, and it's lunch time."
"What have you been doing?"
"I was enjoying the scenery and getting ready to eat lunch."
Report this: LOC-A+O x 4. Patient awake, alert, and oriented times four.

EXAMPLE 2

You approach a scene, and the patient is lying on his side, motionless.
You say, "Hello, I'm Shana, a Wilderness First Responder. Can I help you?"
The patient opens his eyes and responds, "Yes, please save me."
"OK. What's your name?"
"Todd."
"Do you know where you are, Todd?"
"I'm in Lander."
"Do you know what day and time it is?"
"Umm . . . no . . . "
"What have you been doing?"
"Uh . . . I'm not sure."
Report this: LOC-A+O x 2. Patient awake, alert, and oriented times two (oriented only to person and place, not to time or event).

EXAMPLE 3

You approach the scene, and the patient is lying on the ground, eyes open, unable to stand.
You say, "Hello, I'm Shana, a Wilderness First Responder. Can I help you?"
The patient moves a bit, but without purpose. He mumbles something incomprehensible. He does not respond to your question.

You say loudly, "Hello! Are you OK?"

The patient again moves and mumbles, but nothing is understandable.

A few minutes later, you try to logroll the patient onto an insulated pad. He resists the roll to the point you have to restrain him. He continues to mumble incomprehensibly.

Report this: LOC-V. Patient awake, but speech incomprehensible. Patient does not respond to questions, resists movement, and requires restraint.

EXAMPLE 4

You approach a scene, and the patient is lying on his side, motionless.

You say, "Hello, I'm Shana, a Wilderness First Responder. Can I help you?"

The patient does nothing.

You move closer, and in a loud voice say, "Hello! Wake up!"

The patient does nothing. He is not asleep.

Again you say, "Hello! Wake up!"

The patient opens his eyes, but says nothing.

You say loudly, "What's your name?"

The patient mumbles, "Todd."

"Do you know where you are, Todd?"

The patient only stares at you with a blank expression.

"Do you know what day it is?"

Again, all you receive in response is a blank expression.

Report this: LOC-V. Patient not awake, can be aroused with a verbal stimulus. When awake the patient knows his name, but is otherwise disoriented.

EXAMPLE 5

You approach a scene, and the patient is lying on his side, motionless.

You say, "Hello, I'm Shana, a Wilderness First Responder. Can I help you?"

The patient does nothing. He is not asleep.

You move closer, and in a louder voice say, "Hello. Wake up!"

Again, the patient does nothing.

You rub his sternum with your knuckles, applying a painful stimulus.

The patient moans and says something incomprehensible. His eyes remain closed. He moves slightly.

Report this: LOC-P. Patient not awake, can be aroused only with pain. Patient moans, says something incomprehensible. Eyes remain closed.

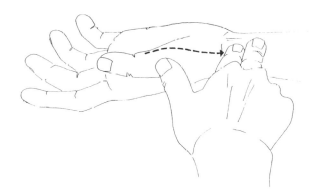

Figure 3-5: *Checking the radial pulse.*

■ HEART RATE

Heart rate may be taken at any place where you can palpate a *pulse*, a pressure wave created by the beating of the heart, a wave of expansion and contraction within an artery as blood rushes through. You can feel a pulse anywhere an artery passes over a bone or near the surface of the body. In the initial assessment you checked for a pulse to determine if the heart was beating with general adequacy. Now you are interested in the *rate, rhythm,* and *quality* of that pulse—more specific information about how well the patient's heart is functioning. At this point in your assessment, it is best to palpate the radial pulse in all patients. The radial pulse is easy to access, and it's more reassuring to a conscious patient to hold a wrist rather than having your fingers pressed into the carotid. It also tells you, if you find a strong radial pulse, that the patient's blood pressure is adequate (see "Blood Pressure," this chapter).

Rate: This is the number of beats per minute. The normal range in an adult can vary greatly, from 50 to as much as 100, although the range usually falls between 60 and 80. Heart rate typically is higher in children, sometimes reaching a norm of 160 in an infant. Counting for 1 full minute is most accurate, but a count for 15 seconds and multiplying by four is acceptable and more often used. In the absence of a watch, take a pulse anyway. At least you can assess the rhythm and quality.

Rhythm: There will be either a clocklike regularity or a sporadic irregularity.

Quality: This refers to the force exerted by the heart on each beat. The force is often best judged by its relation to previous pulse checks. A "normal" quality is strong, a "thready" pulse is weak and an indication of inadequate circulation, and a "bounding" pulse is abnormally forceful.

Note: An absence of a radial pulse is cause for alarm, but delay panicking until you check the other arm. The cause might be an injury to the arm or shoulder.

RESPIRATORY RATE

Breathe in and out once, and you've completed one respiration, or one breath. The *rate, rhythm,* and *quality* of respirations tell you how well a patient is moving life-sustaining air in and out. Breathing is typically effortless; if you notice that a patient is laboring to breathe, let it serve as an indication that something may be wrong. Be aware, however, that patients alerted to the fact that you're checking respirations may voluntarily alter their breathing pattern. Keep your assessment a secret whenever possible by lightly placing one of your hands on the chest and/or upper abdomen (or simply watching) and counting the movements without telling the patient you're counting.

Rate: This is the number of breaths per minute. An adult normally breathes 12 to 20 times per minute. Children and infants breathe faster. Respiratory rates are most accurate if taken for 1 full minute or, at least, for 30 seconds and multiplying by two—but you can count for 15 seconds and multiply by four.

Rhythm: "Normal" breaths are even and regular, with exhalations taking slightly longer than inhalations. Check for irregularities in the pattern, which may indicate a respiratory problem.

Quality: Breathing is usually quiet and effortless, easy and unlabored. Abnormal breathing might include unusual shallowness, unusual depth, labor, pain, noise (snores, wheezes, gurgles, gasps), and a flaring of the nostrils.

SKIN COLOR, TEMPERATURE, MOISTURE

Even though skin is on the outside, away from the body's core where vital life processes are centered, it may give pertinent information on the general well-being of the patient by giving evidence of changes in vital body processes. The skin often reflects the condition of the core.

Color: Pink is the color of skin in the nonpigmented areas of the body—the lining of the eye, inside the mouth, and fingernail beds. It may be difficult to detect subtle changes in darker complexions, but the overall color of the skin is also an indicator. The size of the blood vessels near the skin, in most cases, determines skin color. Abnormal skin appears red, white, blue, or yellow. *Vasodilation* (widening vessels) produces a red, flushed color and indicates problems such as fever or hyperthermia. *Vasoconstriction* (narrowing vessels) produces white, pale, or blotchy skin and may indicate shock or hypothermia. A blue hue, called *cyanosis,* shows a lack of oxygen in the blood. Yellowish skin, called *jaundice,* may indicate liver failure.

Temperature: Skin is relatively warm in a healthy person. Keep in mind that healthy skin may feel cold on a cold day in the wilderness or hot if the patient has been recently exercising. The temperature of the skin will not tell you anything exact about the patient's core temperature, but it may indicate whether or not the temperature is unusually high or low. You will get a more accurate check of skin temperature if you use the back of your hand instead of your palm or fingers. Check skin temperature inside clothing, not on the exposed face or hands.

Moisture: Normal skin, although slightly moist, feels relatively dry. Wetness from excessive sweating may be evident, but if the skin is cool and moist, it may indicate shock. Hot, dry skin may indicate hyperthermia.

Late-Changing Vital Signs

BLOOD PRESSURE

Blood pressure, the pressure of pulsing blood against arterial walls, is one of the most important vital signs, but one in which you will note changes *after* other vital signs have changed. You should know (1) how to get a BP with a sphygmomanometer and stethoscope, (2) how to get a BP with a sphygmomanometer only, (3) how to estimate a BP without instruments, and (4) what that information tells you about the patient.

A **sphygmomanometer,** or cuff, is the instrument used to measure blood pressure. The cuff itself is a rubber bladder inside of a cover. The bladder has two tubes leaving its middle; one connects to a gauge, the other to the rubber bulb that inflates the bladder when squeezed. Bladder sizes vary: adults must have their pressure taken with an adult-sized cuff, children with child-sized cuffs, and infants with baby cuffs, or else the readings will be inaccurate. The cuff is wrapped snugly around the patient's arm, about 1 inch above the elbow, with the two tubes centered over the *brachial* artery, the artery that passes through the region of the elbow on the inside of the arm. Before taking the blood pressure, make sure that the patient's arm is relaxed and that the cuff is approximately on the same level as the patient's heart.

Attached to the cuff is an aneroid gauge, or manometer. The mercury-gravity gauges used in hospitals are more precise but impractical in the field. Inside the aneroid gauge is a metal bellows that expands or collapses with changes in pressure. Those changes are reflected on the dial of the gauge and are calibrated into millimeters of mercury (mmHg) even though mercury is not used in aneroid gauges, so that everyone talks the same language. Advantages of the aneroid gauge include portability; disadvantages include susceptibility to inaccurate readings with major weather changes.

Most commonly, the cuff is used in conjunction with a stethoscope. Many stethoscopes have a two-sided patient end, one flat and called the diaphragm, and the other slightly rounded and smaller and called the bell. Use the diaphragm side when taking a blood pressure, and, if there are two diaphragms, use the larger one. (The bell is usually used for listening for specific respiratory or cardiac sounds.) Be aware that a stethoscope with a diaphragm and a bell only works via one side or the other, and you must test the side you are using by tapping it gently once the ear pieces are in place to know for sure.

With the gauge in view, the rescuer should palpate the radial artery, and then, with a finger on the radial, squeeze the bulb with the other hand and inflate the cuff until the radial pulse disappears plus about 20 mmHg of pressure more. Now the rescuer is ready to take a BP by *auscultation* (listening). Place the diaphragm of the stethoscope over the brachial artery on the skin of the patient. Begin deflating the cuff at a rate of approximately 2 to 4 mmHg per heartbeat—it is important that the needle on the dial drop slowly. The sounds heard through the stethoscope relate the blood pressure. The needle will be pointing to the *systolic pressure* when the first sound is heard. Systolic pressure is the force exerted against the arterial walls by the blood when the heart's left ventricle contracts. A normal systolic is 100 plus the age of the patient, up to 150 mmHg. Blood pressures in women are usually a little lower. A systolic reading above 150 is often considered high blood pressure. When the last sound is heard, the *diastolic pressure* is read. Diastolic pressure is the blood's pressure against the artery when the heart is at rest. A normal diastolic range is 60 to 90 mmHg, with anything above 90 generally considered hypertension, or high blood pressure.

Blood pressure is recorded with the systolic over the diastolic: 120/80. The difference between the systolic number and diastolic number is called the *pulse pressure*. Pulse pressures normally range from 35 to 50 mmHg and are a measure of the pressure of the pulse wave.

Note: Although a blood pressure reading often measures a couple of mmHg higher on the left arm, the difference between left and right arm is insignificant. Take a BP on the arm that's most available. If a second reading is necessary, wait at least 15 seconds before reinflating the cuff. Some patients will be in pain, or nervous, and their apprehensions will alter their pressure. The rescuer needs to be a calming influence.

Without a stethoscope but with a cuff, a systolic pressure can still be obtained by *palpation*. The process is this: Palpate the radial pulse, inflate the cuff, maintain contact with the point of the radial pulse during deflation, and note the position of the needle on the dial at the first palpable return of pulse. That's the systolic pressure. This offers a big advantage in a cold environment since a relatively accurate BP can be obtained without removing clothing. You cannot obtain a diastolic pressure without a stethoscope. Systolic blood pressure by palpation is recorded as the number over P: 120/P.

Without a stethoscope or cuff, as often occurs in the wilderness, it is possible to guess blood pressures by palpating pulses. This may be called blood pressure by estimation. See the chart of suggested estimations.

Suggested Blood Pressure Estimations by Palpating Pulse

If pulse is present at:	Systolic BP is greater than:
Carotid (neck)	50 mmHg
Femoral (groin)	60 mmHg
Brachial (arm)	70 mmHg
Radial (wrist)	80 mmHg
Pedal (foot)	90 mmHg

It is almost always falling blood pressure that is of greatest concern in the wild outdoors. When the BP drops below 70 mmHg, the kidneys lose their ability to remove waste products from the blood. As these waste products accumulate, the blood becomes toxic. When the BP drops below 50 mmHg, the other internal organs start to fail, including the brain.

If a patient loses his or her brachial pulse, the BP is so low that he or she is in trouble. When the carotid pulse is lost, the patient is in serious trouble—or dead.

Other indications of a dropping blood pressure may include a decline in level of consciousness and dizziness or light-headedness if the patient sits or stands up (see Chapter 7: Shock).

Note: Since an adequate blood pressure in most patients is indicated by an alert level of consciousness, a strong radial pulse, and relatively normal skin color, temperature, and moisture, as a general rule in wilderness medicine, LOC=A+O x 4 *plus* a strong radial pulse *plus* normal skin *equals* adequate BP. Without a means to measure BP, it may be recorded on a SOAP note as "strong radial pulse" (see "Write a SOAP Note," this chapter).

■ PUPILS

Normal pupils are equal in size and round in shape, and they have a normal *pupillary response* to light: They *constrict*, or grow smaller, when exposed to light, and they *dilate*, or grow larger, when the light is reduced. You can check a patient's pupils by shining a light, such as a flashlight, briefly into each eye, one at a time, and checking the response. In sunlight, you can cover an eye with your hand, wait a moment for the pupil to dilate, then uncover the eye and observe the response. Normal pupils may be recorded as pupils-equal-round-and-reactive-to-light: PERRL. Pupils that do not respond and pupils that respond unequally when one is compared with the other are significant discoveries (see Chapter 9: Head Injuries).

■ TEMPERATURE

First and foremost, serious changes in core temperature are indicated by signs (other than the number on a thermometer) and symptoms that make an actual measurement low on the priority list for the WFR. And even with a thermometer you will often get inaccurate results. In a patient suffering from hypothermia, for instance, oral thermometers sometimes register temperatures *lower* than the patient's core temperature, due to a cold mouth from breathing cold air. In a heat-stressed patient who has been exercising vigorously, a rectal thermometer reading may be substantially *higher* than the core, due to the heat generated by muscular activity. But, in case you want to use a thermometer, remember you are looking for changes from the normal core temperature of around 98.6°F (37°C). Three areas of the body—the mouth, armpit, and rectum—provide relatively convenient access when you are using a thermometer.

Mouth: Oral thermometers provide the easiest access to an actual measurement. Your patient does not have to expose any body parts to the environment, and most patients are comfortable having the thermometer in place. Standard glass thermometers have found a home in many first-aid kits, but they break easily. Exposure to excesses of heat or cold can alter the accuracy of a standard glass thermometer, and cheap ones can lose their accuracy over time even under controlled conditions. Electronic digital thermometers are more durable and probably more accurate, but they only work as long as the battery does—and they are not a practical addition to most wilderness medical kits.

It is often suggested, sometimes strongly, that your thermometer should be one that reads low temperatures, sometimes called a hypothermia thermometer, because many standard instruments do not register in the hypothermia range. There are, however, many indications of hypothermia far easier to check and far more reliable than a temperature measurement (see Chapter 16: Cold-Induced Emergencies). Thermometers are more useful when checking for rises in core temperature.

The thermometer needs to be held in the mouth, under the tongue, for 3 to 5 minutes. The patient needs to remain still.

Armpit: The same glass thermometers used in the mouth can be used in the armpit, with the same pros and cons. The thermometer may need to be left in place for up to 10 minutes. Axial (armpit) temperatures, in the wilderness especially, are at best a poor guess at core temperature and are not recommended.

Rectum: Anywhere you can stick a regular glass thermometer, you can stick a rectal thermometer, and vice versa. Glass rectal thermometers are typically made of thicker material, which means you'll have to leave them in a bit longer to get an accurate measurement. Digital thermometers with long flexible probes work well for taking a rectal temperature. Other than the heat-stressed patient mentioned earlier, you should get a fairly accurate estimate of core temperature with a rectal measurement. As you can easily imagine, a substantial amount of your patient will get exposed to the environment, and most patients will be made uncomfortable by the procedure. If you take a rectal measurement, the thermometer need only go in 2 to 4 inches.

Note: Temperatures can be taken in the ear near the eardrum with a special thermometer made for that purpose. These tympanic measurements are generally considered unreliable in the wilderness.

Vital Signs (Within Normal Limits) for an Adult

LOC	A+O x 4
HR	50–100, regular, strong
RR	12–20, regular, unlabored
SCTM	pink, warm, dry
BP	90–150/60–90
P	PERRL
T	98.6° F (37° C)

Take the Patient's Medical History

As you practice, you will gain pride in your ability to perform a patient exam and take a set of vital signs, but the tricky part of assessment, and as much as 80 percent of your final verdict, will come from information gathered as you interview your patient. Even with modern medical technology at its current height, physicians often base their assessments on a careful patient history.

Approach the interview with the same calm, confident, competent, compassionate attitude with which you initiated contact with the patient during the initial assessment. An aura of calmness surrounding the scene will often do more for the sick and injured than all the splints and aspirin you can throw at patients.

If you haven't already, establish a relationship with your patient. It is more effective to say sincerely, "I know you must be afraid and in pain, but we'll make you as comfortable as possible," as opposed to panting, "You'll be OK. You'll be fine."

Create a positive situation. Say, "Are you more comfortable sitting up or lying down?" Don't say, "Which hurts more?"

Be enthusiastic without being a cheerleader. Be kind without being nauseating. Be honest without saying everything you're thinking. Beware of the tone of your voice. Beware of your patient's tendency to take the slightest offhand comment as truth. Tell your patient what you know, but avoid speculation and absolutes. Discuss with your patient the possible plans of action.

SAMPLE

A critical mnemonic for gathering patient information is **SAMPLE:**

Symptoms: "Describe what you're feeling. Pain? Headache? Dizziness? Nausea? Stomachache? Hot? Cold?" You need to know the sensations the patient is feeling.

Allergies: "Are you allergic to anything you know of? Foods? Drugs? Animals? Pollens?" You need to know if the patient contacted something that could cause an allergic reaction, and you need to prevent the patient from contacting something that could cause an allergic reaction. You also need to know how serious the allergic reaction could be.

Medications: "Are you taking anything currently? Over-the-counter drugs? Prescription drugs? Recreational drugs? Alcoholic beverages? Herbal remedies? Did you skip taking your medication? Did you take more than usual?" You need to know if medications are causing or could later cause a problem.

Pertinent medical history: "Is there anything possibly relevant I should know about your health? Are you seeing a doctor for anything? Heart problems? Lung problems? Stomach problems? Seizures? Diabetes? Has anything like this ever happened before?" You need to know a patient's medical conditions that could be causing or may later cause a problem.

Last intake/output: "When did you eat or drink last? What? How much? When did you urinate last? Was it unusually yellow? When was your last bowel movement? Did you notice blood or anything unusual? Have you vomited?" You need to beware of what is currently in a patient's stomach, and you need to be aware of a patient's current and future needs for food and water. And you need to recognize when urination and/or bowel movements are indicative of something unusual.

Events: "What caused or preceded this illness or injury? Have you been feeling OK the last few days? Has anything unusual happened that could be related to this event?" For example, when you know the patient fell out of a tree, find out why the fall occurred. And finally, "Is there anything else I should know?" You need to know if events leading to the problem are a part of the problem, and you need to leave no wilderness medicine stone of information unturned.

OPQRST

When the patient has specific symptoms—sensations felt by the patient and not signs you measure—especially if the patient is ill, another mnemonic for gathering information is **OPQRST.** **Note:** When the patient is ill, the MOI is sometimes referred to as the history of present illness (HPI).

Onset: "What initiated the chief complaint? Did it come on suddenly or gradually?"

Provokes or palliates: "Does anything, such as changing position or taking a deep breath, make it better or worse?"

Quality: "Describe the pain or discomfort? For example, is it sharp or dull, constant or erratic."

Radiates or refers, region: "Does the discomfort radiate or refer (move from one body part to another)? What is the region [location] of the pain?"

Severity: "How bad is the pain on a scale of 1 to 10 with 10 being the worst pain imaginable?"

Time: "How long has the pain or discomfort been going on? When did it start?"

Note: If your patient is unconscious, you have to become even more of a detective in your search for clues. Is the patient wearing any medical alert tags on the neck, wrists, or ankles? Is there any information in his or her pockets? What can you learn from witnesses? Is there anyone around who knows the patient? What evidence does the scene hold for you? You need to figure out, if possible, why the patient is unconscious (see Chapter 25: Neurological Emergencies).

STOP! DOCUMENT AND REPORT THE EVENT

Documentation should never interfere with adequate patient care. Documentation, however, is critical. Most importantly, it helps you provide better patient care because you will forget things. The patient deserves to have any physician who later takes over know the details of what happened and what was done. It helps a program, should you be working for one, plot accidents and illnesses and take preventive steps in the future. And it supports you in case of a legal issue (see Chapter 2: Legal Issues in Wilderness Medicine). Even if you're not sure that what you're doing is absolutely correct, you want to record all your efforts on the patient's behalf. It shows you did your best. Documentation also provides an essential message to others if you decide to send out for help. Do not fear documenting too much. Seven years from now you should be able to recreate the event accurately from your document.

With the proliferation of cell phones, radios, and other means of electronic communication, it is likely you will have to give your documentation verbally. In the following sample SOAP note, you can see what a verbal report might "sound" like—but remember to send no specific patient names over a phone or radio.

Write a SOAP Note

When time allows, or if there are spare hands at the scene, all the information gathered during the assessment, and what was done for treatment, and what will be done for the patient, should be *recorded on paper*. The **SOAP** format is a convenient and widely acceptable way to document the event.

Subjective: What happened (MOI/HPI) to whom (include name, age, and sex), where, and when? What is the chief complaint? OPQRST? It is best, as much as possible, to put this information in the patient's words.

Objective: How was the patient found, in what position? What did the physical exam reveal? What are the vital signs, and how did they change over time? What is the pertinent medical history? SAMPLE? Are there pertinent negatives, such as "Patient exam revealed no pain or tenderness."

Assessment: What are the possible problems?

Plan: What are you going to do? Do you have a plan for every assessment you've listed? You may also wish to list problems that you anticipate. What changes, in other words, might you expect over time?

STOP! MONITOR THE PATIENT'S CONDITION

Patient assessment is an ongoing process. You should continue to monitor, to reassess, until the patient is no longer in your care.

Conclusion

You notice a backpack standing open near the patient sprawled on the banks of the Green River and a pair of hiking boots near the pack. Your eyes turn to the young man's feet—he is wearing climbing shoes. The scene appears to be without immediate hazards. You dig your protective gloves out of your first-aid kit.

As you approach the patient, he tries to sit up. You ask him to lie still, and you place a hand on his head. A few questions, and you learn his fall from the sandstone cliff was short, perhaps no more than 6 feet. His chief complaint is pain in his left ankle. He bit his lip, thus the smear of blood. The patient denies striking his head. Other than the left ankle and lip, your head-to-toe examination reveals nothing else. With support to the ankle, he will probably be able to walk out on his injury. Vital signs are all within normal limits. There is no medical history immediately pertinent to this patient other than your discovery that he has had nothing to drink for several hours.

Despite the absence of Mr. Holmes, your Sherlockian approach to assessing the young man has allowed you to deduce that no serious injury has occurred. Indeed, by the time your assessment has ended, and he has downed a liter of water, he chats amiably about how fine most of this wilderness trip has been.

What if more serious problems had been discovered during your assessment? As you work your way through this book, you will gain knowledge and skills necessary to deal with specific emergencies.

A Sample SOAP Note

April 1, 2004

Patient: John Doe

SUBJECTIVE

I have a 22-year-old male who fell approximately 20 feet while rock climbing on Cathedral Ledge. Patient states: "I pendulumed back first into the rock wall." Patient appears reliable. Patient denies wearing a helmet. Patient denies loss of consciousness. Patient states: "My head hurts." Pain described as a constant "5" on a scale of 1 to 10.

OBJECTIVE

Patient was found lying on his back at the base of a route called "They Died Laughing." Physical exam revealed pain on palpation in back of head, neck, and upper back of rib cage near the backbone. No bruising, deformities, or wounds noticed. Good circulation, sensation, and motion noted in all four extremities. No other injuries were found.

Time	6:05 PM
LOC	A+O x 4
HR	84, reg, str
RR	14, reg, easy
SCTM	pink/warm/wet
BP	str. radial pulse
P	PERRL
T	did not check
Symptoms	Pain described as dull ache "like a bad headache" with no radiation.
Allergies	Patient states: "I'm allergic to penicillin. I almost died the last time I took it." Last penicillin reaction was "more than 10 years ago." Patient denies taking penicillin today.
Medications	Patient denies.
Pertinent medical history	Patient denies.
Last intake/output	Patient claims granola bar, cheese, one-half liter of water at noon approx. Urination and defecation are "normal" according to patient. Patient denies vomiting.
Events	Patient states: "I slipped while climbing because a bee scared me."

ASSESSMENT

Possible head and/or neck injury. Possible chest and/or spine injury. Possible loss of body temp.

PLAN

Full manual head and spine immobilization on foamlite pad in sleeping bag. Improvise a cervical collar. Monitor closely for changes. Send a team of "runners" with full documentation and location to get help.

CHAPTER FOUR

Airway and Breathing

You Should Be Able To:

1. Describe the basic anatomy of the human airway.
2. Demonstrate ways to open an airway, including head tilt/chin-lift, jaw thrust, and tongue-jaw lift.
3. Demonstrate how to clear an obstructed airway for an adult.
4. Demonstrate rescue breathing for an adult.

It Could Happen to You

Few places on Earth can compare to the beauty and peace of your campsite near Grave Lake in the Wind River Mountains of Wyoming. You've stepped away from the rest of the group, the group you've been hired to lead, to spend a few minutes alone near the water while dinner preparations near completion. You hear their laughter drifting from beneath the trees. Some of those guys are real jokesters. The faded yellow sun dissolves into rose and gray, and your contentment deepens . . . then a scream from camp sends the hairs up on the back of your neck.

Rushing to the campfire, you find a member of the group lying unconscious, not breathing.

You hear a voice sob, "He choked on something!"

Nothing will strike more fear into a patient than the inability to breathe adequately. Nothing will strike more fear into you than a patient who is not breathing. Of all the skills performed by a first responder, few, if any, will ever compare in importance to those involving a human's airway. If you can open an airway, clear an airway, maintain an airway, and breathe for someone not breathing, you possess skills that may make the difference between life and death.

Air moves in and out of an adult human body, a process called *breathing* or *respiration,* at an average rate of once every 5 seconds. The process has two parts: *inspiration,* or inhaling, and *expiration,* or exhaling. Numerous incidents can cause the breathing process to cease, or *respiratory arrest.* The more common causes are upper airway obstruction, drowning, electric shock, suffocation, heart failure, head injury, seizures, drug overdoses, and severe allergic reactions.

Take away the process of breathing a few minutes, and the heart begins to weaken. Without a return of breathing, the heart will sputter and stop—the moment of *cardiac arrest,* which is also the moment of *clinical death.* If the heart does not start again in an average of 4 to 6 minutes, irreversible brain damage occurs and the brain shuts down, the beginning of *biological death.* Although short time spans separate the cessation of breathing and cardiac arrest and the moment of clinical death and biological death (see Chapter 5: Cardiopulmonary Resuscitation), the first responder must intervene within these small windows of opportunity.

Basic Anatomy of the Airway

Air enters a human body either through the nose into the *nasopharynx* (the upper pharynx) or through the mouth into the *oropharynx* (the central portion of the pharynx). The *pharynx,* the air passageway from nose to *larynx,* or "voice box," at the top of the *trachea* (windpipe), also provides a passageway for food to the *esophagus,* or "food tube." The trachea, visible at the front of the neck, is held open by C-shaped cartilage rings, with the "C" opening toward the back of the neck while the esophagus lies flat until something is swallowed. When something is swallowed, a flap of tissue called the *epiglottis* closes the opening to the trachea, or *glottis,* to prevent solid matter from descending the trachea. Behind the *sternum* (breastbone), the trachea splits into two *bronchi,* one going left and one going right. The bronchi split into *bronchioles* of ever-decreasing size until each one comes to a dead end at a grapelike cluster of *alveoli,* the air cells of the lungs where oxygen from the air is exchanged for carbon dioxide from the blood. For an airway to be open, it must be open from the mouth and/or nose all the way down to the alveoli.

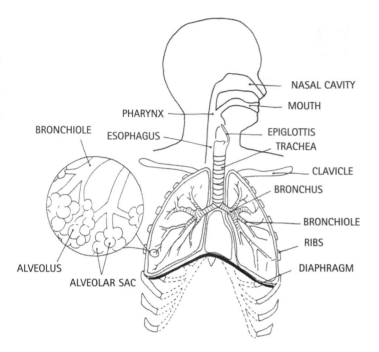

Figure 4-1: *Anatomy of the airway.*

Assessing the Airway

The scene is safe, and you have taken appropriate body substance isolation precautions. The patient appears unconscious. You are now at the A for Airway part of the ABCDE of an initial assessment. Ask the patient, "Are you OK?" or something similar, in a loud voice. If there is no response to your voice, attempt to get a response to pain (see Chapter 3: Patient Assessment). You are not after a specific response *per se,* but after *some* response that gives you an immediate idea of the patient's level of consciousness—a response such as a word, a groan, a movement, or an opening of the eyes.

Remember, with any suspicion of head or spine injury, place your hand gently but firmly on the patient's head prior to asking for a response. **Do not shake a patient who may have a head or spine injury** (see Chapter 8: Spine Injuries). Unless otherwise indicated, the rest of this chapter assumes the patient has *no* head or spine injury.

With no response, you should now call for help. In the urban environment, you would be dialing 911. In the wilderness, you might attract another hiker or your tent mate.

Opening the Airway

With any patient who is not breathing or a patient with labored and/or noisy breathing, you must do all you can to acquire an open airway. Usually, this is best accomplished if the patient is in the supine position, on the back with face up and head in line with the back. A patient in a side position or in the prone (on the stomach) position who requires airway management will need to be rolled **as a unit** into the supine position. Rolling as a unit means to roll the patient's head, shoulders, and trunk simultaneously (see Chapter 8: Spine Injuries).

In an unconscious patient already on his or her back, the most common cause of airway blockage is the tongue. The tongue, a muscle, can relax and fall back far enough to prevent the passage of air. Sometimes simply moving the head and neck into normal anatomical alignment will open the airway enough to save a life. If you move the head into normal anatomical alignment—the head in line with the midline of the body—do so slowly and gently, without forcing past resistance.

Because the tongue is attached to the lower jaw, one of three maneuvers can be used to open an airway that remains closed after aligning the head and neck, three maneuvers that lift the lower jaw—and that move the tongue enough to allow air to pass.

■ HEAD-TILT/CHIN-LIFT MANEUVER

If you do *not* suspect head or neck injury, use the **head-tilt/chin-lift** to open an airway. Kneel beside the patient's head. With one hand on the patient's forehead, apply enough pressure to tilt the head back. At the same time lift the patient's chin with the fingers of your other hand. *Be careful to not push in on the soft tissues under the chin*—doing so could further obstruct the airway. Continue to tilt and lift until the head is in an extended position, the chin pointed toward the sky. With a small child or infant, tilt the head only into a neutral position, the neck only slightly extended. *Full extension could partially close the airway of a small child or infant.*

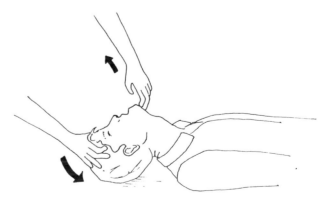

Figure 4-2: *Head-tilt/chin-lift maneuver.*

■ JAW-THRUST MANEUVER

If you *do* suspect head or neck injury, use the jaw thrust to open an airway. One way to do a jaw-thrust starts by kneeling near the top of the patient's head, facing the patient's feet. With your elbows on the ground, place your hands on the sides of the patient's head, your thumbs on the cheekbones, and two or three fingers of each hand at the corner of the patient's jaw where it angles between chin and ear. Lift the jaw with your fingers, using counter-pressure from your thumbs on the cheekbones.

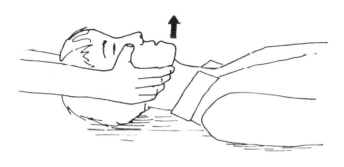

Figure 4-3: *Jaw-thrust maneuver.*

■ TONGUE-JAW-LIFT MANEUVER

To open an airway to remove an object in the mouth, the tongue-jaw lift may be used. Grasping the tongue and lower jaw between your thumb and finger, lift the jaw up and out.

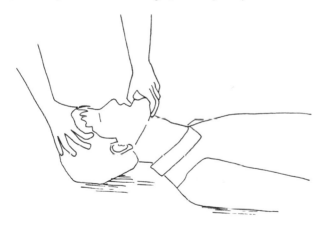

Figure 4-4: *Tongue-jaw-lift maneuver.*

Assessing Breathing

With the airway open, look inside and check for visible obstructions—and remove any you find. Then look, listen, and feel for air movement with your ear and cheek near the patient's mouth. Your head should be facing down the midline of the patient's body, allowing you to watch the chest for movement. Check for about 10 seconds. Adequate breathing will be indicated by the rise and fall of the chest and/or upper abdomen, the sound of air

moving through the mouth, and the feeling of air from the mouth and nose blowing against your cheek. A breathless patient requires either removal of a nonvisible foreign-body airway obstruction, or rescue breathing, or both.

Foreign-Body Airway Obstruction: Conscious Adult

The most common cause of airway obstruction in a conscious patient is food, with meat causing the blockage most often. The obstructing object, whatever it is, usually lodges in the area of the glottis, and may form a partial or a complete obstruction.

A patient with a *partial airway obstruction* can get air past the obstruction and cough in attempts to clear the airway. This person requires no first aid and, indeed, should be allowed to cough his or her way to an open airway. This person, however, should not be left alone in case the obstruction shifts and becomes complete.

A patient with a *complete airway obstruction* cannot get air past the obstruction, cannot speak or cough, and cannot breathe. A patient with a complete obstruction may make weak breathing sounds and/or high pitched wheezing sounds but will not be able to adequately move life-sustaining air in and out. This person typically appears extremely panicked, and he or she may be grasping the throat, involuntarily displaying the "universal distress signal" of choking. This person requires immediate, life-saving help in the form of *abdominal thrusts*, sometimes called the Heimlich maneuver.

Before beginning abdominal thrusts, quickly identify yourself to the patient as someone who can help. Ask, "Are you choking?" If the answer is a nod in the affirmative, yes, then ask, "Can you speak?" If a shake of the head indicates no, then do the following:

If the patient is standing or sitting:

1. Stand behind the patient.
2. Wrap your arms around the waist of the patient, keeping your elbows out and away from the patient's ribs.
3. Make a fist and place it, thumb in, on the midline of the patient's abdomen, just above the navel, well below the *xiphoid* process, the point of cartilage extending below the bottom of the sternum.
4. Grab your fist with your other hand.
5. Pull quickly in and upward with a powerful motion, a motion intended to force the object out of the airway.
6. Repeat the abdominal thrust until the airway clears or the patient goes unconscious.

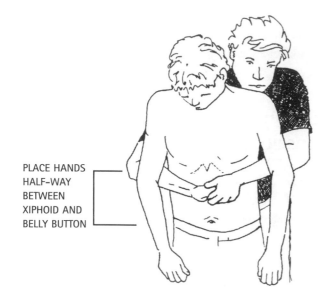

PLACE HANDS
HALF-WAY
BETWEEN
XIPHOID AND
BELLY BUTTON

Figure 4-5: *Heimlich maneuver.*

If the patient is obese or pregnant:

For obese or pregnant choking persons, follow the above directions for the abdominal thrust *with important differences:* Perform the thrust against the patient's chest with your hands on the *middle* of the sternum, and direct the thrust straight back and not in and up.

Figure 4-6: *Chest thrust on pregnant patient.*

If you are choking and alone:

If you suffer an upper airway obstruction while alone, perform the abdominal thrust on yourself. If this fails, position your abdomen against a firm object, such as backpack frame, log, and drop your weight against the object.

If the patient goes unconscious:

When doing abdominal thrusts on a conscious adult, position yourself during the thrusts—legs spread, one leg back and the other leg between the patient's legs—to better control the patient's body weight should he or she lose consciousness. If a patient goes unconscious during abdominal thrusts, you need to lower the patient to the ground and into a supine position. When you feel the patient go limp, brace yourself. Let the patient sag into your arms, lowering the lower body to the ground, then guiding the upper body to the ground while protecting the head. (In an urban situation, you would now ask someone to call 911.)

1. Open the airway with a tongue-jaw lift and look for an obstruction. If you see it, remove it.
2. Perform a head-tilt/chin-lift and attempt to ventilate the patient by sealing your mouth over the patient's mouth, pinching the patient's nostrils firmly closed, and breathing in with enough force to cause the patient's chest to rise (see "Rescue Breathing," below). If air doesn't go in, you will feel resistance, and the chest will not rise. Reposition the head and make a second attempt to ventilate.
3. If the airway remains blocked, expose the patient's chest and begin the chest compressions of CPR (see Chapter 5: Cardiopulmonary Resuscitation). Perform 15 compressions.
4. Repeat steps 1 through 3 until the airway has been successfully opened.
5. If the patient does not resume normal breathing, begin rescue breathing (see "Rescue Breathing," below). If the patient does resume normal breathing, place the patient in the recovery position (see "Recovery Position," this chapter).

Rescue Breathing

The air you inhale contains approximately 21 percent oxygen, and the air you exhale approximately 16 percent oxygen. So with rescue breathing you can provide adequate ventilation for a patient who is not breathing. The WFR may be able to save a life with rescue breathing if it is performed correctly.

Note: Use of a barrier device to prevent direct contact with a patient's mouth and body fluids is strongly recommended. Numerous commercial barrier devices are available. Become familiar with the use of any device you carry *before* you need to use it. Most devices are used similar to the way mouth-to-mouth rescue breathing is delivered.

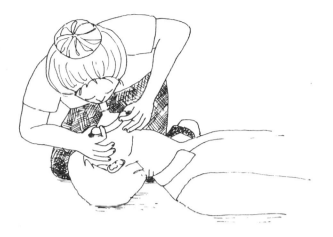

Figure 4-7: *Mouth-to-mask rescue breathing.*

Mouth-to-Mouth Rescue Breathing

After you have determined the patient is not breathing, use the thumb and first finger of your hand holding the patient's forehead to firmly pinch closed the patient's nostrils. If the nostrils are not pinched closed, the air you breathe in will escape through the nose. Hold the airway open with your other hand. Failure to hold the airway completely open will cause you to fail to adequately breathe for the patient.

1. Take a deep breath. This will maximize the oxygen and minimize the carbon dioxide in your lungs.
2. Open your mouth wide and seal it entirely over the patient's mouth.
3. Deliver two initial **full, slow** breaths. Each of your ventilations should take about 2 seconds. Blow only forcefully enough to fill the patient's lungs. **The rise of the patient's chest is the indicator you have delivered a full breath.**

Figure 4-8: *Mouth-to-mouth rescue breathing.*

Too much air blown in too hard will force air into the patient's stomach, causing *gastric distention* (see "Gastric Distention and Vomit," this chapter).

Note: If you're using the jaw-thrust maneuver to open the airway, you'll have to position yourself at the patient's side and seal the patient's nose with your cheek, hold the patient's lower lip down with your thumbs and/or first fingers, and seal your mouth over the patient's mouth. Between breaths be sure to remove your mouth from the patient's mouth, and refill your lungs.

4. If your breath will not go in, reposition the patient's head and attempt to ventilate a second time. A second failed attempt may indicate a foreign-body airway obstruction.
5. Check for a carotid pulse or other signs of circulation (breathing, coughing, movement). If circulation is present, continue rescue breathing at the rate of approximately one breath every 5 seconds for an adult. This rate delivers approximately 12 breaths per minute. If circulation is not present, initiate CPR (see Chapter 5: Cardiopulmonary Resuscitation).

Mouth-to-Nose Rescue Breathing

If the patient's mouth is injured or for some reason cannot be adequately opened, mouth-to-nose rescue breathing can provide adequate ventilation.

1. Perform the head-tilt/chin-lift maneuver, but hold the patient's lips closed.
2. Seal your mouth over the patient's nose.
3. Deliver two full, slow breaths, taking about 2 seconds per breath, and removing your mouth from the nose between breaths. If the patient does not exhale easily through the nose, you may open the mouth to let air escape the lungs.
4. Check for a carotid pulse or other signs of circulation (breathing, coughing, movement). If circulation is present, continue rescue breathing at the rate of approximately one breath every 5 seconds for an adult. This rate delivers approximately 12 breaths per minute. If circulation is not present, initiate CPR (see Chapter 5: Cardiopulmonary Resuscitation).

Mouth-to-Stoma Rescue Breathing

You may have a patient who has had part or all of his or her larynx surgically removed. These people have a permanent opening inserted in the trachea, an opening called a **stoma**. Patients with a partial laryngectomy breathe through the stoma and through the nose and mouth. Patients with a full laryngectomy breathe entirely through the stoma.

Rescue breathing for patients with a stoma needs to be performed through the stoma. If the chest does not rise, suspect a partial laryngectomy, and use one of your hands to hold the patient's mouth and nose closed.

Gastric Distension and Vomit

Severe gastric distension (air inflating the stomach) during rescue breathing can cause vomiting and/or put enough pressure on the diaphragm to limit the air you can breathe into the lungs. Pressing on the stomach to release the trapped air is almost never recommended because it almost always causes vomiting. According to the January 2002 edition of the Journal of Emergency Medical Services (JEMS), there is a 85 to 90 percent mortality associated with aspiration of gastric contents. If, however, adequate rescue breathing is prevented, decompression may be the patient's only hope.

When the patient vomits, whether the gastric area is distended or not, immediately roll him or her into a side position to prevent aspiration. The side position allows the vomit to flow out of the mouth instead of into the lungs, where severe complications such as pneumonia may erupt. When the vomiting ends, wipe out the mouth—a T-shirt or bandanna will work for wiping—then roll the patient once again into a supine position, and continue resuscitation efforts.

Prevent gastric distension by (1) maintaining an adequately open airway, (2) blowing air in slowly and fully but with just enough force to see the chest rise, and (3) allowing adequate time for the patient to exhale between your rescue breaths.

Supplemental Oxygen

Rescue breathing will be more effective if supplemental oxygen, more often a wish than a fact in the wilderness, can be delivered. Several ways of delivering supplemental oxygen exist (see Appendix A: Oxygen and Mechanical Aids to Breathing).

Recovery Position

Since every unconscious patient left on his or her back may lose an airway to a relaxed tongue and/or vomit, every unconscious patient should either receive constant attention from a rescuer and/or be rolled into the recovery position. Even one rescuer can move a patient with a suspected spine injury cautiously and successfully into the recovery position.

1. Kneel beside the patient at waist level.
2. Move the patient's arm on the same side of his or her body you're on until it is fully stretched out above the head, the reverse position to hanging straight down from the shoulder.

The palm should be facing up.
3. Move the patient's arm on the side away from you across the patient's chest with the fingers pointing toward the opposite shoulder.
4. Cross the patient's legs with the away leg on top.
5. With your other hand support the patient's head and neck.
6. Roll the patient, as a unit, assuring the patient's head stays in contact with the down arm. Bring the upper leg forward, like a kickstand on a bike, to maintain the recovery position.
7. Make sure the patient's nose points "downhill" to keep the airway open.
8. If the patient must be left alone, stabilize him or her in the recovery position with whatever is available, such as backpacks, rocks, or logs.

Evacuation Guidelines

Evacuate all patients who have required artificial respirations.

Conclusion

Dropping to your knees beside your client in the Wind River Mountains, you attempt to get a response. Failing to establish responsiveness, you perform a head-tilt/chin-lift maneuver to open his airway. Looking, listening, and feeling for breathing, you detect none. Sealing your mouth over his, you pinch his nostrils closed and attempt a ventilation. It does not go in. Repositioning his head, you attempt a second ventilation without success. Exposing your client's chest, you begin to perform the chest compressions of CPR. With the fifth chest thrust, the client coughs out a chunk of dried meat. You open his airway and give two full breaths, watching as his chest rises and falls. He then spontaneously gasps out a few breaths before his respirations subside into a normal rhythm. The tension eases slightly between your shoulder blades.

Within a few minutes, his return to consciousness appears complete. Although he seems well, you decide to leave the wilderness early to have your client assessed by a physician. He is your responsibility, and you need to know he is truly OK.

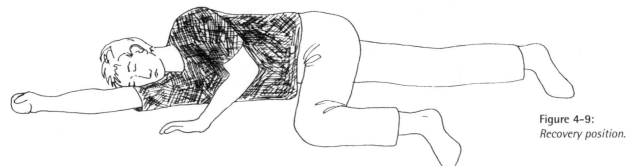

Figure 4-9:
Recovery position.

CHAPTER FIVE

Cardiopulmonary Resuscitation

You Should Be Able To:

1. Describe the basic anatomy of the heart.
2. Demonstrate cardiopulmonary resuscitation (CPR) on an adult.
3. Describe the complications that may occur during CPR.
4. List the criteria for stopping CPR.
5. Describe the special considerations concerning CPR in the wilderness.

It Could Happen to You

Colorado's San Isabel National Forest has attracted its usual large number of elk hunters, men and women from all over the United States who spend an October week in the Rockies. A few of those people will need help, and, as a volunteer on the local search-and-rescue team, you're ready to respond 24 hours a day. You're not surprised when a call comes during dinner.

This one is a little different: a man in his early fifties, not lost, but unable to walk the last half mile to his car due to chest pain and the inability to catch his breath. His hunting partner hiked out and made the call.

You find the patient sitting against a tree beside a well-used trail, his orange jacket reflecting the beams of the headlamps of the rescue team. As you approach, he attempts to stand but collapses face first to the ground.

Cardiac arrest, the cessation of heart muscle activity, will end the lives of approximately a half million humans suffering from heart disease in the United States over the next year (see Chapter 23: Cardiac Emergencies). About 50 percent of those people will die within 1 hour of the onset of signs and symptoms. Most of the patients who are saved receive not only immediate CPR but also more definitive medical treatment, such as defibrillation (see Appendix B: Automated External Defibrillation). In a wilderness environment patients who go into cardiac arrest as a result of long-standing disease will rarely survive, even if your CPR skills are excellent.

Cardiac arrest, however, can be caused by other incidents, most commonly in the wilderness by lightning strikes and submersions in water. In these cases you may be able to save a life with CPR initiated without delay and performed correctly.

Basic Anatomy of the Heart

The heart is a muscular organ lying beneath the lower half of the sternum with the bottom "tipped" slightly to the left side of the body. Your heart is approximately the size of your fist. The heart is divided into four hollow chambers: two upper chambers, the right and left *atria*, two lower chambers, the right and left *ventricles*. These chambers are connected by one-way valves.

Blood from all over the body circulates into the right atrium via two large veins, the *inferior vena cava* and *superior vena cava*. Passing through the *tricuspid valve*, blood enters the right ventricle, from where it is pumped through the *pulmonary valve* into the *pulmonary artery* and out to the five lobes of the lungs—three lobes on the right, two lobes on the left. From the pulmonary artery, blood squeezes through smaller and smaller *arterioles* until it enters capillaries that wrap around the *alveolar sacs* in the lungs. Alveolar sacs have microscopically thin walls, as do pulmonary capillaries. Here, on the alveolar level, gases are exchanged across the thin walls. After exchanging carbon dioxide for oxygen at the alveoli, blood travels through *venules* of increasing size until it reenters the heart at the left atrium via the *pulmonary veins*. It might be of interest to note that all veins in the body carry deoxygenated blood except the pulmonary veins, and all arteries carry oxygenated blood except the pulmonary arteries. All vessels carrying blood toward the heart are veins, and all vessels carrying blood away from the heart are arteries. From the left atrium, blood passes through the *mitral valve* into the left ventricle. A hardy squeeze by the left ventricle sends blood through the *aortic valve* into the *aorta* and on to all parts of the body.

The pumping action of the heart keeps blood circulating under pressure, and each beat sends a pulse through the entire arterial system. Pulses are most easily palpated where large arteries run near the surface of the body. Two of those places are the carotid pulse in the neck and the radial pulse at the wrist.

The heart and the lungs, the *cardiopulmonary system,* work together with the brain as an ultimate team, a perfect triumvirate, each precisely supporting the efforts of the other. If one of the team fails, the other two are soon to follow. Brain arrest, if you will accept that term, is irreversible. Respiratory arrest (see Chapter 4: Airway and Breathing) and cardiac arrest, however, may be reversed.

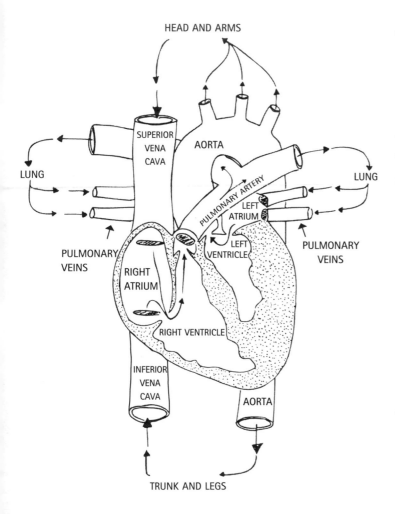

Figure 5-1: *Basic anatomy of the heart.*

Cardiac Arrest and CPR

Respiratory arrest is the cessation of spontaneous breathing. Cardiac arrest is the cessation of heart function. A patient in respiratory *and* cardiac arrest is clinically dead: unresponsive, breathless, pulseless. After 4 to 6 minutes of cardiac arrest, the typical patient will have suffered irreparable brain-cell death to the point where he or she is biologically dead. If CPR is initiated soon enough, however, done well enough, and maintained long enough, the patient may eventually be resuscitated.

CPR consists of a rhythmic combination of rescue breathing and chest compressions. Rescue breathing keeps the patient's blood oxygenated. Chest compressions either directly compress the heart, pushing blood to the brain, or build up pressure in the chest cavity, causing the blood to circulate with enough pressure to maintain life in the brain.

The initiation of CPR may be summarized with the first three letters of the alphabet—ABC—the first three letters of the initial assessment:

A. Check for responsiveness. If unresponsive, open the **airway.**

B. Check for **breathing.** If breathless, give two full, slow breaths.

C. Check for signs of **circulation.** If pulseless, initiate chest compressions.

Cardiopulmonary Resuscitation

The American Heart Association has established the following sequence (Heartsaver CPR) for the rescue of an adult patient who is unresponsive:

1. Assess the patient to be sure he or she needs help.

2. Activate the emergency medical services (EMS) system by dialing 911.

3. Begin resuscitation.

Obviously access to 911 will not be available in the wilderness. For that reason, the step of "activating the EMS system" is left out of sections of this chapter. You can, however—and you are strongly encouraged to—call for help. Anyone who comes running may be able to assist you.

Note: Anyone receiving CPR or who has been resuscitated via CPR should be evacuated to definitive medical care as soon as possible.

Step 1: Check for Responsiveness

If the patient does not respond to an initial check for responsiveness, immediately open the airway with the head-tilt/chin-lift maneuver or the jaw-thrust maneuver (see Chapter 4: Airway and Breathing). As soon as you determine you have an unresponsive patient, you should call for help.

Step 2: Check for Breathing

If a 10-second look, listen, and feel for breathing indicates the patient is breathless, give 2 full, slow breaths—about 2 seconds per breath. If you feel resistance and the patient's chest does not rise with your first attempted ventilation, reposition the patient's head and attempt to ventilate again (see Chapter 4: Airway and Breathing). Once you get the first full breath in, always follow it with a second full breath.

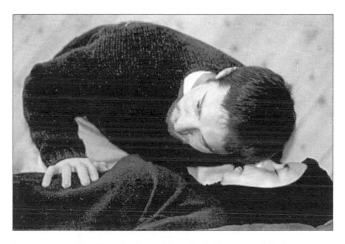

Figure 5-2: *Checking for breathing: look, listen, and feel.*

Step 3: Check for Signs of Circulation

Place two of your fingers on the patient's larynx (the "Adam's apple") and slide your fingers off to the side of the neck you are on. Do not reach across the trachea. In the valley between the larynx and the large neck muscle you will find the carotid pulse. Check for about 10 seconds before declaring you cannot find a pulse. Look also for other signs of circulation that include breathing, coughing, and movement. You may not find a pulse yet see other signs of circulation.

If you find signs of circulation, continue rescue breathing. If you do not find signs of circulation, prepare to do chest compressions, **but** before starting compressions:

1. Place the patient face up on a firm surface, such as the ground.
2. Expose the patient's chest. Time is of the essence: Rip open or pull up shirts or blouses, cut off or slip up bras.
3. Kneel at the patient's shoulder with your knees approximately as far apart as your shoulders.
4. Place the heel of one hand on the patient's sternum at the nipple line. The heel of your hand will be over the lower half of the patient's sternum. Place the heel of your other hand over the hand that's now on the lower half of the sternum. The lower half of the sternum is supported by cartilage flexible enough to allow compressions. If you place your hands too high, you will fail to compress the chest enough. If you place your hands too low, you will soon break off the xiphoid, break ribs, or even break the sternum.

5. Straighten your arms and position your shoulders over the patient's chest.

Chest compressions should be fluid downward motions on the sternum followed by a release of pressure from the sternum to allow blood to flow back into the patient's heart and chest. Compression time and release time should be equal. Avoid jerky, jabbing motions. An adult patient's chest should be compressed 1.5 to 2 inches. Use the weight of your upper body instead of the strength of your arms and shoulders.

You will be performing 15 compressions to every 2 full breaths. The cycle ratio of 15:2 should be done in approximately 15 seconds. To perform at the speed required, you will need to do chest compressions at the rate of 100 per minute. You are not actually doing 100 compressions in 1 minute, but your rate should be 100 per minute. In 1 minute you will actually have done 60 compressions and 8 breaths. Counting softly out loud—"One, and two, and three"—will help you perfect your CPR compressions. *Practice, practice, practice.* Even the most excellent chest compressions will provide no more than one-third of the patient's normal circulation.

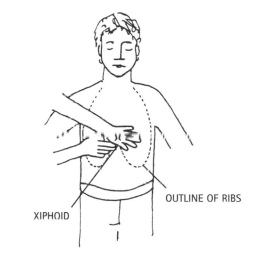

OUTLINE OF RIBS

XIPHOID

Figure 5-3: *Hand placement for adult CPR.*

Adult One-Rescuer CPR

1. Check for responsiveness.
2. With no response, call for help and open the airway. Check for breathing for about 10 seconds.
3. With no breathing, deliver 2 slow, full breaths. Watch chest rise. Allow chest to deflate.
4. Check for a carotid pulse and other signs of circulation—breathing, coughing, movement—for about 10 seconds.
5. With no signs of circulation, start cycles of 15 compressions to two breaths.
6. After four cycles of 15:2 (approximately 1 minute), recheck for signs of circulation. With no signs of circulation, continue cycles of 15:2 beginning with chest compressions.

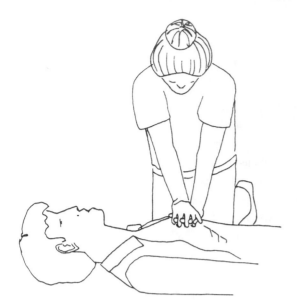

Figure 5-4: *One-rescuer chest compressions.*

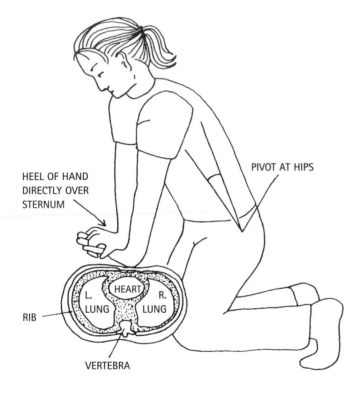

HEEL OF HAND DIRECTLY OVER STERNUM

PIVOT AT HIPS

L. LUNG   HEART   R. LUNG

RIB

VERTEBRA

Figure 5-5: *Cross section of chest during CPR.*

CPR without Artificial Respiration

If you are resistant to doing mouth-to-mouth artificial respirations, the American Heart Association supports studies indicating there is enough residual oxygen in a patient's system to make chest compressions alone effective without artificial respiration for the first 6 minutes after cardiac arrest.

Complications Caused by CPR

1. Even properly performed, CPR may break ribs and/or the sternum. If you feel a fracture occur, check your hand position and continue CPR.
2. The lungs may be bruised. Fractured ribs could cause lacerations of the lungs and the possibility of a pneumothorax and/or a hemothorax (see Chapter 10: Chest Injuries). It is unlikely you will know what has happened internally—so continue CPR.
3. Patients, whether resuscitated or not, tend to vomit during CPR. In case of vomiting **immediately** roll the patient into a side position and open the airway to allow the vomit to flow out of the mouth. When vomit has ceased to flow, wipe out the mouth—a T-shirt or bandanna will serve as a wipe—then roll the patient back into a supine position and continue CPR.
4. Gastric distension may be caused by the rescue breathing phase of CPR (see Chapter 4: Airway and Breathing).

Criteria for Stopping CPR

According to criteria established by the American Heart Association, CPR should be initiated as soon as possible after cardiac arrest and continued until:

1. You are exhausted and unable to continue.
2. You turn the patient over to rescuers or medical professionals of equal or higher training.
3. You resuscitate the patient.
4. The patient is declared dead by someone authorized to do so.
5. You find yourself in imminent danger.

Special Considerations for the Wilderness

Guidelines for the general use of CPR are well defined, regularly updated, and widely distributed. Because the wilderness may impose circumstances that require special considerations in CPR, the following guidelines have been developed by the Wilderness Medical Society. Some of these guidelines may be relevant in urban situations as well.

1. *Contraindications to CPR in the wilderness:* There is no reason to initiate CPR if you find (1) any sign of life in the patient, (2) danger to rescuers, (3) *dependent lividity* (discoloration of the skin, as from a bruise, where noncirculating blood has settled via gravity), (4) *rigor mortis* (stiffness occurring in dead bodies), (5) obvious lethal injury (such as decapitation), (6) a well-defined and written Do Not Resuscitate (DNR) status, or (7) a patient who has a frozen chest.

2. *Discontinuation of CPR in the wilderness:* Once initiated, CPR should be continued until (1) resuscitation is successful, (2) rescuers are exhausted, (3) rescuers are placed in danger, (4) the patient is turned over to more definitive care, or (5) the patient does not respond to prolonged resuscitative efforts (30 minutes or more).

Specific Wilderness Situations

Hypothermia

A cold, rigid, apparently pulseless and breathless patient is not necessarily a dead patient. If you fail to find breathing, rescue breathing should be initiated immediately. The patient needs oxygen, and there is *no* danger to the patient from supplemental oxygen and/or rescue breathing.

Failure to find signs of circulation in a cold patient should not necessarily lead to chest compressions. Apparent pulselessness may be due to hypothermia and the resulting tissue rigidity in combination with a very slow heart rate. Initiation of chest compressions may cause a cold, slow-beating heart to stop.

The initial check for signs of circulation should take 60 seconds instead of 10 seconds. With no signs of circulation, rescue breathing should continue for 3 to 15 minutes—and then a second 60-second check for circulation should be performed.

If the patient is still without signs of circulation, chest compressions should not be initiated on the patient who is rigid from the cold. If the chest is compressible, however, and evacuation to a medical facility is under way—such as by helicopter—begin chest compressions. Otherwise, keep up the rescue breathing and package the patient for long-term management of a cold person (see Chapter 16: Cold-Induced Emergencies).

Avalanche Burial

Breathless and pulseless victims of avalanches are almost always dead from suffocation and/or blunt trauma. Check for snow in the patient's airway. It could block rescue breathing. Hypothermia is sometimes a confounding factor but, in most cases, you will still start and maintain CPR for at least 30 minutes.

Cold-Water Submersion

Near-drowning patients may be successfully resuscitated after prolonged (over 1 hour) submersion in cold water (70°F or less). The younger the patient, the cleaner the water, the colder the water, and the shorter the duration of submersion, the greater the chance for success. CPR should be initiated immediately and continued, if possible, until definitive care is available (see Chapter 19: Immersion and Submersion Incidents).

Lightning Strike

CPR should be initiated immediately on all pulseless, breathless victims of lightning strike. Following a severe electrical shock, respiratory paralysis may persist long after cardiac activity returns. Rescuers should be prepared to provide prolonged rescue breathing (see Chapter 20: Lightning Injuries).

Evacuation Guidelines

Anyone receiving CPR or who has been resuscitated via CPR should be evacuated to definitive medical care as soon as possible.

Conclusion

Kneeling beside the hunter in the San Isabel National Forest, you check for breathing and find none. You grab your rescue mask from the outside pocket of your pack and give the patient 2 full breaths, watching the patient's chest rise and fall, rise and fall. A check reveals no signs of circulation. Baring the chest of the patient, you begin CPR—15 compressions to 2 breaths. The routine continues.

Other rescuers assemble a break-apart litter the team carried in, preparing padding and straps for a carry. Tension stands thick in the dark forest. Beams from headlamps flicker back and forth. Orders are passed around in urgent whispers.

Gurgling in the patient's throat gives too short a warning of impending vomit, and stomach contents splatter the inside of your rescue mask. You are momentarily grateful for the one-way valve in the mask you've been blowing through. Quickly rolling the patient toward your partner, you remind him to keep the airway open. Vomits runs into the duff. You clean your mask with a wad of gauze someone stuffed in your hand. With the mouth wiped out by your partner, you roll the patient onto his back and continue CPR.

Glancing at a fellow rescuer, you see a trickle of sweat beginning to run down his cheek, and you notice the dampness growing across your back beneath your heavy pile sweater. Suddenly, it all seems so hopeless, but you inhale to ventilate once more. . . .

CHAPTER SIX

Bleeding

You Should Be Able To:

1. Describe the importance of blood and the dangers of bleeding.
2. Demonstrate how to control bleeding using direct pressure, elevation, pressure bandages, pressure points, and a tourniquet.

It Could Happen to You

The knife slips out of a chunk of cheddar cheese and into her left hand. It's a Swiss Army knife, new and razor-edged, the kind of blade that seems to first emerge from its red handle with a lust for blood. The cheese is cold and hard, her effort great, and the blade slices cleanly through her lower palm and deep into her upper wrist.

You, her backpacking companion, see little or no evidence of pain. There's a soft squeak of alarm, and a look of surprise on her face. But the knife has nicked her radial artery, and blood begins to spurt with each beat of her heart. Her look of surprise changes to anxiety and confusion. The pulse of escaping blood speeds up. Her skin color fades to pale. Her chest heaves in shallow gasps for breath. She stares at her wrist, and so do you, with a lack of comprehension. In moments blood has splattered everywhere.

The life of every cell in the human body depends on a continuous flow of oxygenated, nourishing blood. Blood is made up mostly of red blood cells, white blood cells, platelets, and plasma. Red blood cells carry oxygen to body tissues and carbon dioxide away from body tissues. White blood cells attack invading microorganisms and produce antibodies that prevent infection. Platelets form clots, which are essential if a wound is ever to stop bleeding. Plasma is the aqueous transportation medium for blood cells, platelets, and nutrients—and it is mostly water.

Although amounts vary depending on body size, the average adult male has approximately 5 to 6 liters of blood (1 liter is slightly more than 2 pints). The smaller the individual, the lower the blood volume. An adolescent of about 100 pounds body weight has approximately 3 liters of blood. Newborns usually have less than 0.5 liter of blood.

An understanding of bleeding and, more importantly, how to control life-threatening bleeding in an injured patient is a critical intervention technique the Wilderness First Responder must learn. All bleeding stops—eventually—but you want to make sure it stops before the patient does.

Types of Bleeding

Hemorrhage is a word that means "bleeding from a wound." Hemorrhaging can be external or internal, and it can be further classified by the types of vessels that have been damaged. *Capillary bleeding* is slow, oozing from these small vessels, and usually bright red in color, although the blood from external wounds is often difficult to distinguish by color. *Venous bleeding* usually comes in a steady flow from these larger vessels, and it can be a heavy flow if the wound is significant. Venous blood tends to be a dark maroon color, due to its lack of oxygen. *Arterial bleeding,* under higher pressure, is usually rapid, often spurting each time the patient's heart beats. Arterial blood tends to be bright red, brighter than capillary blood.

Body Response to Bleeding

When blood vessels are damaged, the body almost immediately begins an involuntary process to stop the blood loss and start the healing. The walls of the damaged vessels constrict, helping to slow the flow of blood. Circulating platelets begin to stick to the tear in the vessel wall and start forming a clot. A wide variety of *clotting factors*—some circulating in the blood, some released from platelets, some released from the damaged cells—interact at the site of the injury. The local action of these factors is the formation of stable threads called *fibrin* that interweave to stabilize the clot and form a matrix that becomes the foundation over which new cells grow to create, eventually, new skin.

Most capillary and small venous hemorrhaging will form a clot and stop bleeding without your assistance, except in a few cases where disease conditions inhibit clotting. Although completely severed arteries will often constrict enough to seal off bleeding, arterial spurting and heavy venous bleeding will usually require outside control to stop the bleeding before serious harm is done.

Control of External Bleeding

The most immediate concern is not how much blood has been lost but how fast it is being lost. Severe blood loss must be stopped as soon as possible, taking precedence over almost everything else during the first few moments with a patient. The arrest of blood loss is known as *hemostasis*.

Despite the urgency of the situation, it is recommended that you first put on a pair of protective gloves, perhaps sunglasses if the blood is splashing, before touching blood or other body fluids to protect yourself against the spread of germs that cause illness (see Chapter 3: Patient Assessment, "Isolate Body Substances"). While locating your gloves, encourage reliable, helpful patients to apply direct pressure and elevation to their own wound, if possible, to facilitate control of the bleeding.

Direct Pressure

The first and best method for control of bleeding is direct pressure. Pressure should be applied directly to the wound with the heel of your gloved hand or the flat part of your fingers (for smaller wounds)—preferably, with a sterile absorbent dressing under your gloved hand. Without a sterile dressing, use the cleanest material available, such as a shirt or bandanna. Most blood loss will slow appreciably or stop with 10 to 20 minutes of direct pressure.

If the absorbent dressing soaks through with blood, in most cases you will want to apply a second dressing and maintain pressure. You may also choose to remove the first dressing carefully, and start fresh with a second dressing, taking care to ensure your pressure is exactly over the bleeding wound. Patience is almost always rewarded with success.

Huge wounds, those for which your hands cannot cover the injury site, may require that you first pack the wound with absorbent material before applying pressure.

Highly vascular areas, such as the scalp, may require prolonged pressure with a bulky dressing. Scalp wounds should be assessed for stability of the skull before applying pressure (see Chapter 9: Head Injuries).

Elevation

As soon as possible and whenever possible, the wound should be elevated above the patient's heart. This, in addition to direct pressure, will further slow blood loss and encourage clotting. Elevation should be performed with extreme discretion if there is the possibility it will make the injury worse, such as wounds associated with severe fractures and dislocations.

Figure 6-1: *Direct pressure and elevation.*

Pressure Points

A pressure point is a place where an artery lies close to a bone and simultaneously close to the surface of the body. By pressing with your fingers or hand on a pressure point, you can slow the flow of arterial blood. Pressure points should not be used in place of direct pressure but as an additional measure in stopping blood loss.

There are two pressure points immediately and easily available: the brachial and the femoral. The *brachial artery* lies on the inside of the upper arm between the armpit and the elbow. By holding the patient's arm with your thumb on the outside and your fingers over the artery, you can press the artery against the *humerus* (upper arm bone). The *femoral artery* crosses the lower pelvis in the groin area. With the heel of your hand, you can press on the crease between the pelvis and leg when the patient is lying down. The crease appears where the leg bends near the groin area.

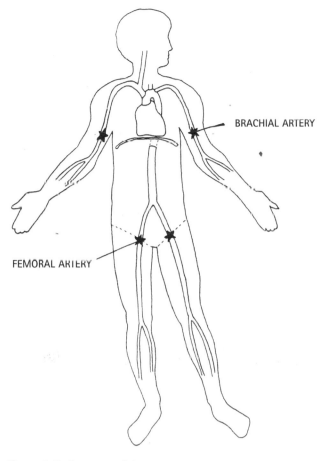

Figure 6-2: *Pressure points.*

BRACHIAL ARTERY

FEMORAL ARTERY

Pressure Bandage

If bleeding persists after 20 minutes and/or if you need your hands free for another job, a pressure bandage should be applied. A pressure bandage is one that holds direct pressure on a wound. With a bulky dressing in place on the wound, conforming roller gauze, elastic wrap, or a clean strip of cloth is wrapped securely around the extremity and tied with the knot over the wound.

The pressure bandage should not be so tight that it causes a disruption in blood flow to the *distal* side of the bandage, the side farthest from the midline of the body (and from the heart). Distal blood flow should be assessed on a regular basis by checking distal pulses and sensation.

Tourniquet

Use of tourniquets in urban situations, where transport is imminent, remains controversial, seldom required, and recommended only as a last resort when the life of the patient is threatened. Using a tourniquet to control bleeding may be extremely hazardous because it interrupts the blood supply and risks causing irreversible injury to the distal extremity due to lack of adequate *perfusion*, the bathing of cells with oxygenated blood under proper pressure.

In the wilderness tourniquets, although rarely necessary, may be of use more often than in urban situations. On upper arms and

upper legs (the only places tourniquets should be used), they can be left in place for 2 hours or more—less is better—before irreversible damage occurs. During that time the tourniquet may be useful in allowing you to stop blood flow long enough to find exactly where the torn vessels are located so that you can increase the effectiveness of direct pressure. A tourniquet also may be useful in encouraging clot formation in severe wounds.

Apply the tourniquet close to the *proximal* (closest to the midline of the body) side of the wound for upper arms or legs, or just above the elbow or knee for lower arm or leg wounds. Use a wide band and tie it snugly around the arm or leg with an overhand knot. (A narrow band, such as a rope, can cause localized neurovascular damage in minutes.) Use a second overhand knot to tie a stick (or anything rigid) into the band—and turn the stick to squeeze off blood flow. Tighten the tourniquet just enough to prevent any leakage from the wound and no more. A loose tourniquet, however, may actually increase bleeding by allowing arterial flow past the constriction but impeding venous return. Carefully monitor the time and your patient. You should loosen the tourniquet periodically to see if it's still needed—and remove it as soon as it's not. **If you apply a tourniquet and leave it applied, you have determined to sacrifice the limb.**

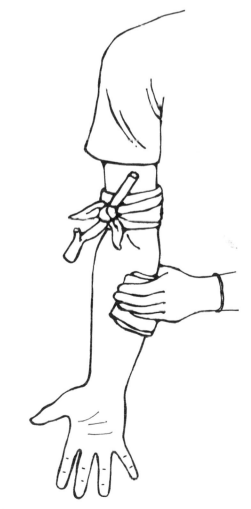

Figure 6-3: *Tourniquet.*

Internal Bleeding

Internal bleeding also can become life-threatening. Beware, for example, when it takes place inside a body cavity (chest, abdomen) and if one or both femurs (upper leg bones) are fractured. A threat to life may develop before you are even aware the patient is bleeding—and without a single drop of blood dripping onto the ground. Anticipate internal bleeding by careful assessment of the patient.

Given the mechanism of injury—for example, a fall on or blow to the abdomen—the signs and symptoms of shock without another obvious cause are usually attributable to internal bleeding. Look for external bruising over a blood-rich internal organ. Watch for guarding, the patient's instinct to physically protect an internal injury. Later signs of internal bleeding may be an increasingly rigid abdomen as the abdominal muscles spasm to protect the damaged area, and an extended abdomen as blood fills the cavity.

Unfortunately, there is little that the WFR can do for internal bleeding other than stabilization of known injuries, treatment for shock, and rapid evacuation (see Chapter 7: Shock).

Evacuation Guidelines

Any patient who has bled enough to exhibit the signs and symptoms of compensatory shock and who does not improve rapidly with treatment, and any patient who has bled enough to exhibit the signs and symptoms of decompensatory shock, should be evacuated. Any patient exhibiting the signs and symptoms of irreversible shock should be evacuated as soon as possible (see Chapter 7: Shock).

Conclusion

With sudden realization, you grab your bandanna from your pocket. Pressing it firmly over the gash in your friend's wrist, you squeeze down with enough force to stop the blood flow. Speaking quietly, your renewed calmness helps calm the patient. You lift her arm, elevating the injury above the level of her heart. A half hour later, when you gently remove the bandanna to check, the wound has ceased to bleed. Judging from her vital signs, you determine her blood loss is well within her physiological ability to compensate, and you begin preparations to treat the laceration (see Chapter 15: Wilderness Wound Management).

CHAPTER SEVEN

Shock

You Should Be Able To:

1. Describe the basic anatomy and physiology of the cardiovascular system.
2. Describe specific causes of shock.
3. Describe the stages of shock with the accompanying signs and symptoms.
4. Demonstrate the management of a patient in shock.
5. Describe the differences between managing shock in an urban environment and in the wilderness.

It Could Happen to You

Snow on the south slopes facing the midwinter sun has turned rock hard. The backcountry skiing, fast and furious, leaves the group exhilarated and exhausted, and the brown and gold tents of camp are a welcome sight. With dinner bubbling on stoves, you're surprised to see the look of concern on a young skier's face when she tells you her tent mate "seems really out of it."

Kneeling in the tent at the patient's side, you immediately notice her skin is pale, cool, and damp with sweat. Her breaths are rapid and shallow. A check of her pulse reveals to you a heart rate of 100, regular and strong.

You ask, "Can you tell me what's going on?"

The patient looks at you as if she didn't understand your question. Her face reveals a high level of anxiety. "I'm not sure," she finally answers, "I feel terrible."

To function properly, the delicate and sensitive brain requires narrowly defined limits of sugar, temperature, oxygen, and blood pressure. When the body senses a drop in blood pressure, it initiates a series of coping mechanisms intended to supply an uninterrupted supply of oxygen-rich blood to the brain. These measures include complicated changes in blood chemistry, shifts in the location of body fluids, and other physiological adjustments. In response to minor problems, such as minor injuries, the system may maintain adequate blood circulation in the brain with little effort. Serious problems, such as injuries that involve major blood loss, require stronger compensatory actions. Attempts at regulation become increasingly more drastic until adequate pressure in the brain is restored, or until the brain dies from *hypoxia* (deficiency of oxygen). Sometimes, even if the person does not die from the initial injury or illness that caused the compensatory mechanisms to shift into high gear, the body is unable to recover—some forms of shock, in other words, defy the body's ability to compensate.

Perfusion is the constant bathing of cells with life-sustaining fluid under proper pressure. *Shock* is inadequate perfusion, a result of inadequate levels of oxygen getting to the cells. Shock can have numerous causes and may be thought of as the "trigger" for the compensatory actions of the body.

As a Wilderness First Responder **you must vigilantly anticipate shock in all patients** because (1) seemingly minor injuries and illnesses may trigger this life-threatening condition; (2) by the time you notice the first signs, the downward spiral of shock may have already begun; (3) as shock progresses, it becomes more difficult to treat; (4) you will not be able to predict how rapidly shock will progress; and (5) appropriate early intervention may save lives.

Basic Anatomy and Physiology of the Cardiovascular System

The cardiovascular system (heart, vessels, and blood) is really two circulatory systems joined at the heart. The right side of the heart pumps blood to and from the lungs. The left side pushes oxygen-rich blood to and from all other parts of the body. Both circuits work simultaneously. *Cardiac output,* the amount of blood pumped out by the heart every minute, is normally enough to meet the needs of all the cells of your body.

Blood leaves the heart through *arteries,* which branch and narrow to *arterioles,* which branch and narrow to *capillaries.* At the capillaries, the oxygen (along with nutrients picked up from the digestive system) is diffused across cell walls and exchanged for carbon dioxide and the waste products of metabolism. The waste-carrying blood moves out the other end of the capillaries and into *venules* that come together to form *veins.* The veins return blood to the heart. Veins are more numerous than arteries and can hold more blood. There is less pressure in the veins, and they have little one-way "doors" to keep the blood from backing up. All vessels can dilate or constrict to meet varying demands from the body.

Shock

The term *shock,* once again, describes the body's condition and reactions when tissues are not being adequately perfused with oxygenated blood. Adequate perfusion requires a certain minimum blood pressure (approximately 90 mmHg in an adult male) necessary for life-sustaining exchanges of oxygen and carbon dioxide to occur. The circulatory system's three main components—the heart, the blood, and the blood vessels—work in concert to maintain adequate pressure.

Heart: The heart must contract at a rate fast enough and strong enough to keep a steady stream of blood flowing through the system. Think of it as a water pump situated at the base of a hill below a cabin. If the pump does not supply water pressure great enough for the occupant of the cabin to take a shower, the occupant can build pressure by increasing the pump's rate, the number of strokes the pump makes per minute. Similarly, the body monitors blood pressure at sites called *baroreceptors* (nerve endings sensitive to pressure changes). When a baroreceptor detects low pressure, it sends a message that signals the pump, in this case the heart, to beat faster.

Blood: There must be enough blood in the system to keep the blood vessels filled. Using the cabin and water pump as a model, you can easily see that if there is not enough fluid in the system of pipes, no amount of rapid pumping will cause the water to rise all the way up to the cabin, and the occupant is left high and dry. Likewise, if the body is low on blood, through either dehydration or hemorrhage, the heart will not be able to work hard enough to maintain an adequate pressure. The only remedy to this problem is to add more blood to the system. The body will do this on its own, but the process requires time and raw materials: water and nutrients.

Vessels: The highly elastic arteries and veins form a container for the blood and must be precisely the proper size if the pressurized system is to function at its optimum. In the cabin example the occupant may determine that the only way to get suitable pressure to the cabin is to narrow the diameter of the pipes that lead there. You can see that given a certain volume

and a limited pumping capacity, forcing the same liquid through a smaller pipe will increase the pressure at the end of the run.

The human body can narrow, or *vasoconstrict,* its "pipes" very efficiently. A healthy body involuntarily makes minute adjustments in the functioning of the circulatory system that keep the body well pressurized. For example, when an adult male donates a half liter of his approximately 6 liters of blood during the local blood drive, his vessels make up for the loss by constricting—shrinking the container, you might say—until his body is able to replenish the supply. Stress, trauma, and disease, however, are all capable of interfering with perfusion and initiating shock.

Types and Causes of Shock

All shock results from a failure of one or more of the major components of the cardiovascular system. With numerous specific causes, all shock is classified generally by the component of the cardiovascular system that is failing.

Cardiogenic Shock: The term *cardiogenic* describes shock stemming from failure of the heart. A heart attack, in which the muscle itself has sustained damage from *ischemia* (deficient blood supply), may result in cardiogenic shock. Trauma, too, may cause damage, such as a rupture of one or more of the heart's chambers, making it an ineffective pump.

Hypovolemic (Low-Volume) Shock: The term *hypovolemic* describes shock that occurs from low fluid volume within the cardiovascular system. Bleeding secondary to trauma is a common cause of shock seen by WFRs. It can often be easy to determine the relationship between large amounts of blood lost outside the body and the signs of shock in a trauma patient. Always remember, however, that internal bleeding, whether caused by trauma or disease, can result in shock as well. Patients can die from the amount of blood lost into the abdominal cavity. A ruptured spleen is one of the many causes of massive abdominal bleeding. As another example, a broken femur, when it perforates the femoral artery, may cause serious internal loss of blood directly into the thigh (see Chapter 6: Bleeding).

Another common wilderness cause of hypovolemic shock is dehydration. Excessive sweating, lack of adequate water intake, and gastrointestinal illnesses that result in diarrhea and/or vomiting are all culprits. When the general fluid level in the body drops, water is absorbed from the blood into other areas of the body to make up the deficit.

Burns are another major cause of hypovolemic shock. In addition to any blood loss associated with burns, the plasma that continues to leak from damaged tissue can account for significant fluid loss. The blisters that result from partial-thickness burns serve as visible reminders that fluid is outside the body, rather than inside where it is needed. Usually much more serious, however, are full-thickness burns, which, having eliminated the protective barrier of skin, allow fluid to escape the body at a rapid rate.

Vasogenic (Low-Resistance) Shock: The term *vasogenic* describes shock resulting from failure of the blood vessels to maintain sufficient resistance to blood flow. Spine injuries, for instance, may sever, pinch, or otherwise damage nerves that control muscles in the veins and arteries. When these muscles relax, the vessels dilate, and the container becomes too large for the amount of blood in the system, a condition sometimes called *neurogenic shock*. An overdose of some drugs that affect the nervous system may produce the same problem.

Septic shock, a form of low-resistance shock, is the name given to shock stemming from infection. When bacteria, viruses, or some other agents of illness proliferate rapidly in the system, the colony produces waste products. If these toxins are numerous enough, they damage blood vessels, causing them to leak and to lose muscle tone. The resulting vascular dilation can initiate shock.

A severe reaction caused by oversensitivity to an allergen may result in *anaphylactic shock*, another form of low-resistance shock. Stings from honeybees and other *Hymenoptera* (wasps, hornets, fire ants), and the ingestion of certain drugs, particularly penicillin-based antibiotics, are common causes of anaphylactic shock. This reaction causes the vessels to dilate and the pressure to drop, but patients are more likely to die from airway obstruction caused by swelling of the airway (see Chapter 28: Allergic Reactions and Anaphylaxis).

Because both septic and anaphylactic shock allow blood to leak into spaces between cells, called *interstitial spaces*, patients with either condition may look somewhat different than patients in other forms of shock. Their skin may be red or mottled and warm rather than the characteristic pale, cool, and clammy. Their bodies may be trying hard to get rid of the excess fluid, making them prone to vomiting and diarrhea.

Stages of Shock

Whatever the type of shock, if the injury or illness is severe enough, patients may progress down through stages of shock that lead to death. Sometimes patients may stabilize on their own—but they may, however, need proper care to survive.

Stage 1: Compensatory Shock

When blood pressure begins to drop, the body will respond in ways that work to keep the pressure up. During this stage, the patient typically appears anxious and/or confused. Heart rate increases. **Please note that the activity of the heart is the first and primary indicator of shock.** Respiratory rate increases. Skin typically grows pale or chalky white or ashen gray as peripheral blood vessels constrict to draw blood into the core. The patient usually begins to sweat. The redistribution of blood away from the abdomen may lead to nausea and vomiting, and the patient may be thirsty. If these actions maintain enough blood pressure to allow for adequate perfusion, this stage of shock is called *compensatory shock*.

You may have experienced a temporary reaction very similar to compensatory shock the last time a loud noise outside your tent startled you in the middle of the night or you suddenly realized a bear was following you down the trail. When the brain perceives some threat, it prepares the body for immediate action by directing the adrenal gland to release its hormones. This adrenaline rush, the "fight or flight" response, induces the same physiologic changes as those in compensatory shock. This stimulation prepares you to make your best physical response to the situation. Once the threat subsides, a healthy body will return to a normal resting state. These "stress reactions" or "alert responses" to possible damage usually last no more than 10 to 20 minutes if no real damage to the patient has occurred.

Psychogenic Shock

Sometimes fainting is referred to as *psychogenic shock*. In what is thought to be the mind's attempt to avoid an unpleasant situation, it will dilate the blood vessels, causing pressure in the brain to drop suddenly. Fortunately the condition is almost always self-limiting because the skeletal muscles relax, too, and permit the body to drop into a horizontal position. Provided the person who faints is not strapped into a seat and chest harness while dangling in a crevasse or standing on a narrow rock ledge following a 5.9 lead, the body will slump to a level where the heart and brain are on the same plane, enabling blood to easily flow back into the brain and restore consciousness.

Signs and Symptoms of Compensatory Shock

1. Anxiety; confusion.
2. Rapid, strong pulse.
3. Shallow, rapid respirations.
4. Cool, clammy skin.
5. Normal to pale or chalky white or ashen gray complexion.
6. Blood pressure within normal limits.
7. Pupils are PERRL (pupils-equal-round-and-reactive-to-light).
8. Thirst; nausea; vomiting.
9. Dizziness, lightheadedness, and/or weakness.

COMPENSATORY SHOCK VIA BLOOD LOSS

When you donate blood, the amount taken is usually one unit (0.5 liter, about 1 pint) or roughly 8 to 10 percent of your total blood volume—and it's no problem. Blood volume loss of up to 15 percent usually is well tolerated by a patient. As a crude estimate, this amount of blood will just about fill a 1-liter water bottle, the size carried by most people on wilderness trips.

Loss of 15 to 30 percent of an individual's blood volume, however, may be considered serious. But with adequate treatment, the compensatory blood shunting usually maintains a normal, or even slightly elevated, blood pressure.

ORTHOSTATIC CHANGES IN VITAL SIGNS

Orthostatic hypotension (positional low blood pressure) is another indicator that a patient may be in shock. Orthostatic hypotension is further defined as a fall in blood pressure, typically greater than 20/10 mmHg, on assuming the upright posture. A significant orthostatic change in a patient's pulse may be defined as a change equal to or greater than 15 to 20 beats per minute. To check for orthostatic changes, assist the patient to either sit up or stand, wait for approximately 2 to 3 minutes, then check for differences in the vital signs from when the patient was lying down or sitting. Orthostatic changes may also cause the patient to complain of dizziness, light-headedness, or nausea.

SKIN COLOR CHANGES

During compensatory shock the patient's skin typically turns pale or chalky white. People with dark skin are more likely to appear ashen gray when in shock. One way to assess *all* patients for perfusion—and perhaps the best check for African-Americans or other people of color—is to note the color of their mucous membranes. Pull down an eyelid or look at the inside of the lips. Reddish or bright pink is normal. Light pink or pale is a sign of diminished circulation.

Stage 2: Decompensatory Shock

Sometimes compensatory mechanisms are ineffective. Perhaps there is too much damage for compensation. Blood pressure starts to drop. Now the body enters *decompensatory shock,* sometimes called *progressive shock.* The patient may become obsessively anxious and spiral down through progressively deteriorating levels of consciousness. The heart rate continues to increase. There is a greater increase in the respiratory rate and a decrease in the depth of respirations. The lesser amount of oxygen reaching the tissues changes the skin from pale to cyanotic (blue), evidence of lack of oxygen. *Diaphoresis* (profuse sweating) will begin and may soak the body.

Signs and Symptoms of Decompensatory Shock

1. Altered level of consciousness.
2. Rapid, weak pulse.
3. Rapid, shallow respirations.
4. Cold clammy skin; diaphoresis.
5. Very pale to cyanotic or mottled complexion.
6. Blood pressure dropping, with weakening radial pulse.
7. Pupils sluggish to react.

DECOMPENSATORY SHOCK VIA BLOOD LOSS

With a loss of 30 to 40 percent in blood volume, death may occur, if the loss is rapid. The patient's body can no longer compensate for the reduction in blood volume. But, even at this stage, the patient may be saved with aggressive treatment for shock and rapid evacuation.

Stage 3: Irreversible Shock

Eventually the body runs out of options and begins to sink into *irreversible shock.* In this stage, hypoxia suffocates the brain, heart, liver, kidneys, and other vital organs, and accumulating waste products poison the entire system. The patient becomes unresponsive. An extremely rapid heart rate is strongly indicative. Respirations may slow down and may become labored. The skin grows cold and mottled. Blood pressure drops to the point of being undetectable by palpating a pulse.

Sometime during this final stage, the body releases an overwhelming surge of energy, a final all-out attempt to recover. This "bounce back," as it is sometimes called, may cause a return of blood pressure and consciousness. But the effect is short-lived, and unfortunately so is the patient.

Signs and Symptoms of Irreversible Shock

1. Unresponsive.
2. Extremely rapid heart rate.
3. Slow, agonal respirations.
4. Cold and damp or cold and dry skin.
5. Cyanotic or mottled appearance.
6. Blood pressure perhaps undetectable, with no pulses palpable.
7. Pupils very slow to react or fixed and dilated.

■ IRREVERSIBLE SHOCK VIA BLOOD LOSS

Loss of more than 40 percent of the blood volume and the resulting inadequate perfusion are typically fatal. Only definitive medical intervention will preserve the life of this patient.

Risk Factors

Bleeding and/or trauma are obvious conditions that predispose a patient to shock. Dehydration alone can lead to shock. The dehydrated person who is also a patient with trauma is doubly at risk. Children and young adolescents have dangerously deceptive reactions in shock. They compensate very well for a long period of time, often masking the severity of the condition. Then, with very little warning, they may decline through the progressive stage quite rapidly. The elderly and the chronically or acutely ill will often have weakened systems, impairing their ability to compensate. Pregnant women are supplying blood to two organisms, and, because the body diverts blood from the abdomen in shock, the fetus is placed in great jeopardy.

Management of Shock

Because of the extended transport time associated with a wilderness environment, the downward spiral of shock must always be on the mind of the WFR. Anticipating shock in someone far from a hospital is so critical that the first act of management is actually an act of prevention—and proper hydration is the key. Be sure that all members of your party are maintaining an adequate water intake, generally indicated by clear and copious urine output. Then, whenever an injury or illness suggests, do not wait for the signs and symptoms of shock to appear in the patient. **Anticipate and treat for shock until you unequivocally rule it out.**

Early intervention will slow down the spiral and extend the amount of time a patient can remain viable. On the other hand, rough handling and anxious attention contribute to the general decline of the patient's condition. Handle the patient gently and speak in soothing, reassuring tones. Keep the patient informed of what you are doing and encourage a positive attitude. You will have comparatively long periods to spend with your patient. Use that time to attend rigorously to the details of patient care. Act deliberately and with confidence. Avoid panic or, at least, keep it to a quiet minimum. Move your patient only when necessary and strive to make her or him as comfortable as possible. Anxiety can exacerbate shock by making the heart and lungs work harder. Therefore it's important to keep the patient calm. **Your calm attention may be all that keeps your patient alive.**

As a WFR, another priority is to prevent further decline of the patient, and, therefore, whenever possible, it is important to treat the underlying cause of shock. During the initial survey on a trauma patient, for example, you should detect any gross venous and arterial bleeding, and stop this bleeding immediately (see Chapter 6: Bleeding). Applying traction to a broken femur, as another example, will reduce pain and may slow down hemorrhaging. In fact, splinting any fractures will lessen pain and anxiety associated with those injuries and should contribute to stabilizing your patient's condition (see Chapter 12: Fractures).

Remember, shock indicates that cells are not receiving an adequate supply of oxygen. Supplemental oxygen, although typically not available in the wilderness, may be of great benefit. High-flow/high-concentration oxygen would best be administered via a non-rebreather mask (see Appendix A: Oxygen and Mechanical Aids to Breathing).

You can further assist the body's compensatory efforts by placing the patient in a supine position (horizontal and face up) and raising the legs slightly, no more than 10 to 12 inches, a maneuver that allows whatever blood is circulating to flow more easily to the brain. You can position a shock patient in the wilderness by simply placing him or her on an insulating pad with the legs flexed at the knees and placed over a backpack or canoe bag.

Beware: Do not overelevate the feet, which puts pressure on the opening of the stomach and may encourage nausea and vomiting.

Beware: Do not raise the legs if your assessment is cardiogenic shock (see Chapter 23: Cardiac Emergencies).

Note: Pregnant women do better in the *left lateral recumbent position* (on the left side with right knee drawn toward body and elevated on a pillow). In the supine position the heavy mass of the fetus may press on the inferior vena cava, reducing blood return to the heart.

The body's attempts at temperature regulation, a function normally carried out by the circulatory system, virtually cease during shock. Therefore, insulating your patient from the cold and preventing heat gain in an extremely hot environment may be important.

Monitor vital signs often, preferably every 5 minutes or so. Pay particular attention to the patient's heart rate and mental status. They will be your primary and secondary indicators, respectively, of how he or she is doing. An increasing heart rate and increasing levels of anxiety and restlessness indicate progressive shock. An accurate chart of changing signs informs professional rescue and emergency room personnel how quickly they need to act and will assist them in making critical decisions in incidents involving multiple casualties.

"Nothing by mouth" is the standard of care for shock patients in the urban environment. But you must be concerned with the oral hydration of the shock patient in the wilderness, as

Figure 7-1: *Shock treatment position.*

long as he or she is conscious, can accept a container of water, and can swallow. You may administer liquids, preferably cool liquids in small amounts. Lightly salted water may be beneficial—the salt should not be detectable by taste. Avoid any drink that tastes sweet. What you do not want to do is encourage vomiting with hydration.

General Shock Management

1. Assess and monitor the ABCs-airway, breathing, and circulation.
2. Keep the patient physically and emotionally calm.
3. Treat any treatable causes (stop serious bleeding, splint fractures).
4. Administer oxygen (high-flow/high-concentration) if possible.
5. Handle gently.
6. Elevate legs, but no more than 10 to 12 inches.
7. Maintain body temperature.
8. Monitor vital signs.
9. Orally rehydrate, when appropriate.
10. Consider an evacuation of the patient.

Evacuation Guidelines

Initiate a rapid evacuation of all patients who exhibit sustained or progressive deterioration—the compensatory shock patient who does not improve, and any patient whose vital signs continue to worsen. If the shock patient is not getting better, the shock patient should be getting out as soon as possible.

Conclusion

Although no cause is immediately evident, the skier-turned-patient shows signs of shock. Calmly and gently, you pad beneath her legs, raising them approximately 10 inches. You cover her with her tent mate's sleeping bag since she is already in her own.

Proceeding through a focused history and examination, you discover she took a hard fall on the last downhill run of the day. She landed, she says, on her left side. Your physical exam, performed reaching beneath the sleeping bags, elicits pain when you press on her left side, low on her rib cage. Taking a look, you see bruising approximately over the spleen. Perhaps internal bleeding is a probable cause.

A second, and then a third, set of vital signs show increasing heart rate and increasing respiratory rate. She shows an increasing level of anxiety. Picking a team of your fastest skiers, you begin preparations to send out a request for a rapid evacuation.

Ken Thompson, EMT, contributed his expertise to this chapter.

CHAPTER EIGHT

Spine Injuries

You Should Be Able To:

1. Describe the basic anatomy of the spinal column.
2. Describe the most common mechanisms of injury to the spine.
3. Demonstrate proper assessment for spine and spinal cord injury.
4. Demonstrate proper emergency care for the spine-injured patient.
5. Describe the special wilderness considerations for a focused spine assessment that may allow a decision to discontinue spinal treatment.

It Could Happen to You

April's sun-washed warmth only hints at the blast furnace heat of summer to come. Mountain biking the Kokopeli Trail, your group passed the Colorado border miles back. A lot of Utah stretches out dry and sandy ahead, and your party stretches out over a quarter mile. The silence breaks to a faint call of your name, drifting to you from somewhere behind. You stop and look back over the rough track you've just completed, a section that included a dicey piece of sloping sandstone. You see arms waving overhead in the distance. There's nothing to do but retrace your hard-earned wheel turns.

You find Steve flat on his back in soft sand at the bottom of the sandstone slope, his bike approximately 10 feet away. The bike appears intact. His helmet is attached by its strap to the back of the bike. A nasty abrasion on Steve's forehead oozes blood. You hear him groan as you approach.

"He went over the handlebars," someone says.

Injuries to the spine rank among the most intricate and potentially devastating problems a Wilderness First Responder has to deal with. Spinal cord damage can produce paralysis or death as a result of the initial mechanism or of mishandling by well-meaning rescuers. The problems of proper management are compounded by the lack of equipment and the remoteness of wilderness environments. Skilled handling and adequate equipment, some of which may need to be improvised, are keys to successful patient care.

Basic Anatomy of the Spine

The spinal column is made up of thirty-three vertebrae. The top seven are called *cervical vertebrae* (C1–C7). They form the neck, the most flexible and least protected portion of the spine and, therefore, the portion most prone to serious injury. The topmost cervical vertebra, C1, sometimes called the *atlas*, supports the weight of the head, as the mythical Titan named Atlas supports the weight of the world on his shoulders. C2 is sometimes called the *axis*, most of the movement of the head on the spine occurs there. A small spike of bone, the *odontoid process*, rises from C2 into C1 to form a peg on which the head rotates. The twelve *thoracic vertebrae* (T1–T12) are attached to the ribs and protected by back muscles, making them relatively rigid and secure. The lower back is composed of five *lumbar vertebrae* (L1–L5). They are large and strong, but the muscles supporting them are highly susceptible to injury and pain from improper lifting and moving techniques. The five *sacral vertebrae* are fused together to form the *sacrum*, where the spine is attached to the pelvic girdle. The final four *coccygeal vertebrae* are fused together to form the *coccyx*, the lower tip of the vertebral column—the tailbone.

The nonfused vertebrae of the neck and back are joined to each other by ligaments attached to their *anterior* (front), *posterior* (back), and *lateral* (side) surfaces. Between these vertebrae are discs of cartilage that allow some movement between bones and that serve as shock absorbers.

Down through an opening in each of the vertebrae—an opening called the *vertebral foramen*, or spinal canal—passes the spinal cord. The spinal cord is a continuous bundle of nerves running from the brain to the juncture of L1 and L2 in the lumbar region, approximately the level of the navel. Below L1 the spinal cord branches into loose fibers called the *cauda equina* and resembling, somewhat, a horse's tail. The spinal cord is protected by the bones of the spine and by a continuation of the same meninges and cerebrospinal fluid that protect the brain (see Chapter 9: Head Injuries). Spinal nerves branch out from the spinal cord between the vertebrae, sending countless peripheral nerves to the body. This is the connection of the brain to all parts of the body.

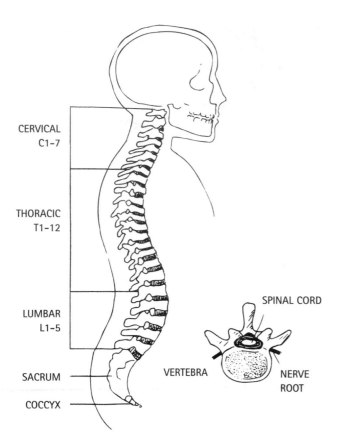

Figure 8-1: *Basic anatomy of the spine.*

Mechanisms of Injury

Most injuries to the spinal cord occur as a result of injury to the spinal vertebrae. Although spinal column injuries usually heal well, injuries to the nerve tissue of the cord are often irreparable. Spinal injury is produced by one or more of the following mechanisms:

1. Excessive flexion (chin forced toward chest, such as when a hiker tumbles backward down a hill). Flexion is especially dangerous when the head is axially loaded (such as when a diver hits the bottom of shallow pond).
2. Excessive extension (head thrown backward, such as when a driver is thrown backward when his vehicle is hit from behind by a second vehicle).

3. Compression (vertebrae forcefully driven together, such as when a climber falls and lands sitting).
4. Distraction (vertebrae forcefully pulled apart, such as in hanging).
5. Excessive rotation (vertebrae forcefully twisted, such as when a skier tumbles and the skis fail to release).
6. Excessive lateral bending (vertebrae forced to one side or the other, such as when a stationary skier is hit from the side by a high-speed skier).
7. Penetration injury (such as a stabbing, gunshot wound, or goring).
8. Sudden and violent deceleration (such as a 30-foot unroped fall).

Types of Spinal Cord Injuries

Vertebral damage—breaks in the bones of the back—can occur without spinal cord damage, but the mechanism of injury can cause fragments of bone or disc to cut the cord. If the cord is cut on impact, the damage is almost immediately obvious. Spinal cord damage can also occur without significant damage to bones and discs due to swelling of the cord. An injury that causes swelling of the cord within the spinal canal can produce as serious an injury as a "broken back"—injuries such as *paraplegia*, paralysis of the lower portion of the body, or *quadriplegia*, paralysis of all four extremities and almost always the trunk. Sometimes hours pass before obvious spinal cord damage from swelling shows up. Even though paralysis may occur hours later, some signs and symptoms of spinal damage *will* appear shortly after the injury.

Uninterrupted Spinal Cord

The spinal cord is intact despite the possibility of fractures and ligamental disruption. There are no changes in *circulation, sensation,* and *motion* (**CSM**) in the extremities—no decrease in strength of pulses, no loss of feeling, no loss of motor function, and no unusual weakness in the arms and/or legs. Proper care is critical to ensure the spinal cord stays intact.

Partial Spinal Cord Injury

Normal blood flow to the spinal cord is interrupted with the possibility of damage to cord tissue. There are one or more changes in CSM, evidence of nerve injury—decreases in strength of pulses, loss of sensation, loss of normal motor function in the extremities, and possibly weakness in the arms and/or legs. Early recognition and stabilization in the field are critical to keep the patient's injury from getting worse.

Full Spinal Cord Injury

There is complete interruption of spinal cord function with obvious loss of CSM—decreases in strength of pulses, loss of sensation, loss of motor function, and perhaps paralysis. Patients are often in respiratory distress. Early recognition and stabilization are critical to keep the patient alive.

General Assessment of the Spine

When you suspect possible spine injury, take extra time to be sure your patient assessment is accurate and complete. During your assessment of the scene, attempt to determine, as accurately as possible, the mechanism of injury and the forces involved. The spine may have been injured when

1. The patient suddenly and forcefully stops—such as in motor vehicle accidents, falls from a height (especially if the height is as much or more than twice the patient's height), high-speed skiing accidents, high-speed bicycle accidents, and diving accidents (especially dives into shallow water and dives from a height).
2. The patient hits or lands head first.
3. The patient lands on rocks instead of a softer surface.
4. The patient has a head injury, especially if the injury involved loss of consciousness.
5. The patient has a penetrating injury to the spinal region.

Treat all unconscious trauma patients as if they have a spinal injury until it is proven otherwise, preferably by physical examination in an emergency room.

In your initial assessment, at B for Breathing (see Chapter 3: Patient Assessment), note any difficulty breathing. Breathing difficulty may be a sign of spinal cord damage. Nerve messages to the diaphragm come from between the upper cervical vertebrae, thus the old reminder: Above C4, they breathe no more! Damage between C4 and T1 often shows itself with "belly breathing" in a patient: The chest muscles are not working, so the patient's belly rises and falls, sometimes dramatically.

Vital signs indicating shock may also indicate spinal cord damage. In this case the shock would be neurogenic, with loss of control over the dilation and constriction of blood vessels due to nerve damage. Although neurogenic shock is sometimes difficult to assess, the patient is often vasoconstricted (cool and pale) above the point of spinal cord injury and vasodilated (warm and flushed) below the point of injury.

Neurological deficits in the extremities may indicate spinal cord damage. With cord damage patients may complain of numbness or tingling in the hands and/or feet or unusual sensations of cold or heat in the extremities. Examination may disclose weakness or inability to move the extremities. It is important to note that other injuries (including fractured bones and dislocated joints) may cause altered sensations when the spinal cord is perfectly intact.

During your check of the patient's back—via rolling, if necessary—palpate carefully each and every bone along the spinal column. Pain and/or tenderness anywhere in the spinal column are indicators of possible spine damage. The patient's back should be inspected visually and palpated on the skin and not through thick clothing that could hinder a proper assessment. In an urban environment the clothes are typically cut away. In the wilderness your wish to retain the integrity of the clothing—the patient may need it later—creates a situation requiring some thought. Removing clothing may cause too much movement of the patient. Clothing can usually be loosened—unzipped or unbuttoned—and pushed gently out of the way. A patient on his or her back may be safely logrolled to gain access to the back (see "Lifting and Moving a Patient," this chapter). It is important to note that patients with spine injuries may initially move, even walk around, and still be seriously damaged. **Do not ask the patient to move his or her spine to confirm your assessment.**

Signs and Symptoms of Spinal Column Injury

Injury to the spinal column may produce one and usually more than one of the following signs and symptoms:
1. Difficulty breathing.
2. Signs and symptoms of shock.
3. Altered sensations (numbness, tingling, unusual heat or cold sensations) in the arms and/or legs, especially in the hands and feet.
4. Weakness or paralysis.
5. Pain and/or tenderness along the spinal column.
6. Obvious evidence of injury to the spinal column (cuts, punctures, bruises).
7. Incontinence (loss of bladder and/or bowel control).

Treatment for Suspected Spinal Injury

In general the immediate emergency care for a spinal injury is to immobilize the entire spinal column, including the head and pelvis of the patient, on a long backboard with a cervical collar and head immobilizer in place and then to transport the patient to a medical facility. The rule, in most cases, is to overtreat rather than to undertreat.

Figure 8-2: *Manual stabilization of C-spine.*

Treatment for spinal injury *must* include the following:

1. Manually (hands-on) stabilize the head and cervical spine, assure an adequate airway, and keep the patient as still as possible. This may require gentle alignment of the patient's cervical spine into a neutral position. Realignment should be slow and steady. **Do not attempt to realign the cervical spine if it requires force or causes pain,** in which case the patient must be immobilized as found. If the airway must be opened, use a jaw thrust (see Chapter 4: Airway and Breathing). Ideally, maintain manual stabilization during the entire patient assessment and until mechanical stabilization is secured.

2. Check for adequate circulation, sensation, and motion in the extremities.

3. Apply a cervical collar. Commercial collars should be properly sized for the patient and applied to the patient according to the manufacturer's instructions. In the wilderness cervical collars can be improvised (see "Special Considerations for the Wilderness," this chapter). Collars do not adequately immobilize the cervical spine, but they do serve as an adjunct until full mechanical immobilization is established. Even after application of a cervical collar, commercial or improvised, manual stabilization should be maintained.

4. If supplemental oxygen is available, it should be started as soon as possible.

5. Move the patient to a long backboard or a rigid litter (see "Lifting and Moving a Patient," this chapter).

6. For the patient's comfort, pad the places where the patient will feel pressure, especially beneath the head. Use a thin pad beneath the head to prevent flexion of the cervical spine. Avoid the voids: Fill the spaces, for instance, under the small of the back and beneath the knees and ankles.

7. Secure the patient to the long backboard or litter. Straps should be placed across bones: upper chest and shoulders, pelvis, and upper and lower legs. Conscious patients will prefer to have their arms free if the arms are undamaged. **Do not place the straps where they will interfere with breathing.** Pad the straps in places where they may cause the patient discomfort. Pad any spaces between the straps and patient to prevent shifting of the patient during carrying. **Secure the head last.** If the head is secured first, shifting of the cervical spine may occur during strapping. The patient should be completely immobile when you have finished.

8. Reassess the patient for adequate circulation and normal sensations in the extremities. Inadequate circulation may indicate you have strapped the patient *too* tightly.

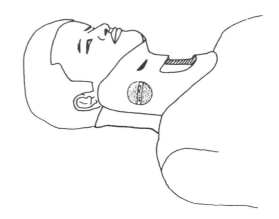

Figure 8-3: *C-collar.*

Lifting and Moving a Patient

Although trepidation may surround movement of a patient with a suspected spinal injury, it must be done for one or more of several reasons, including rolling the patient to assess the back, getting the patient off the cold ground, and lifting the patient into a litter.

Logrolling a Patient

Although one rescuer can successfully logroll a patient into a side position (see Chapter 4: Airway and Breathing), three, or even four, rescuers make the process easier and safer for the patient.

1. One rescuer maintains manual stabilization of the patient's head and neck during the process. This "head" rescuer gives the movement commands.

2. The patient's arms are positioned along the sides of his or her body, if possible.

3. Additional rescuers position themselves on the same side of the patient, reaching across the patient at the shoulder, hip, thigh, and lower leg.

Figure 8-4: *One-rescuer logroll.*

4. On command, rescuers roll the patient toward themselves and onto his or her side.
5. The back may now be assessed. Backboards, pads, sleeping bags, etc., may now be placed under the patient.
6. On command, rescuers roll the patient back into a supine position.

You may find you must logroll the patient onto a backboard. Place the board parallel to the patient and slightly off center, toward the patient's head. Place the patient's arms along the sides of the body (palms toward thighs), and logroll the patient as a unit, maintaining careful manual alignment of the head and spine. Slide the patient as a unit along the longitudinal axis of

his or her body toward the head of the board to center the patient on the board. **Do not shove the patient laterally in order to center the body on the board.** While the patient is in a side position, you can use the time to carefully check the back again.

Lifting a Patient

It is better to lift the patient as a unit while maintaining careful alignment of the head and spine, slide the board or litter underneath, and lower the patient into position on the board or litter. This method requires an adequate number of people. Think of it as being "beamed aboard," a technique for moving people

Figure 8-5: *Three-rescuer logroll.*

Figure 8-6:
Preparing to BEAM.

familiar to *Star Trek* fans. Think of **BEAM** as *body elevation and movement*. With someone manually in control of the head and spine, five or six additional rescuers divide approximately according to strength on either side of the patient. Sliding their hands underneath the patient, the rescuers, using their legs and not their backs, lift the patient on the command of the rescuer at the head, and lower the patient on the command of the rescuer at the head ("On three we go down. Anybody not ready? OK, one, two, three. . . ."). Strong rescuers may even move patients short distances using this technique, such as out of the shallows of a lake onto shore. Proper technique is critical; if you find yourself in charge of a group that has never "beamed" a patient before, practice first, if time allows, on a healthy person.

Standing Takedown

If a patient who must be treated for a potential spine injury is found standing, place the board or rigid litter against the patient's back and lower patient and device to the ground while maintaining careful manual stabilization of the head and spine.

Special Considerations for the Wilderness

Patients suspected of having a spine injury should be rapidly evacuated from a remote environment with full immobilization. Swelling may be taking place, choking off the blood supply to the spinal cord, and rapid evacuation to definitive care may be beneficial. Intravenous drugs exist that could possibly, if administered soon enough, reduce or reverse the effects of some spinal cord injuries.

Packaging the Spine-Injured Patient

In the wilderness adequate packaging of the patient for a long evacuation is extremely important to reduce the chance of increasing the severity of a spine injury. It is difficult to improvise a safe, adequate backboard, especially one that must be carried over rugged terrain, but with attention to detail it is possible to create an improvised litter (see Chapter 36: Wilderness

Transportation of the Sick or Injured). **Improvised litters should be used to evacuate a spine-injured patient in only the most extreme circumstances,** such as when there is *no* possibility that help will arrive with a commercial litter. Commercially available rigid litters provide adequate spine immobilization and are preferable for evacuating a spine-injured patient. During a wait for a litter, the patient should be moved onto a sleeping pad or similar protection from the ground. The patient should be collared, and the patient's head and neck should be kept manually stabilized or secured between padded rocks, logs, backpacks, or other such objects.

When a litter is available, the patient should be packaged with attention and comfort. Padding should be placed under the entire patient, and extra padding should be placed at stress points created by being tied to a rack: behind the knees, in the small of the back, beneath the head and neck. All straps should be adequately padded for comfort. The patient is essentially made *free of any movement* (**FOAM**), but, while being carried, the straps can be tightened for steep terrain and loosened on flatter terrain for increased comfort. In addition, the patient usually requires insulation from the cold and protection from rain or snow.

Most patients are packaged supine, but some patients may require packaging in a side position, especially unconscious patients who require airway management. Adequate padding to prevent patient shifting on the litter is the key to side immobilization, and a firm pad beneath the head is required to provide proper positioning of the cervical spine. Again, **avoid the voids.**

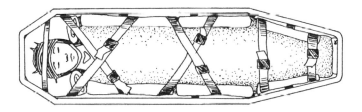

Figure 8-7: *Patient secured in litter.*

Improvisation of a Cervical Collar

An improvised cervical collar should immobilize the cervical spine as much as possible. Some improvised collars provide little immobilization, and some provide a lot, but the collar remains an important aspect of treatment. The fact that something is supporting a patient's neck reminds that person to refrain from moving.

While positioning a collar, make sure the patient is moved as little as possible, but a second rescuer may raise the head slightly to assist. What you use to improvise a collar should especially discourage flexion of the chin toward the chest. A couple of suggestions follow:

1. Foamlite sleeping pads can be cut into the shape of a cervical collar and held in place with tape.
2. Sweaters or parkas can be rolled and placed around the patient's neck using the sleeves to secure the garment in place.

Remember: A collar is *not* a substitute for consistent hands-on stabilization of the head and neck. The standard of care requires a collar, a head immobilizer, a rigid litter, and enough straps to secure the patient from movement.

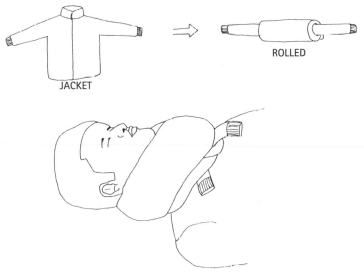

Figure 8-8: *Improvised C-collar.*

The Focused Spine Assessment

First, a disclaimer. There is always a slim chance you will miss a spine injury and cause harm to the patient. If you always want to be certain, every patient with a mechanism of injury (MOI) for spinal injury should be immobilized on a backboard or litter. Wilderness medical training, however, offers a way to do a focused spine assessment that may allow you to decide to discontinue spinal precautions. The focused spine assessment is extremely accurate, and it has been approved by the Wilderness Medical Society. It gives all those patients who do not have a spine injury freedom from hours or even days of spinal immobilization and all those rescuers freedom from carrying a patient miles and miles to have him or her cleared of spine injury minutes after arrival at the hospital.

The patient must receive a full patient assessment—initial assessment, head-to-toe examination, vital signs, history—with his or her spine manually immobilized *prior to* considering the patient a candidate for a focused spine assessment. Do not allow your wish to discontinue spine precautions interrupt the patient's need for a full assessment or create a bias in your decision.

The performance of a focused spine assessment must be made as a separate and distinct assessment, not as a part of the standard patient assessment system. A focused assessment of the spine, therefore, will be the *second* time you are checking for possible spine damage.

The patient must be reliable: alert and oriented, sober, and without distractions. Does "alert and oriented" mean alert and oriented to person, place, time, and event, that is, A+O x 4? (See Chapter 3: Patient Assessment.) That would be best, but if the patient is only A+O x 3 (with immediate loss of vivid memory as to the event), that person could still be considered for a focused spine assessment if the rest of the criteria are met. "Sober" means the patient does not have drugs or alcohol in his or her system in amounts that alter reliability. A reliable patient is also free of distractions, physical and/or psychological, that could block perception of spinal pain. A patient with a grossly angulated fracture, for instance, or a mother in the depths of worry over her child may be too distracted to pay attention to a focused spine assessment. An unreliable patient with an MOI for spine injury must be treated for spine injury.

The patient must be free of altered sensations in the extremities. Altered sensations include numbness on palpation, weakness when checked for strength, tingling (pins and needles), inability to move, and even temperature irregularities such as cold feet without another reason for cold feet. A patient with an MOI for spine injury who has altered sensations in the extremities must be treated for spine injury.

The patient must have no complaints of pain in the neck and back on a second careful palpation of the spinal column. Almost all patients with a spine injury will complain of pain, and those free of pain will almost always have some tenderness or guarding in the neck and/or back when palpated. A patient with an MOI for spine injury who has pain or tenderness in the neck and/or back must be treated for spine injury.

Now the patient may be allowed to move about freely but cautiously, if no other injuries, of course, would hamper moving about. If the patient experiences the onset of numbness or weakness later, or the onset of sharp pain later—not the tightness of muscle spasms—he or she must then be treated for spine injury.

Evacuation Guidelines

Initiate an immediate evacuation for anyone treated for a suspected spine injury. Whenever possible, the evacuation of the patient should be carried out with a commercial spinal immobilization device.

Conclusion

Back in the sand of Utah, Steve moans and tries to sit up. You discourage any movement, placing your hands on his head, lowering him once again to the ground.

A full patient assessment reveals no injuries other than mild abrasions to the forehead and right forearm. Taking a deep breath, you methodically check off the points that must be covered during a focused assessment of the spine: no loss of reliability, no pain or tenderness on careful palpation of the entire spinal column, no altered sensations in the extremities.

The wounds are cleaned and dressed. You remind your friend to let you know immediately if his condition changes, specifically changes in neck or back pain, and the trip continues.

Section Three

Traumatic Injuries

Head Injuries

It Could Happen to You

Ice formed overnight on the rocky slope called Central Gully at the head of Huntington Ravine. It happens that way sometimes in October on New Hampshire's Mount Washington. Two 20-year-old hikers—one male, one female—move off the trail seeking surer footing. She doesn't find it. Slipping on a more technical section of the ravine, she tumbles about 20 feet. Somewhere on the way down her face strikes forcefully against a stone.

Her friend descends carefully to find her sitting up, bleeding heavily from the nose, a nose that has already begun to puff up alarmingly. He has had enough first-aid training to assess the young woman for spine damage. He decides her spine has not been involved. They choose to hike out.

Within 30 minutes, the young man notices his friend beginning to act "weird." She grows increasingly irritable and confused. She can't remember if this is the right trail. Before long she asks, "What happened to me?"

He tells her, but five minutes later she asks the same question again.

Within an hour, she sits down beside the trail, refusing to walk farther. When he tries to help her to her feet, she slaps his hands away and angrily uses words he has seldom heard her utter. As he sighs deeply in frustration, she collapses. He is unable to get any response from her other than moans.

Head injuries range from simple scalp damage to severe brain damage. Serious, potentially life-threatening injury is usually a result of the uncompromising structure of the *cranium* (skull)—if the brain starts to swell, it is squashed because there's no room for the swelling to take place. Head injuries that involve the brain are a major source of disability and death in the wilderness because there is little the Wilderness First Responder can do other than recognize the risk to the patient, offer what small aid is available, and arrange an evacuation as soon as possible. Knowing when and how fast to work toward an evacuation of a head-injured patient is a critical skill for the WFR.

Basic Anatomy of the Head

The bulk of the brain is the *cerebrum*, the gray matter, the center of higher functions such as thought and emotion. In the back of the head, beneath the cerebrum, is the *cerebellum*, where equilibrium and coordination are controlled. The *brain stem* is at the base of the brain and is responsible for vital vegetative functions, including circulation and respiration. Enclosing the brain, and the spinal cord, are three blood-rich layers of tissue called, collectively, the *meninges*. Their names, from the brain outward, are the *pia mater*, *arachnoid*, and *dura mater*. (Together, the meninges may be remembered as a **PAD** for the brain.) *Cerebrospinal fluid*, a clear liquid manufactured at a constant rate in cavities in the brain called *ventricles*, continually circulates through the *subarachnoid space*, the space between the arachnoid and the pia mater meninges.

The brain sits about one millimeter from the inside of the cranium, protected by hair (on many people), the scalp, the thick skull, and the shock absorbency of tissues and the cerebrospinal fluid surrounding the brain. The brain maintains control of almost everything that goes on within the human body as long as levels of oxygen, temperature, blood sugar, and pressure stay adequate and balanced within the head. When brain tissue is damaged, however, it does not repair itself like other tissues in the body. Loss of brain cells is permanent. Loss of too many brain cells adds up to the death of the patient.

Types of Head Injuries

Injuries to the head may or may not involve scalp damage that requires medical attention. Injuries to the head that involve the

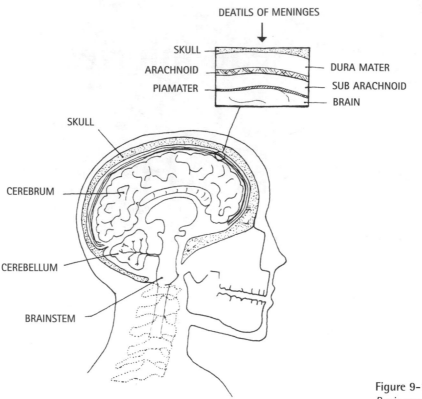

DEATILS OF MENINGES

Figure 9-1:
Basic anatomy of the head.

brain may be closed (the integrity of the skull remains intact) or open (the skull is fractured and/or an object has penetrated the skull). All injuries that may involve the brain require immediate attention.

The single most important sign indicating the possibility of brain damage is a change in the patient's level of consciousness. A normal level of consciousness—alert and oriented to person, place, time, and event—will deteriorate and typically involve alterations in personality as the pressure from swelling increases on a damaged brain.

Scalp Damage

Lacerations and avulsions of the scalp have a habit of bleeding profusely due to the blood-rich nature of the scalp and to the fact that blood vessels in the scalp do not constrict rapidly after injury to help slow bleeding. Scalp tears also tend to reveal the skull, which tends to make rescuers nervous. Fortunately for rescuers and patients, scalp damage is seldom a serious injury, and shock rarely results from scalp bleeding alone, except in small children.

To encourage scalp bleeding to stop, apply direct pressure with a bulky dressing. The recommendation of a bulky dressing is based on consideration of the integrity of the skull beneath the wound. If a blow to the head was forceful enough to fracture the skull, too much direct pressure could press bone fragments into the brain, with devastating results. A bulky dressing disperses the pressure. Blood loss can be further discouraged by pulling the edges of the wound together, perhaps with wound closure strips, although it may be difficult to get the strips to stick to a bloody scalp. Hair from both sides of the cut may be twisted and tied across the cut to hold the wound closed.

Once bleeding has stopped, the wound should be cleaned and dressed (see Chapter 15: Wilderness Wound Management). Hair may be clipped to make wound management easier, but do not shave the scalp. Shaving increases the risk of infection.

There may also be a rather large "goose egg" developing rapidly on the head of the patient after an impact injury to the skull. Types of helmets worn, for instance, by rock climbers or mountain bikers may prevent such lumps but do not necessarily prevent brain injury, even though helmets typically lessen the severity of the impact. Large lumps may be treated with the application of cold and, by themselves, are not a cause for concern. The amount of brain damage, which may be cause for great concern, is usually relative to the forces involved.

Open Head Injury

A skull fracture is an *open head injury*, an injury that involves a break in the integrity of the skull. In addition to decreased level of consciousness, a fracture of the skull may be indicated by obvious signs that include (1) fracture lines visible beneath a tear in the scalp; (2) deformity, often a depression, at the injury site; (3) "raccoon eyes" (black-and-blue discoloration around swollen eyes); (4) Battle's sign (black-and-blue discoloration behind and below the ears); (5) seizures; and (6) cerebrospinal fluid (CSF) or blood mixed with CSF leaking from the nose and/or ears.

Note: You should not attempt to stop the flow of fluid from ears or nose resulting from a head injury. The patient may be relieving some increasing intracranial pressure. A patient with a fractured skull should be evacuated as soon as possible.

Signs and Symptoms of a Skull Fracture

1. Decrease in level of consciousness.
2. Visible fracture lines where the scalp has been torn away.
3. Depression in the skull.
4. Raccoon eyes.
5. Battle's sign.
6. Seizures.
7. Cerebrospinal fluid or blood mixed with CSF from the nose and/or ears.

Unconsciousness

Unconsciousness is a term that often is used incorrectly by the rescuer. Remember that there are various levels of consciousness on the AVPU scale. The levels of consciousness range in descending order from fully alert to totally unresponsive (LOC = A, V, P, U). It is important to note that a patient who is unconscious due to a blow to the head does not wake up, does not become alert, when stimulated by either verbal or painful stimuli. So the phrase "unconscious following a blow to the head" should only be used when the patient is totally unresponsive to any stimuli. In other words, LOC = U.

When an object such as an ice ax, knife, or bullet penetrates the skull, permanent brain damage almost always results. The wound should be covered with bulky sterile dressings. Often the object lodges in the brain, sticking out of the skull. Impaled objects should be stabilized in place with large bulky dressings, *not* removed. Bleeding should be allowed to continue, not stopped with direct pressure, in hopes that some of the pressure within the cranium will self-release. Any penetrating wound to the head requires an immediate evacuation of the patient.

Closed Head Injury

A *closed head injury* involves an immediate and often transient loss of normal brain function—a period of unconsciousness—following a violent impact to the head. No matter how minor the

severity of the closed head injury, there will always be some damage to brain cells. Two immediate factors will help you decide how severe the trauma to the patient's brain has been: the *duration* of unconsciousness and the *depth* of unconsciousness.

In the urban environment any blow to the head resulting in loss of consciousness should send the patient to a hospital for evaluation. In the wilderness environment the decision-making process leading to the evacuation of someone knocked unconscious may be plagued with doubt. If you evacuate anyone knocked unconscious, you will be acting responsibly. But if the loss of consciousness lasts momentarily, and if the patient responds immediately, wakes up to verbal or painful stimuli, and remains without symptoms of brain injury after regaining consciousness, you may sometimes choose to let the patient remain in the wilderness.

Another scenario you may encounter is the successive brain injury. Successive closed head injuries, even minor ones, may cause cumulative brain damage and increase the risk to the patient. Therefore, if the patient reveals that this is not the first time he or she has been knocked unconscious, your eagerness to have the patient evaluated in a hospital should increase.

Keep in mind that many outdoor organizations have a standing protocol to evacuate *anyone* who has been knocked unconscious. Such a protocol definitely eases the burden of decision on the WFR.

A serious injury to the brain may or may not be associated with any obvious signs of damage to the head or face. There does need to be sufficient force to slam the brain around inside the skull. What may happen inside the patient's head is this: Blood flows out of broken blood vessels and, sometimes, blood serum starts to leak out of the damaged area of the brain. Swelling results. As the size of the brain increases, there is less and less room for the flow of life-sustaining blood. *Intracranial pressure* (ICP), the normal pressure inside the skull, starts to rise, and, as a result, the brain may sustain permanent damage or death.

Brain injuries may be categorized into one of several types. A *contusion* (bruise) to the brain may lead to cerebral *edema* (swelling) that could cause an increase in ICP. An *epidural hematoma* (bleeding above the dura) is almost exclusively arterial bleeding and typically causes a rapid rise in ICP and a quick death. A *subdural hematoma* involves bleeding between the dura and the arachnoid and/or bleeding between the arachnoid and the cerebrum. Bleeding between the arachnoid and the cerebrum may be called a *subarachnoid hematoma*. Subdural bleeding is almost exclusively venous bleeding, which causes a slow rise in ICP and may lead to death hours or even days after an accident.

Although the progress of increasing ICP is not absolutely predictable, certain changes in the patient will probably occur, sometimes fast, sometimes slowly. Once regaining consciousness, the patient may become increasingly disoriented and irritable, and, perhaps, combative. These changes in level of consciousness are a result of pressure on the cerebrum.

Remember: Level of consciousness changes are the first and foremost signs of brain damage.

Later, after level of consciousness changes, breathing will tend to be erratic and/or rapid and deep. Heartbeats slow down and begin to bound. Blood pressure rises. If you can monitor blood pressure, you will see a widening of the pulse pressure—the systolic pressure rising faster than the diastolic pressure. Heart rate and blood pressure changes are serious, and both are caused by pressure on the brain stem that is being crushed against the floor of the skull. Skin may appear flushed, especially in the face. Later indications of increasing ICP may include pupils that become distinctly unequal and unusual, rigid body postures.

Serious brain damage may be detectable by other signs and symptoms that include (1) prolonged unconsciousness, (2) a headache that increases in intensity, (3) complaints of blurred vision or other visual disturbances extending over 1 hour, (4) excessive or unusual tiredness or sleepiness, (5) protracted nausea and vomiting, (6) seizures, and (7) *ataxia* (unusual loss of balance).

Without definitive treatment for increasing ICP, the patient's brain may rupture (*herniate*) through the floor of the skull, the only place where there is a hole for the brain to try to squeeze through. Death immediately follows. Even if the brain manages to gain control of the swelling, the loss of circulation often results in permanent loss of the brain's normal function.

Brain Injury vs. Shock

Signs of increasing intracranial pressure should not be mistaken for signs of shock. The triad of changes indicating a rise in ICP are (1) erratic respiratory rate, growing rapid and deep; (2) slowing and bounding heart rate; and (3) rising blood pressure. The triad of changes indicating shock are (1) rapid and shallow respiratory rate; (2) increasing and weakening heart rate; and (3) dropping blood pressure. Although a patient may have sustained injuries that could cause either increasing ICP and shock, the signs and symptoms tell which problem should be treated. In other words, vital signs showing increasing ICP mean treat for increasing ICP, and vital signs showing shock mean treat for shock (see Chapter 7: Shock).

Signs and Symptoms of Increasing Intracranial Pressure

1. Distinct changes in mental status, often described as disoriented to irritable to combative to coma.
2. Slowing and bounding heart rate.
3. Erratic respiratory rate, later becoming rapid and deep.
4. Rise in blood pressure.
5. Flushed skin, especially in the face.
6. Intense headache.
7. Disturbances in vision.
8. Excessive sleepiness.
9. Protracted nausea and vomiting.
10. Ataxia (loss of balance).
11. Seizures.
12. Unequal pupils.
13. Rigid body postures.

Levels of Head Injury

For quick reference, all head injuries can be loosely divided into three levels:

1. **No loss of consciousness** means only an extremely rare chance of serious problems, even if heavy external bleeding and a huge "goose egg" bump may accompany the injury. In the wilderness monitor the patient for at least 24 hours, watching primarily for changes in level of consciousness, and evacuate if the signs and symptoms of serious brain damage develop.

2. **Momentary loss of consciousness** indicates that the patient's brain underwent some transient loss of function and, therefore, some brain cell damage. Keep in mind the patient may not be aware he or she has been unconscious. Ask witnesses, if possible. People knocked unconscious only momentarily are often OK, but they may only appear OK initially and then start to deteriorate. Either evacuate the patient as soon as possible, while he or she can still walk, or monitor closely for 24 hours, awakening the patient several times during the first night—every couple of hours—to evaluate the level of consciousness. Evacuate if the signs and symptoms of serious brain damage develop.

3. **Long-term loss of consciousness** and/or other warning signs and symptoms indicating the brain has been seriously damaged warrant an immediate evacuation.

Treatment for Serious Head Injury

Any force severe enough to cause unconsciousness may have caused damage to the cervical spine. If the patient regains consciousness and reliability, you may proceed with your assessment with the possibility of a focused assessment of the spine (see Chapter 8: Spine Injuries) with the decision to take no spinal precautions. Any patient who remains unreliable must, of course, be treated for spine injury as well as brain injury.

Without spinal damage, and with a patient capable and free of signs and symptoms of brain damage, you may decide to start walking the patient out as a means of evacuation. A patient incapable of walking and/or a patient with obvious signs and symptoms of increasing ICP and/or a skull fracture or penetrating head wound should be evacuated as soon as possible by whatever other means are available.

If the patient must be left alone for any reason, roll her or him carefully into a stable side position, the recovery position, to ensure an airway.

Elevation of the patient's head to decrease ICP is recommended. Some experts believe that head elevation reduces ICP, a "good" thing. Other experts, however, say it reduces cerebral perfusion, a "bad" thing. If local protocols exist, you should follow those protocols. Without protocols, since patients with a serious head injury are almost always being treated for spine injury as well, stabilization of the patient supine with the head in a neutral position is a priority. Beyond that, elevating the patient's upper body at approximately 30 degrees is recommended.

Since oxygen deprivation is the immediate threat to life, supplemental oxygen, when available, would be of great benefit. In the wilderness with a patient in the later stages of brain injury, artificial respirations—mouth-to-mouth/mouth-to-mask—*may* help keep the patient adequately oxygenated.

Do *not* give painkilling medications to head-injured patients with signs and symptoms of brain damage or during an evaluation period while you're waiting to see if signs and symptoms appear. These medications may mask the signs and symptoms you are trying to monitor. Some medications, such as aspirin and ibuprofen, may increase the rate of bleeding in the brain. Any patient who feels a great need for painkilling drugs after a blow to the head is a patient who needs a doctor.

Once in a hospital, even the seriously head-injured patient has a chance of survival. Oxygen therapy, drug therapy, and surgical intervention are available. That's why caregivers keep the patient awake or arouse him or her regularly. If arousal is impossible, the hospital staff knows it's time for radical procedures. In a wilderness situation you should wake the patient periodically to assess the level of consciousness and, thus, the level of brain damage. In the wilderness once you have decided to evacuate a patient for treatment for serious brain damage, you may let him or her sleep.

Evacuation Guidelines

Immediate evacuation is necessary for any patient with signs and symptoms of increasing intracranial pressure and/or a skull fracture or penetrating head wound. It is highly recommended to initiate an evacuation for any patient who does not respond to aggressive attempts at verbal and painful stimulation immediately after a blow to the head—in other words, if you cannot get an alert response after stimulating the patient.

Conclusion

Fortunately for the young woman lying unconscious in New Hampshire, her friend remembers to leave her in the recovery position—to maintain her airway—while he hurries on for help. He leaves his parka covering her as much as possible.

The rescue team arrives to find the ground around her face splattered with vomit. Her respirations are deep and sighing, her heart beats slowly and hard, her face appears deeply flushed. She does not respond to any stimulus. Supplemental oxygen is started at a high flow. She is loaded into a litter with her head and shoulders elevated approximately 30 degrees for the carry to the nearest road.

Long after dark, she lifts off for a quick helicopter flight to Boston. After extensive surgery she is able to return to a full and healthy life.

CHAPTER TEN

Chest Injuries

You Should Be Able To:

1. Describe the basic anatomy of the chest and normal breathing.
2. Define and describe the signs and symptoms of the most common chest injuries, including fractured rib, fractured clavicle, flail chest, pneumothorax, hemothorax, and pericardial tamponade.
3. Describe the treatment for the most common chest injuries.

It Could Happen to You

Posted at the trailhead, a small sign read BEWARE OF BEARS, which explained why Dave straddled a limb high in the ponderosa pine to tie off a bag of food. The knot held, but Dave didn't. Smashing into several branches on the way down, his rush to the ground ended with a gut-wrenching thud and an *ooooff* of forcefully expelled air.

Kneeling close to the baghanger-turned-patient, your first words are, "Don't move." To your assessing eye, clues begin to appear: scrapes on bare legs and arms—ugly, but not serious; respiratory effort—shallow and labored, an immediate concern; a ragged tear on the upper left side of Dave's T-shirt reddening with seeping blood.

Dave, fully alert, denies pain in his head or spine, but you opt on the side of caution, asking another member of your party to maintain manual control of his head and neck. His heartbeat is fast, but strong and regular. You lift his shirt to take a close look at his chest. You know that any injury to the chest may lead to severe respiratory difficulty and a critical patient; you want to know as much about a chest injury as possible, and as soon as possible.

Few traumatic injuries offer Wilderness First Responders as great an opportunity to watch a patient die as a serious chest injury. Although well protected by the construction of the chest, the organs within the chest cavity—lungs, heart, and great blood vessels—once damaged, can quickly create a life-threatening situation. For that reason, **any injury to the chest should be considered an immediate life threat until proven otherwise.**

Basic Anatomy of the Chest and Normal Breathing

Ribs form a bony cage that surrounds the *thorax* (chest cavity). Each of the twelve pairs of ribs attaches to a thoracic vertebrae. The top ten pairs attach to the sternum by way of cartilage. The last two pairs of ribs are shorter and "float" free, with no anterior attachment. Beneath each rib runs an *intercostal nerve, artery,* and *vein.*

Lungs fill both sides of the thorax, and the thorax is lined with a tough membrane called the *parietal pleura.* The lungs are covered with a membrane called the *visceral pleura.* Between the two pleurae exists a potential space, the *pleural space.*

Between the lungs, beneath the lower half of the sternum, lies the heart, surrounded by its own membrane of fibrous connective tissue called the *pericardial sac.* The central area of the thorax also houses the great vessels—the *aorta* and *vena cava*—that carry blood from and to the heart, respectively.

Being spongy and relatively flaccid, the five lobes of the lungs (three on the right side, two on the left) depend on nearby muscles to function. A domed shelf of muscle, the *diaphragm,* relaxes between the chest and abdominal cavities when not in use. Every few seconds it contracts and flattens downward, enlarging the chest cavity. Secondary to the diaphragm, *intercostal muscles,* the ones that lie between the ribs, contract to lift the rib cage and expand the chest cavity. Lungs slide along the chest wall as it moves up and down, slipping easily on a lubricating layer of *synovial fluid.* The fluid keeps the lungs stuck to the chest wall, so the lungs are pulled open as the chest expands. On inhalation, air rushes in to fill the void created by the expansion of the chest cavity—the active phase of breathing. As tension releases from the diaphragm and

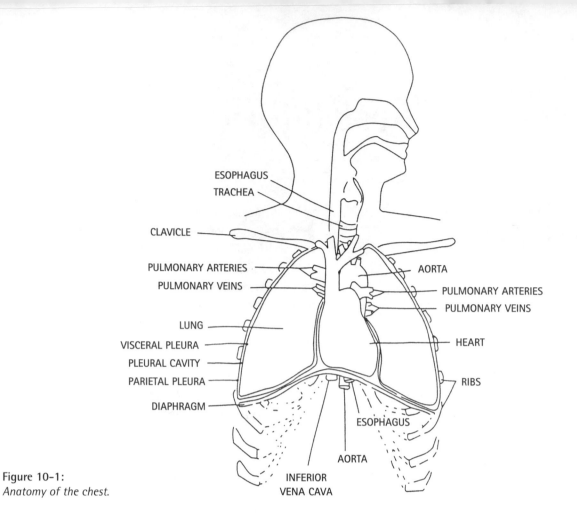

Figure 10-1:
Anatomy of the chest.

Labels in figure: ESOPHAGUS, TRACHEA, CLAVICLE, PULMONARY ARTERIES, PULMONARY VEINS, LUNG, VISCERAL PLEURA, PLEURAL CAVITY, PARIETAL PLEURA, DIAPHRAGM, AORTA, PULMONARY ARTERIES, PULMONARY VEINS, HEART, RIBS, ESOPHAGUS, AORTA, INFERIOR VENA CAVA

intercostal muscles, air leaves the lungs—the passive phase of breathing. The breathing process, all things considered, resembles very much the action of bellows that once hung beside just about every fireplace in the world.

Types of Chest Injuries

Injuries to the chest may be *closed* (the skin of the chest is unbroken) or *open* (an object penetrates or has penetrated the chest wall). These injuries may result from blunt trauma, penetrating trauma, or compression trauma.

Two important signs indicate the possibility of a serious chest injury: changes in respiratory rate and changes in respiratory effort. Normal breathing at 12 to 20 breaths per minute is effortless and painless. A patient with rapid breathing, painful breathing, noisy breathing, and/or difficulty breathing should send you into quick-response mode.

Injuries to Bones

■ FRACTURED RIB/CLAVICLE

Broken ribs and clavicles (collarbones) account, by far, for most chest injuries. Although painful, they rarely create a serious

condition for a patient. Pain may explode in the damaged area whenever the patient inhales, especially in a rib fracture, when the active phase of breathing causes contraction of the muscles around the broken bone. The patient typically takes shallow breaths to ease the pain. Typically, the patient also presses a hand over or squeezes an arm against the site of a fractured rib to help stabilize the broken bone or bones, another pain reducer. For a fractured clavicle the patient often supports the arm on the injured side, taking weight off the injury (see Chapter 12: Fractures).

Signs and Symptoms of a Fractured Rib/Clavicle

1. Pain at the fracture site.
2. Point tenderness at the fracture site.
3. Pain with movement, coughing, and deep breathing.
4. Shallow breathing.
5. Guarding the fracture site.
6. Bruising at the fracture site.

Supplemental oxygen, if available, may ease the pain and anxiety. As long as spinal precautions need not be taken, aid the patient in assuming a semi-reclining position that should make breathing easier. You can help the patient by doing more than he or she is doing to stabilize the fracture. The simplest and most effective treatment involves suspending the arm on the damaged side of the chest in a sling, also known as the *sling-and-swathe technique*. The arm then hangs protectively over the fracture of a rib or supports the arm beneath a fractured clavicle. Secure the arm in place by tying strips of cloth around the chest and the slung arm, or by putting the patient's sweater or jacket on over the arm. If your first-aid kit does not contain sling material, you can improvise with relative ease using extra clothing or even by folding the shirt the patient is wearing up over the arm and securing the material in place with a safety pin (see Chapter 12: Fractures).

The sling-and-swathe technique leaves one arm virtually useless, which is sometimes a disadvantage on the trail. A less protective but functional alternative to help stabilize a rib fracture is to pad the site with a folded cloth, such as a folded T-shirt, and secure it with tape placed about halfway around the chest.

Anti-pain, anti-inflammatory drugs such as ibuprofen may be given, but avoid narcotics, medications that depress the patient's respiratory drive.

Patients with minor fractures to the ribs may need nothing done for them. You should, however, encourage the patient to periodically take deep breaths to keep the lungs clear. The patient's ability to deal with the discomfort will be your best guide. But all people suspected of having a broken rib or clavicle should be watched over the next day or two. Any marked increase in a patient's respiratory effort indicates a need for rapid evacuation. Since patient comfort is also a factor, you may decide to evacuate someone with a suspected fracture simply because he or she is not able to enjoy the wilderness experience.

FLAIL CHEST

If two or more consecutive ribs are broken in two or more places and/or the sternum has broken loose, a free-floating section of chest wall, called a *flail*, sometimes results. A minor flail is often a subtle injury, never assessed until the patient is X-rayed. A major flail, however, is a major threat to life. You may see the flail move in opposition to the rest of the chest wall during breathing, a condition known as *paradoxical respiration*. Paradoxical movement is easiest to see when the patient lies on his or her back, something the patient may resist doing. A severe flail makes it extremely difficult for the patient to breathe adequately, partially due to the paradoxical movement of the chest and, perhaps, even more because of underlying lung trauma. A pneumothorax and/or a hemothorax may develop as well (see "Injuries to Lungs," this page).

Signs and Symptoms of a Flail Chest

1. Chest pain.
2. Difficulty breathing, even at rest.
3. Paradoxical respiration and/or deformity at the fracture site.
4. Bruising at the fracture site.
5. Signs and symptoms of shock.

Taping a bulky dressing securely over the flail to stabilize the fracture site often allows the patient to inhale a little more easily. If the tape runs from midline to midline, halfway around the patient's body, the stabilization is usually improved. If, however, the patient gets worse with the flail secured, remove the tape and dressing. Evacuation of the patient on his or her side, injured side down, sometimes aids breathing, but the patient, other injuries allowing, may prefer to assume a semi-reclining position. Supplemental high-flow/high-concentration oxygen may be of great benefit if it is available.

Injuries to Lungs

PNEUMOTHORAX

A simple fractured rib may not be so simple if a bone fragment stabs inward to puncture a lung. Air escaping the lung on inhalation and collecting in the chest outside the lung in the pleural space creates a condition called a *closed pneumothorax*. Over a period of time determined by the extent of the injury, the "dead air" in the chest cavity grows in size. The patient will experience increased breathing difficulty, especially difficulty taking a deep breath, and a rising level of anxiety as the trapped air takes up more and more breathing room. With a stethoscope, you may hear diminished breath sounds on the affected side. If your hearing is acute, you may be able to hear diminished breath sounds on the affected side by pressing your ear against the patient's chest (see "Breath Sounds," this chapter).

Signs and Symptoms of a Pneumothorax

1. Sharp chest pain.
2. Increased breathing difficulty, with diminished breath sounds on the affected side, even at rest.
3. Signs and symptoms of shock.

Treat the patient as you would a patient with a fractured rib, using when possible supplemental oxygen, semi-reclining position, and stabilization of the fracture site. More often than not, a pneumothorax reaches a point where it gets no worse. But it can worsen until the patient is unable to breath adequately, a condition known as a *tension pneumothorax,* one of the most life-threatening chest injuries. In a tension pneumothorax, air has kept on leaking from the damaged lung, collecting in the pleural space until the lung on the affected side compresses to the size of a tennis ball. Air continues to leak out, compressing the heart, the great vessels, and even the other lung. Neck veins may bulge, sometimes called *jugular vein distension,* or JVD, and the trachea may deviate toward the uninjured side, especially noticeable on inspiration. With a stethoscope, you may hear diminished breath sounds or fail to hear breath sounds on the affected side. When the pressure of trapped air inside the chest reaches maximum tension, breathing stops.

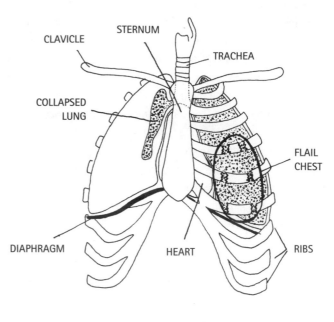

Figure 10-2:
Anatomy of collapsed lung and flail chest.

Signs and Symptoms of a Tension Pneumothorax

1. Sharp chest pain.
2. Increased breathing difficulty, with eventual absence of breath sounds on the affected side.
3. Signs and symptoms of shock.
4. Distended neck veins.
5. Tracheal deviation.

Medics with advanced training are often able to puncture the chest with a small hole that lets the trapped air out. Since most of us cannot do that, and since you never know whether a pneumothorax will go to tension or not, **suspicion of any pneumothorax calls for an immediate and rapid evacuation.** It would be best if the patient is carried, preventing an increased workload on an already damaged lung. You may, however, choose to walk out if that's the speediest evacuation option and if the patient can tolerate it. If you are waiting for transport, remember to aid the patient to sit in a semi-reclining position, if possible, which makes breathing easier.

Note: A pneumothorax can occur spontaneously, without an injury to the chest. Not usually life threatening, a *spontaneous pneumothorax* results when a weak spot on a lung ruptures. Treatment is the same as for any pneumothorax, and the patient must be monitored for the unlikely possibility of the pneumothorax going to tension.

With a chest injury that tears a lung or a bronchial tube, little bubbles of air may move into subcutaneous tissue and appear on the skin of the chest, neck, or face. Termed *subcutaneous emphysema,* the bubbles move around and may make a crackling sound when you push on them. Although not a problem itself, subcutaneous emphysema is another indication of a serious chest injury, most often a pneumothorax.

■ HEMOTHORAX

Broken ribs can tear into blood vessels, causing enough damage to allow an accumulation of blood in the pleural space, a condition known as a *hemothorax.* A small hemothorax can create no signs and symptoms, but just as air may accumulate in a pneumothorax, blood may accumulate to the point at which the lungs cannot expand and breathing ceases to be possible. With a stethoscope, you may hear diminished breath sounds or fail to hear any breath sounds on the affected side, usually at the base of the lungs. If the hemothorax is extensive, you may see *hemoptysis* (blood coughed up). A hemothorax and pneumothorax can occur simultaneously, a condition known as a *pneumohemothorax.*

Shock is possible, and treatment for shock, if it develops, would be appropriate, except that the patient will probably breathe much easier sitting up instead of lying down. Otherwise, the patient with a hemothorax receives the same treatment as a patient with a pneumothorax.

Breath Sounds

Although it is not a "typical" first responder skill, the ability to listen for breath sounds can augment your ability to make an assessment of lung injury. Normal breath sounds are heard best at the end of a deep inspiration, which is why doctors ask patients to "take a deep breath" when they check breathing with their cold stethoscopes. A normal breath sound is like a hollow rush of air through the stethoscope.

There are three places to listen for breath sounds: the apex of the lungs (just below mid-clavicle), the axillary point (under the armpits, about nipple high), and the subscapular point (just below the scapula on the back). You should make a comparison, one side to the other. Report normal breath sounds as "lungs bilaterally clear."

Abnormal breath sounds include (1) wet sounds (rales), a sound like the faint bubbling of a household aquarium through the stethoscope; (2) wheezing, a somewhat high-pitched sigh; (3) diminished or muffled sounds on one side when compared to the other side; and (4) the absence of sounds on one side.

■ SUCKING CHEST WOUND/ OPEN PNEUMOTHORAX

If the chest has been opened by a penetrating object, the resulting hole may bubble and make noise when the patient breathes. When the chest moves, air flows in and sometimes out through the wound as well as through the mouth and nose. This is appropriately called a *sucking chest wound,* and the air being sucked in through the wound does nothing to sustain life.

Immediate action is required. Plug the hole. Your hand, preferably gloved, will work fine, and quick action could prove life saving. Continue treatment by covering the wound with an occlusive dressing—something that lets no air or water pass through. Clean plastic will work. Tape this dressing in place leaving one side, or at least one corner, free in the hope that air collecting under tension in the chest will self-release. If collecting air does not release, the *open pneumothorax* can become a tension pneumothorax.

If the patient's pneumothorax goes to full tension, he will not be able to breathe. Death is imminent. You can try pushing your little finger or some dull-ended object gently into the hole in an attempt to release the trapped air. This is not fun and not advisable in most situations, but it may be your only chance to save the patient.

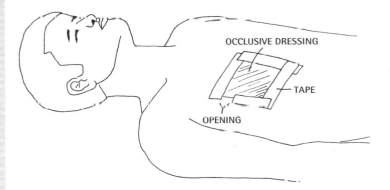

Figure 10-3:
Bandaging of an open chest wound.

■ IMPALED OBJECT

When you inspect the chest and find an object still wedged in a wound, **do not attempt to remove it.** Removal has an excellent chance of making the injury worse by encouraging serious bleeding or creating a sucking chest wound. Pad well around the impaled object with the padding rising at least as high as the object, and tape the padding in place to stabilize the object. Evacuation of the patient, of course, should be high on your list of priorities.

■ PERICARDIAL TAMPONADE

Blunt trauma to the chest wall may cause *pulmonary contusions* and/or *myocardial contusions,* producing shortness of breath or chest pain, respectively. These contusions may, of course, be small or large. Chest trauma, blunt or penetrating, also may cause blood or other fluids to leak from the myocardium (heart) itself into the pericardial sac. The sac will not stretch, so the heart is squeezed by the increasing pressure, a condition known as *pericardial tamponade,* an immediate threat to life. As the heart beats with less and less efficiency, the pulse pressure (the difference between the systolic and diastolic blood pressure) narrows. In other words, the two numbers grow closer and closer to the same number. You should see the pulse growing weaker and shortness of breath increasing. You may see distended neck veins. Shock results. Death is imminent.

There is little to be done in the field for pericardial tamponade other than early recognition, supplemental oxygen, if available, and the most rapid evacuation possible.

General Signs and Symptoms of Lung Trauma

1. Pain, tenderness, and unequal chest expansion on inhalation.
2. Wounds, bruises, deformity, or instability of the chest wall.
3. Increasing shortness of breath, especially with the patient at rest.
4. Vital signs indicative of shock.
5. Subcutaneous emphysema.
6. Diminished breath sounds or wet breath sounds on the injured side.

General Treatment Guidelines

1. Maintain an adequate airway.
2. Give supplemental oxygen, if available.
3. Inspect the bare chest: look, ask, feel.
4. Treat any injury you find: stabilize fractures, close open wounds, stabilize impaled objects.
5. When no other injuries prevent movement, allow the patient to assume a position that provides the most comfort.
6. Avoid pain medications that depress respiratory drive.
7. Plan an evacuation (except for simple fractured rib).

Evacuation Guidelines

Any patient with increasing breathing difficulty while at rest or increasing signs and symptoms of shock and any patient treated for a serious chest injury should be evacuated as soon as possible.

Conclusion

Beneath the tear in Dave's shirt you find a superficial abrasion, the source of the blood. His breathing remains labored, but he appears to be getting no worse. His chest expands equally on both sides when he inhales. A complete examination reveals no serious problem, and you decide, after a focused assessment of the spine, to take no spinal precautions. Sitting up, Dave is able to breathe more easily.

Your careful palpation of the ribs below the abrasion elicits a pain response at a specific point. With encouragement, because it hurts, Dave is able to take a deep breath. Your assessment: a cracked rib.

For the next day the group remains camped at this spot, and Dave wears a sling supporting the arm on his injured side. He reports sleeping "fairly well" through the night. Although extensive bruising appears on his injured side, his ability to breathe without pain increases.

You keep a watchful eye on his condition, but he is able to finish the trip with the rest of the group.

CHAPTER ELEVEN

Abdominal Injuries

You Should Be Able To:

1. Describe the basic anatomy of the abdomen.
2. Describe the general signs and symptoms of abdominal injuries.
3. Describe the general treatment for abdominal injuries.
4. Demonstrate specific treatment for blunt abdominal trauma and penetrating abdominal trauma.

It Could Happen to You

As a novice wrangler, you're happy to have landed a job with a small company that leads horsepacking trips in the Wenaha-Tucannon Wilderness, a vast area shared by Washington and Oregon. One of the reasons you got this job is your medical training. None of the other wranglers on this 10-day ride have more training than an urban-oriented basic first-aid course. It's the evening of the fourth day before your skills are needed.

The dinner bell rings the call to grub, but one of the clients is a no-show. Another client says the missing rider was complaining of a stomachache. He says he saw him lying down by the river after the horses were unsaddled and hobbled for the night. You go to investigate, and most of the group follows.

You find the man on his side with his legs drawn up in a fetal position, his arms crossed over his stomach to guard his abdominal area. He moans softly. His skin appears pale, cool, and clammy. His breathing is rapid and shallow.

"Anybody got an idea about what happened?" you ask.

"Well," offers one of the other wranglers, "he got kicked in the gut by that ol' pack mule at lunch break, but he jumped right up and seemed OK."

Abdominal injuries—trauma to internal structures of the body between the diaphragm and the lower end of the pelvis—may be generally classified in two categories: (1) *blunt trauma*, a closed abdominal injury caused by a forceful blow to the abdomen; and (2) *penetrating trauma*, an open abdominal injury caused by an object that penetrates the abdomen. The extent of injury is often difficult to assess in the wilderness—or anywhere else, for that matter. No other region of the human body has more potential to conceal serious blood loss. The mechanism of injury may seem trivial, and the patient may initially present with signs and symptoms that appear relatively normal before indications of serious injury seem to suddenly manifest themselves.

Serious abdominal trauma in the wilderness carries a high risk for mortality due to two factors: (1) Several body systems may be involved in the injury, and (2) several hours to days may lapse before definitive care can be reached, and there is little treatment that can be provided in the field. The chance of saving a patient with serious abdominal injury depends largely on the Wilderness First Responder's ability to make an early accurate assessment and provide rapid evacuation.

Basic Anatomy of the Abdomen

The abdominal region extends from the vertebral column in back to the abdominal muscles in front, from the diaphragm on top to the pelvis on the bottom. It is important to remember that the diaphragm may rise as high as the fourth rib on exhalation, and injury below that point may involve the abdomen.

The abdominal cavity is divided into four quadrants for assessment purposes by drawing an imaginary vertical line and horizontal line through the *umbilicus* (belly button). The quadrants are named "right" and "left" in reference to the patient's body, not the rescuer's body. The contents of each quadrant are as follows:

Right upper quadrant (RUQ): the right and largest section of the liver, the gallbladder, part of the transverse colon, and the duodenum (first section of the small intestine).

Left upper quadrant (LUQ): the left and smaller section of the liver, the stomach, part of the transverse colon, and the spleen.

Right lower quadrant (RLQ): part of the small intestine, the ascending colon, the appendix, the right ureter, approximately half of the urinary bladder, and, in females, the right ovary, the right fallopian tube, and approximately half of the uterus.

Left lower quadrant (LLQ): part of the small intestine, the descending colon, the sigmoid colon, the left ureter, approximately half of the urinary bladder, and, in females, the left ovary, the left fallopian tube, and approximately half of the uterus.

The kidneys and pancreas, which lies between the kidneys, are situated beneath the lowest two ribs in the *retroperitoneal* space (the space behind the *peritoneum*, the membrane lining the abdominal cavity). If they were within the peritoneum, they would be in the upper right and left quadrants.

Some of the abdominal organs—pancreas, liver, kidneys, and spleen—are solid and are mostly tucked under the rib cage for more protection. Some are hollow—stomach, intestines, gallbladder, urinary bladder, and female reproductive organs. Solid organs can be damaged by penetrating trauma, but they can also rupture from blunt trauma. Blood loss from solid organs can be life threatening, especially from the liver, which receives about 30 percent of heart's output every minute, and the spleen, which receives about 10 percent of the heart's output every minute. Hollow organs are more likely to be damaged by penetrating trauma, but they also can rupture from blunt trauma, especially if they happen to be full of food, urine, or fecal matter, depending on their purpose.

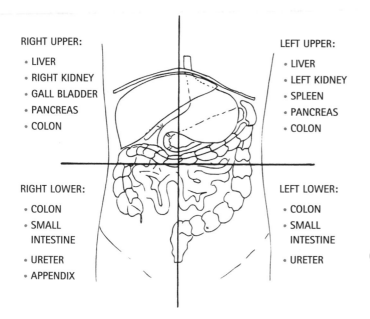

RIGHT UPPER:
- LIVER
- RIGHT KIDNEY
- GALL BLADDER
- PANCREAS
- COLON

LEFT UPPER:
- LIVER
- LEFT KIDNEY
- SPLEEN
- PANCREAS
- COLON

RIGHT LOWER:
- COLON
- SMALL INTESTINE
- URETER
- APPENDIX

LEFT LOWER:
- COLON
- SMALL INTESTINE
- URETER

Figure 11–1: *Anatomy of the abdomen showing quadrants.*

General Abdominal Trauma Assessment

1. Determine the mechanism of injury, and assess if it is indicative of abdominal trauma.

2. Observe the patient's body position. Although pain is not always a reliable indicator of the seriousness of the injury, a patient who lies still with legs drawn up into a fetal position is often assuming a posture that minimizes, as much as it can, serious pain.

3. Look at the abdomen with the patient lying down. A normal abdomen is gently rounded and symmetrical. Look for distension and/or an irregularly shaped abdomen, both of which may indicate a serious injury. Look for bruises, penetrating wounds, or obvious injuries, and don't forget to check the flanks, where damage to the kidneys might be indicated. Find out as much as you can about any object that has penetrated the abdomen. If you see a gunshot entrance wound, check as soon as possible for an exit wound.

4. Feel the abdomen with flat fingers. Press gently in all four quadrants of the abdomen. Watch for a pain response. Normal abdomens are soft and not tender to palpation. Feel for rigid muscles, lumps, and pain specific to a local spot, all of which may be signs of injury.

5. Ask the patient about his or her condition. Has this ever happened before? The OPQRST questions, relative to pain, may be very helpful:

   O (Onset): Did the pain come on suddenly or gradually?

   P (Provokes or palliates): Does anything make the pain worse or better?

   Q (Quality): How does the patient describe the pain? Sharp? Dull?

   R (Radiate or refers, region): Where is the pain, and does it radiate or refer to another region?

   S (Severity): On a scale of 1 to 10, with 10 being the most serious, how does the patient rate the pain?

   T (Time): How long has the pain been there?

6. Listen with an ear or stethoscope pressed against each of the four quadrants. This takes time. In 2 to 3 minutes, within each quadrant, bowel sounds (gurgling noises) should be heard. Absence of noise means something is not working right.

7. Monitor the vital signs for indications of shock.

8. Ask if any blood has appeared in the urine, stool, or vomit.

Blunt Trauma Assessment and Treatment

A patient suffering a severe blow to the abdomen needs to have an adequate history taken because damage on the inside is often not very obvious from the outside. What hit the patient? How fast was it going?

When you look at the abdomen, you may see bruising. When you press gently on the abdomen during your examination, the patient may reflexively "guard"—a sign of a potentially serious abdominal problem. When you palpate the abdomen, you may feel unusual lumps.

Hollow organs, when ruptured, may release substances into the abdomen such as stomach acid, digested food, or bacteria. The contents of hollow organs are highly irritating. *Peritonitis,* an inflammatory reaction in the abdominal cavity, will probably develop, with pain—described as sharp, stabbing, or burning—increasing and spreading throughout the abdomen. Pulse and respirations quicken. The muscles of the abdomen become increasingly rigid and distended. Your ear placed against the patient's abdomen, periodically, reveals a decrease in the gurgles and rumbles of normal bowel activity as the intestines become paralyzed. The patient will lie very still on her or his back or side but demand flexion of the knees to take pressure off the abdominal muscles. Pain is provoked by movement and any attempt to straighten the legs.

Because solid organs bleed when ruptured, and because blood is less irritating than the contents of hollow organs, the signs and symptoms of peritonitis may not appear. Instead, the signs and symptoms of shock may appear. You should also see increasing rigidity and distension and possibly hear the loss of bowel sounds. Pain may increase. Pain from a damaged liver may refer to the right shoulder, and pain from a damaged spleen may refer to the left shoulder. Threat to life may be immediate.

With any abdominal injury anticipate nausea and vomiting. Over time, blood may appear in the vomit (often looking like coffee grounds), in the urine (often appearing pale pink), or in the stool (often described as black and tarry), depending on where the damage occurred. Over time, a fever may develop.

Stay alert to the possibility of vomiting. Generally, treat for shock. Patients suffering blunt trauma should be kept in the position of comfort they choose—if no other injuries prevent this—and kept warm. If you are involved in their evacuation, comfort and warmth should be extended during the carry. In general, nothing should be given to the patient by mouth, but on an extended evacuation sips of water, preferably cool, may be necessary to prevent dehydration.

Penetrating Trauma Assessment and Treatment

The immediate seriousness of any penetrating abdominal injury, as with blunt trauma, is determined by what got damaged inside and how bad it's bleeding. With severe bleeding, shock is imminent, and immediate evacuation is the only chance of salvation. Over time the risk of infection is very high. General assessment of the patient is the same as with a closed injury. General treatment of the patient is the same as for a patient suffering blunt abdominal trauma. Specific treatment varies somewhat, depending on the soft tissue involvement. External bleeding should be controlled. Wounds should be cleaned and bandaged (see Chapter 15: Wilderness Wound Management). Impaled objects, in almost all cases, should be stabilized in place.

An *evisceration* is a specific type of penetrating injury that opens the abdomen and lets intestines protrude. In short-term care, cover the exposed bowels with sterile dressings soaked in disinfected water to prevent drying out. Check the dressings every 2 hours to make sure they stay moist. Cover the moist dressings with thick, dry dressings, and rapidly evacuate the patient. In long-term care, over several hours, the exposed intestines will do better if they are flushed clean with disinfected water and "teased" back inside by gently pulling the wound open. If teasing does not work, you may have to gently push the exposed loops of intestine back inside the abdominal cavity. Then clean and bandage the wound.

Evacuation Guidelines

All patients with serious abdominal injuries require rapid evacuation. The evacuation should be as gentle as possible. Often surgical intervention is all that will eventually prevent death.

Conclusion

Exposing the abdomen of your horsepacking client, you notice substantial bruising over the left upper quadrant. In response to your questions he says, yes, that's right where the mule kicked him. When you press on the left upper quadrant, the patient groans, flexing his abdominal muscles to guard the area. Palpation of the other three quadrants produces unremarkable results. He refuses to roll onto his back. He says when he's curled up on his side the pain eases a little. A first set of vital signs shows a heart rate of 110, strong and regular, and a respiratory rate of 20, shallow and slightly labored.

Organizing members of the party, you lead a carry of the patient to his tent, where he is bedded down as comfortably as possible. A second set of vitals shows a marked increase in the numbers.

You begin immediate preparations to send for help. The trail in was easy, and a group of competent riders, led by one of the other wranglers, will travel through the night, carrying full documentation and location information.

CHAPTER TWELVE

Fractures

You Should Be Able To:

1. Describe the basic anatomy of the musculoskeletal system.
2. Describe the signs and symptoms of upper and lower body fractures.
3. Demonstrate how to treat simple upper and lower body fractures.
4. Demonstrate how to treat complex upper and lower body fractures, including femur fractures, angulated fractures, and open fractures.

It Could Happen to You

You've set a camp near Minnesota's Blue Mounds because it's a pleasant spot and one of the few places in the Land of 10,000 Lakes where you can rock climb. Early-summer mosquitoes and ticks make you glad you brought the tent. To the west the first hint of color announces the beginning of the end of the day, but the routes here are short. You and your partner agree there's ample light for an hour or so of climbing.

Halfway up a straightforward 30-foot crack, you've broken a mild sweat. Safely belayed by your partner, you ease off a moment and look to your left. A solo climber has started up an open face about 50 feet away. He's only about 8 feet off the ground. As you watch, he reaches high and both his feet come off their precarious steps. He drops to the broken rock at the foot of the face, landing on his feet.

It doesn't look like much of a fall, and you return your attention to the route you're on when the solo climber's scream of pain jerks your head back in his direction. Even from up here you can now see the weird angle at which his lower right leg lies.

Injuries to bones, muscles, tendons, and ligaments—known as *the musculoskeletal system*—are, undoubtedly, among the most common emergencies dealt with by the Wilderness First Responder. The types of injuries to this system, or at least the ones the WFR should be ready to assess and treat, are fractures, dislocations, and athletic injuries (strains, sprains, and tendinitis). Your quick and adequate response can prevent further injury, reduce pain, and, sometimes, make the difference between discomfort and permanent disability.

Basic Anatomy of the Musculoskeletal System

The two principal parts of the musculoskeletal system are *muscles* and *bones*. Muscles are specialized tissues that contract, or shorten, when stimulated, providing power and motion. Bones—the skeletal system—give muscles attachment points for locomotion (movement from one place to another) to occur. Muscles and bones also provide support and protection for vital organs. The rest of the musculoskeletal system includes *tendons*, connective tissues holding muscles to bones; *ligaments*, connective tissues holding bones to bones; and *cartilage*, which is tough, elastic tissue that forms protective pads where bone meets bone.

Types of Fractures

Any break in the normal continuity of a bone is a *fracture*. These breaks can be caused by force being applied directly to a bone, such as a skier's leg striking a tree, or they can be caused by indirect force, such as falling on an outstretched hand and breaking the *clavicle* (the collarbone) or the *radius* (the larger lower arm bone) near the elbow. Bones can break along interesting lines with interesting names—transverse, oblique, spiral, comminuted, and greenstick, for instance—but that matters little or not at all to the WFR. Fractures may be *closed*, when the skin is intact, or *open*, when the skin is broken over the site of the fracture. Bone does not have to be visible through the opening in the skin for the fracture to be open. What matters to you as a WFR is (1) determining if the injury could be a fracture and (2) managing the injury properly.

General Assessment of a Fracture

The general principles of any assessment, and thus of fracture assessment, involve *looking, asking,* and *feeling* (**LAF**).

Looking requires careful removal or cutting away of the patient's clothing. You cannot treat an injury that you cannot see. But cut away clothing with discrimination because it might be needed before an evacuation is over. Much can be gained by comparing the injured and uninjured sides. Look for bruising, bleeding, deformity, and lack of symmetry. Look for guarding, overprotectiveness of the injury site by the patient.

Asking starts with finding out what happened and what forces were involved. The greater the forces involved, the greater the chance of a significant injury. Ask about pain—where and how much. Ask about sounds—snaps or pops heard when the injury occurred.

Feeling means palpating the injury site for signs of a fracture. Do you feel muscles spasming around the injury? Does it feel unstable? Do you feel (or hear) bony *crepitus,* the grating of bone against bone? Is there "point tenderness," a certain spot that hurts when touched? Is there adequate circulation (pulses), sensation (feeling), and motion (the ability to move distal to the injury site)?

In most cases fractures are determined by X-rays, not by field assessments, unless there are obvious signs, such as gross deformity—two knees, as it were, on the same leg. You are looking, asking, and feeling for signs and symptoms of a fracture, but, in the end, a great deal of your wilderness assessment of a possible fracture will include the willingness of the patient to use the injured body part. If you and the patient decide the injured part is not usable, you will, in most cases, splint the injury as if it were a fracture and arrange an evacuation.

Signs and Symptoms of a Fracture

1. Pain specific to the injury site.
2. Swelling and/or bruising.
3. Deformity.
4. Tenderness and/or point tenderness.
5. Sounds, such as snaps, pops, or crepitus.
6. Loss of circulation, sensation, and/or motion (CSM).
7. Wounds at the site with or without protruding bone.

General Principles of Fracture Treatment

Managing fractures in remote environments requires common sense and sensitivity to the needs of the patient. For example, with a badly sprained ankle, where a fracture is a possibility, you would ideally immobilize the body part and put the patient at rest with instructions for elevation and ice. In the wilderness, however, you must weigh other factors: the desire of the patient to ambulate on a suspicious ankle injury; the availability of people to transport the patient; the type of terrain involved in transport; the severity of the environment; and the patient's need or desire to continue the trip. Even though the best medical judgment would preclude using the injured ankle, the best decision in a remote environment might be, if the patient can use the injury, to immobilize the ankle in a splint that allows the patient to hobble along on his good ankle using an ice ax, ski pole, or wooden stick as a crutch. This could be the safest and most reasonable decision based on the situation.

The basic and best treatment for all suspected fractures,

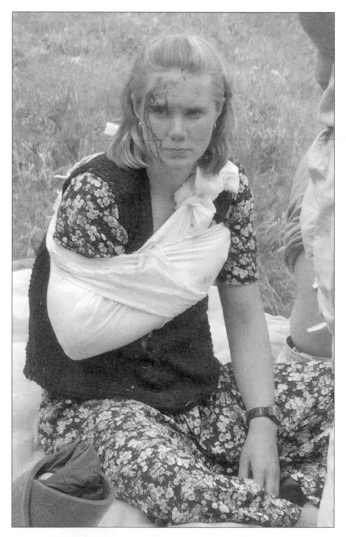

Figure 12-1: *Sling and swathe.*

however, is the same: an immobilizing splint. A patient with a bone that might be broken should have circulation, sensation, and motion (CSM) assessed beyond the injury point before and after splinting. CSM should be regularly monitored as long as the patient is in your care. Any loss of CSM should be immediately investigated, in case you can do something to reestablish healthy blood flow, such as by loosening the splint.

Suspected fractures, in most cases, should be splinted in a position of function, which means arms flexed at approximately 90 degrees at the elbow, hands slightly curled, legs flexed at approximately 5 to 10 degrees at the knee, feet approximately perpendicular to the legs.

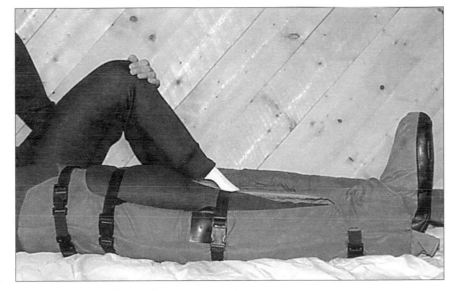

Figure 12-2: *Leg splinted with Crazy Creek chair.*

A splint should involve plenty of padding for support and comfort, and it should incorporate something rigid for additional support. Sufficient padding can almost always be found in the wilderness in the form of extra clothing. When extra clothing is in short supply, you can bulk up what you have by stuffing in forest duff, leaves, and such. The padding should be thick enough to provide comfort and arranged in a manner that eliminates voids within the completed splint. Voids allow the possibility of painful shifting. Avoid the voids! Padding, however, should not be so thick that patients find themselves too bulky to move. A big, fat arm splint, for instance, might make it difficult to walk comfortably. Padding is essential, but bulk is not.

Rigid splinting materials also typically abound in the form of stays from packs, Crazy Creek chairs, sticks, sleeping pads, and the like. Therm-A-Rest pads offer an advantage: They can be deflated, secured in place, then inflated for additional rigidity. Commercial splints of practical value for some broken bones in wilderness situations include the SAM Splint and foldable wire splints.

A splint should immobilize the joints above and below a long bone injury or the bones above and below a joint injury. Most upper extremity fractures can be adequately immobilized with a *sling and swathe*. Both sling and swathe may be created from triangular bandages, but many improvisational techniques can also be used successfully. A long-sleeved shirt, for instance, can be formed into a sling using the arms of the shirt to go around the patient's neck. Cut a few strips from the bottom of a T-shirt to use as swathes. After you've cut a few swathes from the bottom of a T-shirt, the rest of the shirt can be used as a sling by putting the patient's head through the head-hole on the shirt and using the arm-holes to hold the patient's arm. As a minimal sling and swathe, you can fold up the bottom of the shirt the patient is wearing and safety-pin it in place for a sling; zip a jacket closed over the torso as a swathe.

Most lower extremity fractures can be immobilized with sleeping pads held firmly in place with any material that can be wrapped around the splint and secured in place, such as elastic wraps, triangular bandages, belts, rope, and strips of cloth. In the absence of sleeping pads, lower extremities can be splinted with paddles, ski poles, ice axes, sticks, or other such objects plus adequate padding.

With a reliable patient ask him or her to let you know if there are any changes in how the splinted body part feels—such as pain, tingling, or numbness—in case something shows up in between your periodic CSM checks

Remember: There should be adequate circulation, sensation, and motion before *and* after a splint is applied.

Almost all patients with a fracture will benefit from **HIRICE:** hydration, ibuprofen, rest, ice, compression, elevation (see Chapter 14: Athletic Injuries). If your protocols allow, give medications for pain, medications such as ibuprofen or another nonsteroidal anti-inflammatory drug. If your protocols allow, severe pain from a fracture may be treated with stronger pain medications.

General Principles of Splinting

1. Check CSM—circulation, sensation, and motion.
2. Splint the injury in a position of function.
3. Surround the injury site with padding.
4. Support the injury site with something rigid.
5. Immobilize the joints above and below a possible fractured bone or the bones above and below an injured joint.
6. Check CSM.

Specific Treatment for Upper Body Fractures

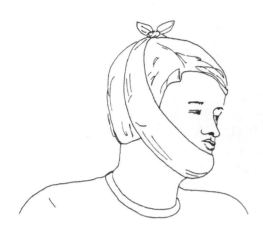

Figure 12-3: *Stabilized jaw.*

The Mandible (Jaw)

Secure the jaw in place with a wide wrap that goes around the head. Be sure to provide for an adequate airway and a way for the patient to quickly remove the wrap should he or she feel like vomiting (which often happens after a severe blow to the jaw).

The Scapula (Shoulder Blade)

Fractures of the *scapula* are quite often stable and require nothing more than sling-and-swathe immobilization. Immobilizing the arm against the body wall is nature's best splint, and many upper extremity injuries can be very adequately padded and immobilized in this manner with a sling and swathe. With any sling and swathe, the hand and wrist should be accessible so that pulses may be monitored.

The Clavicle (Collarbone)

The *clavicle* is a small bone that is easily fractured by a direct blow, by falling on the shoulder, or by falling on an outstretched arm where the force is transmitted up to the clavicle. Patients with a fractured clavicle often support the weight of the arm on the affected side with their opposite hand. A fracture of the clavicle may be treated with a sling and swathe, which should support the weight of the arm, taking pressure off the clavicle. Monitor the patient for increasingly difficult breathing, an indication of possible lung injury (see Chapter 10: Chest Injuries).

The Humerus (Upper Arm)

It is worth noting that the *humeral shaft* may be palpable on the *medial* (inner) side throughout its entire length (from armpit to elbow). Therefore, when a fracture of the *humerus* is suspected, beginning either proximal or distal to the patient's area of com-

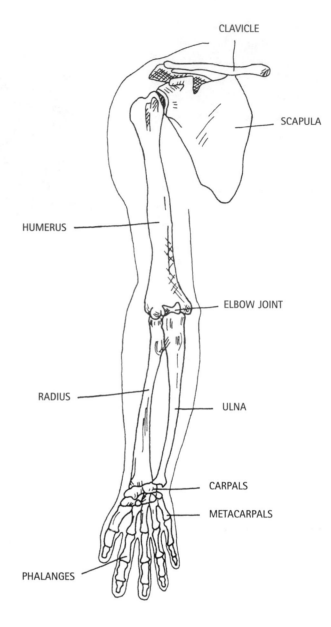

Figure 12-4: *Bones of shoulder and arm.*

plaint, palpate the entire shaft. In this way, very small, nondisplaced fractures may be identified. A fractured humerus may be immobilized with a sling and swathe. It may be beneficial to place a pad over the outside of the humerus to protect the fracture. With an unstable humerus fracture, the entire arm can be immobilized, including the wrist and lower arm secure in a splint, against the chest wall with the elbow not cupped in the sling. This allows the arm to hang down so that gravity can pull gentle traction on the fracture, which may increase patient comfort.

The Lower Arm

For suspected fractures in the region of the elbow, the *radius* (large lower arm bone), *ulna* (small lower arm bone), and wrist, create an adequate splint that incorporates the joints above and below. Adequate padding must be placed around the elbow to

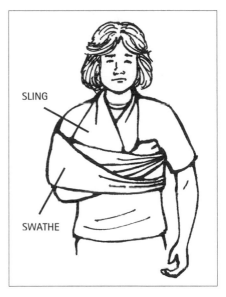

Figure 12-5:
Sling and swathe.

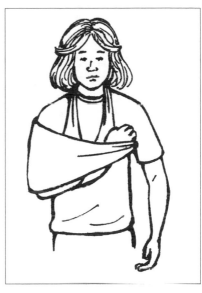

Figure 12-6:
Sling and swathe with elbow free.

Figure 12-7:
Improvised sling from a shirt.

prevent point pressure from the splint. If at all possible, splint the elbow at approximately 90 degrees of flexion (the position of function) to elevate the forearm and hand and, thus, reduce swelling. The stability provided by a solid splint is worth the effort, especially in a long and difficult transport. Place the hand in a position of function with a rolled-up glove, sock, or other soft material tucked in the palm. Then immobilize the hand, wrist, and forearm in a splint.

The Hand and Fingers

When splinting the entire hand is advisable, a suitable splint may be made by placing the entire hand in a functional position with a rolled elastic bandage—or a suitable wad of something similar in size—in the palm of the hand. Then wrap the entire hand with an elastic wrap or roller gauze before securing it to something rigid. Fractures of the upper *phalanges* (fingers) should be splinted in a curled position of function, not in an extended (straight) position. The use of adjacent digits for splinting ("buddy splinting") may be appropriate at times, depending on the severity of the injury. When buddy splinting, place something soft and absorbent, such as gauze, in between the fingers for greater comfort.

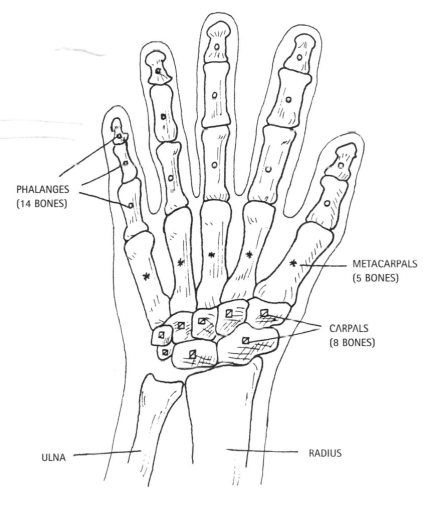

Figure 12-8: *Bones of the hand.*

Specific Treatment for Lower Body Fractures

Fractures of the hip, pelvis, and femoral shaft present major problems in wilderness medicine, due foremost to the potential for shock, the forces involved that also create a possibility of spinal injury, the inevitable transport problem, and, in the case of a femoral shaft fracture, the need to maintain adequate traction for extended periods of time.

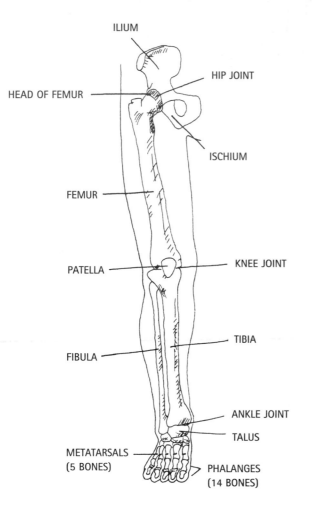

ILIUM
HIP JOINT
HEAD OF FEMUR
ISCHIUM
FEMUR
KNEE JOINT
PATELLA
TIBIA
FIBULA
ANKLE JOINT
TALUS
METATARSALS (5 BONES)
PHALANGES (14 BONES)

Figure 12-9: *Bones of the hip and leg.*

The Hip

The *hip joint* is the proximal end of the femur and its socket in the pelvis. In fractures of the hip, the typical position of external rotation and shortening may or may not be present. The fracture may be one of two types: an *impacted femoral neck* type or an *acetabular* fracture (a fracture of the socket portion of the hip). Making the assessment might be difficult. As a general guideline, if a patient has sustained significant trauma and has painful motion in the region of the hip plus pain with weight bearing,

anticipate carrying him or her out well padded and secure on a litter, sled, or backboard. The patient should be carefully assessed for the possibility of spine injury (see Chapter 8: Spine Injuries). It is usually beneficial to pad between the patient's legs and to secure the legs together. Suspected fractures in the hip area should not be placed in traction, an unnecessary technique that could reduce healthy blood flow to the joint.

The Pelvis

In suspected fractures of the *pelvis*, it is imperative that the patient be observed for progressive shock due to the significant blood loss often associated with these fractures (see Chapter 7: Shock). The patient should be carefully evaluated for the possibility of spine injury and transported in a well-padded litter during evacuation. Securing the pelvis by carefully wrapping a sleeping pad around the pelvic region and tying the pad snugly in place may provide some relief for the patient. Inflatable sleeping pads may be secured around the pelvis prior to being inflated, and then inflated for additional support. The support around the pelvis should not go above the belly button, allowing access (later) to the abdomen by a physician who might not want to remove the splint but who might want to access the abdomen for definitive treatment. Padding between the patient's leg is also recommended.

The Femur (Thigh Bone)

Fractures of the *femoral shaft* require traction, except for fractures of the upper end (hip region) and lower end (knee region) of the femur. Although upper- and lower-end femur fractures are technically femur fractures, it is *not* acceptable to place such fractures in traction. Traction may cause decirculation injury to the hip or knee.

In a major expedition or in an extended trek in remote regions, the WFR in charge should plan ahead for the type of traction that will be used if a femoral shaft fracture should occur. There are commercial traction devices available that are lightweight and easy to apply, such as the Kendrick Traction Device, or KTD. Improvising femoral traction can be satisfactory but should definitely be practiced prior to the actual event.

The fracture of the femoral shaft must be treated in traction for many important reasons. The most important is that traction reestablishes normal length and conformity of the musculature, and this tends to slow the bleeding that often occurs in the thigh. A patient with a closed femoral shaft fracture can easily lose a liter of blood into the thigh, and, if the fracture is movable and the thigh unstable, this bleeding can continue. With bilateral fractured femurs, bleeding may constitute a threat to life. If fracture fragments exist, they may cause further vascular damage if they move. Other reasons for femoral traction include relief of pain, prevention of converting a closed fracture to an open one, and control of further soft tissue damage.

Once a fractured femur has been assessed, someone should be assigned to apply manual traction to the extremity until a mechanical traction device is in place. **In an ideal world, manual traction will not ever be released until mechanical traction is in place.** A general rule of how much traction to apply is 10 percent of the patient's body weight—but a practical rule is pull until the pain of the muscle spasm is relieved and/or the patient's legs are the same length.

A patient with a fractured femur requires evacuation on a rigid litter to keep the risk of further injury to a minimum.

AN IMPROVISED FEMUR TRACTION SPLINT

Many improvisation techniques have been devised for creating a femur traction splint. More wait to be invented. Here is one that has proven successful:

1. Secure a fixation splint, such as a sleeping pad, firmly to the injured leg. This provides support for the fracture and patient comfort both while the traction device is being improvised and after the device is in place. Extend the fixation splint beyond the knee and place padding behind the knee to create 5 to 10 percent of knee flexion; this will make the extremity much more comfortable than if the knee is fully extended into traction.

2. Create an anchor at the sole of the foot, a point from which traction will be mechanically applied. This is often referred to as an *ankle hitch.* An ankle hitch can be made from rope, webbing, belts, cloth, duct tape, or similar material. Numerous techniques can be used to create an ankle hitch: (1) With a single piece of material, form a "Z" over the patient's lower leg. Run the tips of the material under the leg and back through the loops of the "Z." Tighten the system around the patient's leg, and make an anchor at the sole of the foot by tying the ends together. (2) With two pieces of material, place a loop of one over the patient's leg, and a loop of the second piece under the patient's leg. Pass the ends

back through the loops. Tie the ends together to form an anchor at the sole of the foot. (3) With no other means to create an ankle hitch, consider removing the boot from the injured leg. Cut holes in both sides of the boot near the sole in line with the ankle bones. Run a short piece of material through the holes and tie the ends below the boot. Replace the boot and secure it to the patient's foot. **This movement may cause a lot of pain for the patient and should not be your first choice.**

Is the patient better off with boot left on or the boot taken off? The boot and sock provide padding for the ankle hitch. If you do leave the boot and sock off, you'll need to pad well beneath the ankle hitch. With the boot on, a gross determination of sensation and skin warmth can be made by palpation. The patient can relate the sensation of numbness or tingling in the toes, if this should occur. In cold weather, however, the patient may not be able to maintain adequate foot warmth with the boot on, depending on the type of boot. Foot warmth may be easier to maintain with the boot off and adequate insulation added to the foot. The determination to remove footwear should be left to the person in charge and should be based on consideration of all pertinent factors.

3. Find a shaft about 12 to 18 inches longer than the injured leg—measured from the groin or the hip. This can be a stick, a ski pole, a tent pole, a paddle, or some similar object. If the shaft is too long, it can get in the way when the patient is moved.

4. Secure one end of the shaft to the patient's upper thigh on the injured side.
   Note: With only shorter shafts available, you can well pad the groin area of the patient and secure the shaft between the patient's legs. Although this tends to be less comfortable, it is acceptable.

5. Find a piece of rope, cord, or strong cloth to make a trucker's hitch. Traction by the trucker's hitch is pulled from the loops at the sole of the foot and the end of the shaft until the traction of the device is equal to or greater than the manual traction.

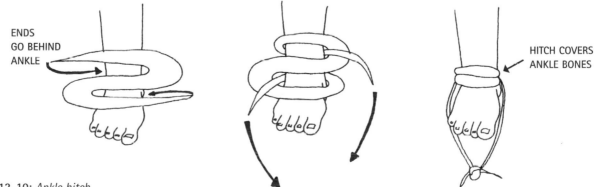

ENDS
GO BEHIND
ANKLE

HITCH COVERS
ANKLE BONES

Figure 12-10: *Ankle hitch.*

6. When mechanical traction is stable, use wraps of cloth, elastic bandages, or some similar material, to secure the shaft firmly in place. The injured leg can be secured to the uninjured leg for additional support.

Note: If you find yourself alone with a patient suffering a fractured femur, you will have to forego manual traction, gather the materials, and carefully apply mechanical traction when everything is in place.

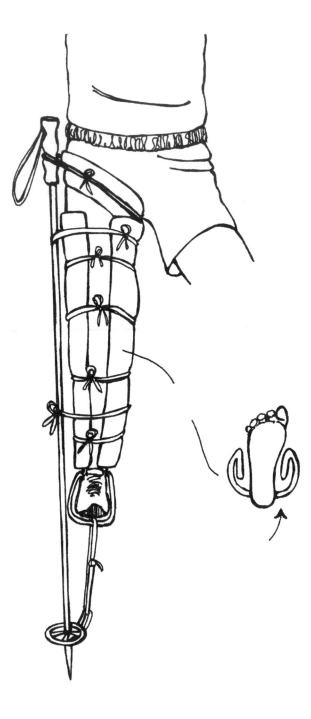

Figure 12-11: *Improvised traction splint.*

The Patella (Kneecap)

Fractures of the *patella* may result from a fall directly on the knee and may be difficult to differentiate from a severe contusion unless there is an obvious deformity. Therefore, a patient with such an injury should be immobilized in a splint that extends from well above the knee to well below the knee. The patient may be permitted to walk with assistance if terrain and other factors dictate that this is the best course of action. A sleeping pad rolled up from both ends simultaneously and secured around the knee, leaving the kneecap free, works well to immobilize injuries in the knee area.

The Tibia and Fibula (Lower Leg Bones)

Fractures of the *tibia* (large lower leg bone) and *fibula* (small lower leg bone) require adequate fixation splinting. The knee, ankle, and both lower leg bones need to be stabilized.

Remember: Pad behind the knee to keep the knee comfortably flexed at about 10 degrees, and secure the foot at a 90-degree angle to the lower leg.

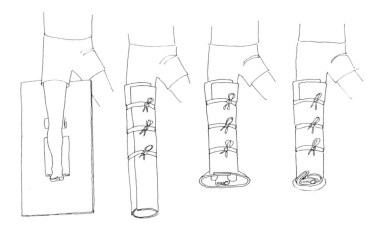

Figure 12-12: *Splinting of lower leg.*

The Ankle

Fractures of the ankle are often difficult to assess, but you will have a patient unwilling to use the ankle, and that means a splint will be needed. Ankle fractures also are often associated with dislocations, leaving the ankle deformed. In fracture-dislocations of the ankle, early gentle reduction of the deformity is extremely important and most often quite easy to do. Simply holding the foot and applying gentle traction at the heel can greatly improve the deformity. The sooner this is done, the better. Basically, the damage has already been done by the fracture-dislocation and, therefore, any attempt at improving the position will have a beneficial effect on circulation (see Chapter 13: Dislocations).

This also makes splinting more stable and comfortable. Ankle splints need to hold the foot approximately perpendicular to the leg, and they provide better support if they extend almost up to the knee.

The Foot and Toes

Fractures of the foot may be splinted similarly to fractures of the ankle. Fractures of the lower phalanges (toes) seldom require anything more than walking carefully and swallowing painkillers. Additional support may be gained by padding between the fractured toe and neighboring toes and "buddy splinting" the injured toe to an uninjured toe.

Complicated Fractures

Angulated Fractures

CSM!

In *deformed* or *angulated* fractures, when angles exist in bones where no angle should exist, gentle traction to obtain better alignment for stable splinting is the appropriate and preferred treatment. Be sure to assess CSM before and after reduction of angulations. Generally, no harm will be done by gently realigning a long bone fracture to a more anatomically aligned position. Applying a splint to a badly angulated fracture is difficult and, most often, unstable. Gentle traction, with an assistant applying stabilization to the proximal side of the extremity, results in an overall improvement in pain, circulation, and splint stability with a negligible risk of creating further vascular or neurological damage.

Traction should first be applied *in line*—in the direction the bone presents. Then, when the patient has relaxed, movement toward normal alignment should progress slowly. Patient relaxation is often signaled by a significant reduction in pain after traction-in-line is applied. Two signals will indicate that you should stop your attempt to reposition the fracture: (1) a need to use force, indicating, possibly, that the bone ends are somehow trapped out of alignment; and (2) a marked increase in pain, indicating, possibly, that nerves and/or blood vessels are caught at the fracture site. **Do not use force. Do not cause patient a marked increase in pain.** Once the bone has been returned to normal anatomical alignment, splinting may proceed without any need to maintain traction except in the case of a fractured femur.

Open Fractures

In an open fracture without visible bone ends, the wound should be thoroughly cleaned via irrigation and bandaged prior to splinting (see Chapter 15: Wilderness Wound Management). The rigid part of the splint should not press on the wound.

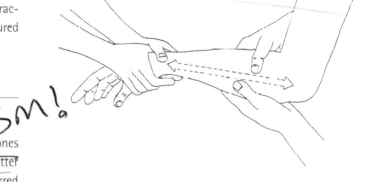

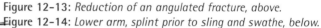

Figure 12-13: *Reduction of an angulated fracture, above.*
Figure 12-14: *Lower arm, splint prior to sling and swathe, below.*

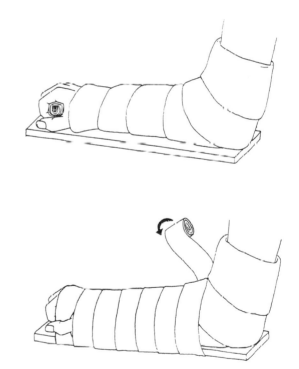

In case of an open fracture with visible bone ends, the best approach is to gently irrigate the wound and the protruding bone ends with water. Bone ends should be irrigated and/or flushed and *not* scrubbed. In a wilderness environment, gentle traction-in-line should be applied, a technique that almost always results in the bone ends slipping back under the skin. A sterile bandage or the cleanest available material should be placed over the wound and the fracture then splinted. The rigid part of the splint should not press on the wound. Broad spectrum antibiotic therapy is indicated, if available and if your protocols allow it. The point here is that correction of gross angular deformity and adequate splint immobilization takes precedence over concern regarding protruding bone ends. Whether bone ends slide back under the skin during realignment should *not* be a major concern during treatment of an open fracture in the wilderness. If, however, the bone ends fail to slip back under the skin, they should be covered with a moist sterile dressing prior to splinting.

Long-Term Fracture Care

Over extended periods of time, extremity splints may become unbearably uncomfortable. You may need to periodically and gently remove splints to apply cold to injury sites, massage parts of the body not fractured but contained within the splint, or just give the fractured extremity a "breather." Although you cannot remove traction from a fractured femur, you may need to take over with manual traction and release the pressure of the ankle hitch periodically.

Evacuation Guidelines

Almost all suspected fractures, especially those that are unstable, should be evacuated to definitive medical care. There is no great rush except in the case of open fractures, fractures with decreased CSM distal to the injury, and fractures of the femur, hip, or pelvis.

Conclusion

After a careful, belayed descent, you hurry to the side of the injured climber. His face is a grimace of pain, and he breathes between clinched teeth. Your initial assessment reveals no immediate threat to the patient's life, and your focused assessment uncovers no injuries other than an angulated fracture not far above the right ankle.

With your climbing partner holding counter-traction well above the injury site, you take careful hold of the patient's foot and pull to create traction in line. After a moment of increased pain, the patient remarks that it "feels better now, please don't stop pulling." With traction, you slowly return the leg to normal anatomical alignment.

Releasing traction but holding the leg firmly in position, you ask your partner to return to your tent for a sleeping pad and the first-aid kit. When he returns, you create a well-padded and rigid splint that encloses the lower leg, stabilizing the knee and ankle. With the injury splinted, you and your partner lift and carry the patient to a relatively comfortable nearby spot. You intend to stay with him while your partner goes for help.

CHAPTER THIRTEEN

Dislocations

You Should Be Able To:

1. Describe the difference between long-term (wilderness) care and short-term (urban) care of dislocations and the importance of wilderness intervention.
2. Describe the signs and symptoms of upper and lower extremity dislocations.
3. Demonstrate how to treat upper and lower extremity dislocations.

It Could Happen to You

A period of melt-and-freeze temperatures in the Old Cascades region of Oregon has left the snowy surface hard, but not hard enough—just enough of a crust to hold weight a teasing moment before dropping snowshoers into the softer underlying whiteness. For additional support everyone in the group you're leading is using two ski poles. It could have been frustrating, but the day's weather is pleasant, the group congenial, and spirits are high. The walk up the ridge goes quickly.

On the descent, the trip goes even faster for Deborah. Her lighter weight kept her from crunching through on an especially dense crust in an opening on the steep, forested hillside. She "skies" away from the group on her snowshoes. Planting a pole instinctively as a brake, she slips to the end of her arm's reach, and levers her right shoulder out of joint. Deborah plops into a sitting position with a look of surprise that changes rapidly into one of intense pain. She sits huddled in the snow with her right arm cradled in her left.

Any joint of the body can come out of joint, out of its normal bone-to-bone relationship. A *dislocation* is a complete or partial disruption of the normal relationship of a joint. The mechanism of injury can be indirect (as when a stuck ski pole levers out a shoulder or the high brace of a kayaker pries out a shoulder) or direct (as when a fall on a shoulder forcefully knocks it out). Direct mechanisms tend to cause more damage to the joint because of the forces involved. Dislocations are sometimes associated with fractures, the probability being higher if the forces involved were direct.

Our concern is primarily with the more obvious complete dislocations of joints. These are more easily identified and more incapacitating to the patient. In an urban situation the treatment for a dislocation is relatively simple: The injury should be stabilized—splinted—in the position found, and the patient should be expeditiously transported to a medical facility. In the unique situation of a wilderness accident, however, it is extremely important to be able to make an accurate assessment and attempt a *reduction*—a return to a normal bone-to-bone relationship—of a dislocation as quickly as possible after it occurs. Discretion must obviously be used when evacuation to a nearby medical facility can be easily accomplished, such as when the patient is approximately an hour or less from definitive medical care. In that case, medical protocol requires you to leave the reduction to hospital medics.

There are some major advantages to early reduction of dislocations, including the following:

1. Reduction is easier immediately after injury, before swelling and muscle spasm have developed.
2. Transport of the patient is easier after reduction.
3. Reduction most often results in dramatic relief of pain.
4. Immobilization of the injured joint is easier to accomplish and more stable after reduction.
5. The safety of the entire party may be jeopardized during the evacuation of a patient with an unreduced major joint dislocation.
6. Early reduction reduces the circulatory and neurological risks to the extremity.

Disclaimer: The reduction of dislocations falls outside the scope of practice for Wilderness First Responders unless the WFR is acting under specific protocols established by and managed by a physician adviser.

General Assessment and Treatment of Dislocations

As with any musculoskeletal injury, have a good LAF—look, ask, and feel—at the patient (see Chapter 12: Fractures). There are several signs that may be helpful in identifying a dislocation. When a joint is dislocated, there is nearly always restriction of motion through its normal range. There is often obvious deformity in comparison with the uninjured side. Bony crepitus, the grating of bone on bone, is almost always absent. There is often a typical, identifiable posture of the dislocated joint, which the patient will maintain to minimize the pain. Obtaining a history of the mechanism of injury (MOI) can also be helpful and will be discussed for individual joints in this chapter.

Assess for normal CSM—circulation, sensation, and motion—distal to the injured joint. Any vascular or neurological deficits distal to the injury should be noted and should initiate an immediate attempt to reduce the dislocation to prevent permanent disability.

Note: Do not be concerned about causing additional damage to any fracture associated with a dislocation. These fractures are most often improved in alignment with the reduction of the joint. The same is true of vessel or nerve impairment associated with a dislocation—reduction reduces the impingement to these structures as well.

When a fracture of the femur or humerus accompanies a dislocation at the hip or shoulder, respectively, the dislocation may not even be assessed in view of the more apparent major fracture. In these cases, splinting of the fracture is the main concern. The dislocation, for all practical purposes, is a secondary issue and usually not amenable to reduction by ordinary means.

When reducing dislocations, try to have as many factors in your favor as possible. Use other party members where necessary; communicate with the patient very openly about what you are going to do; be very positive and deliberate in your approach. The use of intramuscular or intravenous drugs can make the job much easier, should they and a person licensed to use them be available. Gentle (but steady) and persistent traction is the rule rather than forceful jerking or ballistic maneuvers to reduce the joint. Remember: You and the patient have much to gain.

Work quickly but calmly. Act fast, but go slowly. The sooner an attempt to relocate, or reduce the angle of dislocation, is made, the easier it will be for you and the patient. You need to get the patient to relax as much as possible. Suggest that the patient take deep breaths and let all the tension go from the painful area. Lightly touch the dislocation, reminding the patient where to concentrate attempts to relax.

Place yourself comfortably in a position that allows you to pull gentle traction in line. Take a firm grip of the arm or leg distal to the injury, pulling *in line* with the way the patient is holding the extremity. Pull steadily and increase pull slowly—and remember to avoid jerking.

Note: Do not use force. Stop if you cause a marked increase in pain.

A patient may feel pain when you attempt reductions. Do not be discouraged by this, despite the fact that your patient might. Avoid marked increases in pain, such as those that cause a patient's voice to rise an octave or more.

If the dislocation does not relocate after you have pulled traction in line for a minute or so, gently move the extremity toward normal alignment. Maintain traction during the movement. What you are trying to do is manually relax the muscular spasms and encourage the bones into normal alignment. You can move the bones toward a more normal alignment as long as (1) the patient does not complain of markedly increasing pain and (2) you are not forcing movement. The muscles, tendons, and ligaments want to be back where they belong. You are encouraging them to move back. Most patients can feel the relocation take place and can tell you when relocation has occurred.

Patients with a history of chronic dislocations may have a preferred reduction technique, a technique they want you to use. These patients are usually ready, willing, and able to tell you what to do. It would be wise to follow their advice.

Occasionally a dislocation will be complicated enough to prevent relocation. You will have to splint the extremity in the position the patient is holding it. Make the splint as comfortable as possible. Evacuate this patient as soon as you can.

Signs and Symptoms of a Dislocation

1. Pain and/or tenderness.
2. Loss of normal range of motion.
3. Deformity (loss of normal symmetry), including swelling.

Specific Treatment for Upper Body Dislocations

The Mandible (Jaw)

Mandibles can be dislocated by direct force (a blow) or indirect force (such as biting aggressively on a very hard apple). After wrapping something soft and protective around your thumbs, reach gently into the patient's mouth, gripping the back molars. Pull down and forward until relocation occurs. The mandible may be secured with a wide wrap that holds the lower jaw to the upper jaw by extending entirely around the head from bottom to top. The wrap needs to be easily removable by the patient if he or she feels nausea on the verge of vomiting. The patient may require a soft diet for several days. Without protocols an evacuation of the patient is not required unless the reduction fails.

The Shoulder

Outdoor travelers, particularly paddlers and skiers, often exert sudden force on an already extended arm, forcing the head of the humerus forward and downward. These *anterior-interior* dislocations of the shoulder joint are the most common, accounting for over 90 percent of shoulder dislocations. The patient usually stabilizes the shoulder in the most comfortable position and cannot bring the involved extremity across his or her chest to a position of rest. The upper arm typically is held away from the body in various positions and cannot be brought into a sling-type position. This differentiates a dislocation from a fracture of the humerus, in which instance the patient is invariably splinting the upper arm against the chest wall for comfort. Check circulation, motor, and sensory function to the hand, as well as sensory function along the outer aspect of the shoulder, and document your findings.

Posterior dislocations of the shoulder are rare. In this instance, the upper arm and forearm is held across the anterior chest wall and attempts at externally rotating the upper arm away from the chest are restricted and painful. The assessment is often difficult to make.

The simplest and easiest method for reduction of a dislocated shoulder begins with asking the patient to lie down. With the patient supine, pull gentle traction straightaway from the body in a line with the direction the patient presents the arm to you. One of your hands pulling at the elbow is often enough. A second rescuer is not needed. Under traction, turn the arm into *external rotation* (the upper arm turned away from the body as if the patient was going to wave "hello" to someone). External rotation is extremely important. Without a reduction, try to move the arm, still under traction, up to 90 degrees from shoulder. At 90 degrees, the arm looks sort of like the patient is ready to throw a baseball. Reaching 90 degrees from the shoulder is not critical, but moving the upper arm into external rotation (the ready-to-throw position) is critical. If the shoulder does not reduce, move the elbow to the belly button with the upper arm held in external rotation. Then move the patient's hand, on the affected side, to the opposite shoulder. Traction should be maintained until reduction is achieved, or until rescuer and patient tire to the point at which traction can

no longer be maintained. Take as much time as you need, and make as many attempts as you are able.

A second method is placing the patient prone (face down) on, say, a large rock or log, and letting the arm hang down toward the ground with 15 to 20 pounds of weight secured to the hand. This method may be slow and relaxation is critically important, but the muscles generally fatigue in time. Manual assistance by manipulation of the shoulder is helpful.

Unreduced shoulders should be splinted across the chest with padding between the arm and the chest. After a successful reduction, immobilize the arm with a sling and swathe. The sling supports the forearm horizontally in front of the body. One

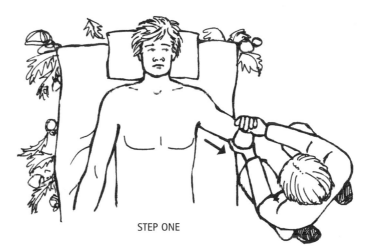

STEP ONE

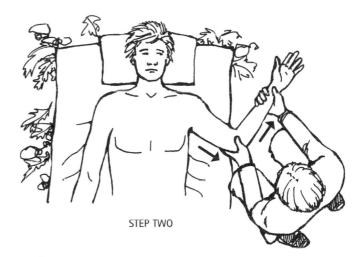

STEP TWO

Figure 13-1: *Aerial view of patient lying on ground with arm under traction and then into external rotation for shoulder reduction.*

swathe secures the arm to the chest wall with a wrap around the body, and the other secures the elbow to the chest wall. In both cases, evacuation of the patient is recommended.

The Elbow

Look for obvious deformity when compared with the uninjured elbow and for restricted flexion and extension of the joint. Most commonly the *olecranon* (bony upper end of the ulna) dislocates toward the rear, and you can see the resultant bony prominence of the olecranon posterior to the elbow.

For elbows, reduction requires two people: One pulls up on the humerus, and the other pulls out on the forearm while pushing down on the forearm. Apply a slow, steady pull. The patient's ability to flex with elbow to 90 degrees is a sign of reduction. Immobilize the injured elbow in a sling and swathe, as described for the shoulder. If reduction is not possible, splint in the position in which the elbow is found. Evacuation of the patient is recommended in both cases.

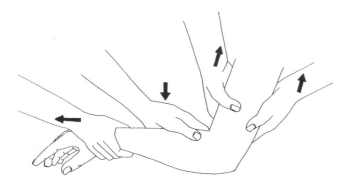

Figure 13-2: *Elbow reduction.*

The Wrist

Wrist dislocations are often difficult to differentiate from a fracture and often difficult to reduce. Therefore, immobilization in a splint is the treatment of choice. Circulation and neurological function to the hand are usually not compromised, but, if so, reduction should be attempted with traction in line and movement toward normal alignment. The most common neurological deficit is median nerve involvement, resulting in numbness over the palmar surfaces of the thumb, index, and middle fingers, and one half of the ring finger. A dislocated wrist may be reduced by grasping the patient's hand as if you were shaking hands while holding the patient's arm above the dislocated wrist. Then pull traction while moving the wrist toward normal anatomical alignment. An evacuation is required.

Upper Phalanges (Fingers)

Obvious deformity and limited function are the main assessment factors in dislocated finger joints. Reduction of the common dislocations of middle or distal finger joints is best accomplished by maintaining the digit in partial flexion. The weight of the patient's arm will apply counter-traction. You actually may have to push the dislocated base of the joint back in place while traction is being applied to the partially flexed digit. This is much more successful than attempting to apply traction to a straight digit. If the finger seems "locked out," you should increase the deformity of a dislocated finger a bit, with traction, prior to movement toward reduction.

There are two hand dislocations in which reduction is difficult, if not impossible, without surgery. The first and most common is dislocation of the *metacarpophalangeal joint* (where the finger joins the hand) of the index finger, and the second is dislocation of the metacarpophalangeal joint of the thumb. The thumb is sometimes reducible closed, but the index metacarpal rarely, if ever. Make one attempt and then quit, immobilizing the joint in a functional position rather than persisting with multiple attempts.

Finger dislocations may be associated with skin disruption. It is recommended to cleanse the open wound as thoroughly as possible and accomplish the reduction. The patient would benefit from being placed on oral antibiotics, if available and allowable by medical control.

Do not splint a reduced finger to a rigid splint. Tape the injured finger to an uninjured neighboring finger and encourage the patient to flex the damaged finger regularly to promote healing. An evacuation for simple dislocations of fingers that are reduced is not required, unless you have a protocol that dictates otherwise.

Specific Treatment for Lower Body Dislocations

The Hip

The majority of hip dislocations are posterior. The hip will be moderately flexed, internally rotated, and *adducted* (brought across the midline toward the opposite leg). Any attempt to extend the hip for splinting or easier transport will be resisted by the patient and is mechanically nearly impossible to accomplish. The mechanism of injury is usually a fall in which the hip is flexed and the impact is transmitted longitudinally through the knee, driving the femoral head posteriorly from the *acetabulum* (hip socket). The mechanism would be similar to that sustained in a dashboard injury in a motor vehicle accident. A climber, for instance, may dislocate a hip in a fall as he or she makes contact with a rock wall after reaching maximum rope length and swinging back into the rock face. The typical posture of the posteriorly dislocated hip often makes it easy to differentiate from a fracture of the proximal femur, which leaves the leg externally rotated.

Anterior dislocations of the hip result in a posture of extension, external rotation, and *abduction* (the thigh is held away from the midline of the body). Again, attempting to extend the hip to a neutral position would be most difficult, if not impossible, and resisted by the patient.

The reduction of a hip dislocation requires two people, ideally, with one applying counter-traction to the pelvis as the patient lies in a supine position on the ground. The involved hip and knee are flexed to 90 degrees with the rescuer straddling the patient, preferably straddling both legs, and applying traction in an upward direction. If traction straight up from the patient fails to work, try varying the pull to lateral. If only one person is available to attempt the reduction, the patient can be placed prone over a log, rock, or bench, and the traction applied downward with the hip and knee flexed 90 degrees. Once reduced, the injured hip must be splinted and immobilized to the uninjured extremity and the patient transported in a supine position.

Make every attempt to reduce a dislocated hip. Without reduction, joint death can occur in as little as 6 hours. If the hip fails to reduce, evacuation must be expeditious.

Figure 13-3: *Hip reduction.*

The Patella (Kneecap)

The patella is most often displaced laterally (to the outside) with the knee held in flexion by the patient for comfort. Dislocations of the patella are often recurrent and the result of a pivoting type of injury with a partially flexed knee. The patella, you will find, is not movable and obviously out of place.

Apply gentle traction to extend the leg. Massaging the large thigh muscles often helps in extending the leg and may be a critical part of reduction. You can place a small pad under the patient's heel to hold the leg in a slightly overextended position. As the knee is extended, the patella will usually reduce itself, although you may have to wait for relaxation and reduction. Keep massaging the thigh. Occasionally a slight nudge of the

patella toward normal alignment with your hand can be used to shorten the wait. Immobilize the extremity in a splint that reaches well above and well below the knee. With the knee extended and immobilized in a splint, the patient will most often be able to walk well enough in this "walking knee splint" without further risk to the joint. To ease walking, create a crutch, using a stick, ice ax, ski pole, or some other device.

The Knee

A frank dislocation of the knee results from radical tearing of the joint out of its normal alignment and presents a disaster at best. Major ligamentous disruption is the rule. The knee may not be dislocated at the time of exam (it may pop out then pop back in on its own), but gross instability is the major clue and vascular impairment is a major symptom. Check pulses and motor function in the ankle and foot.

Observe the joint closely and apply gentle traction to realign the joint as well as possible. Reduction is seldom difficult. As with all dislocations, the sooner realignment is established, the better. With a knee, it is often even more important to act quickly since circulation distal to the knee can be compromised. Splint as securely as possible, making sure not to compromise circulation to the foot. This patient must be carried out as soon as possible.

The Ankle

Dislocated ankles are most commonly associated with fractures and obvious deformity. Bony crepitus is possible with this dislocation. Vascular impairment to the foot must be assessed by checking CSM.

It is important to reduce the degree of deformity as much as possible. Ordinarily this is not difficult due to the gross instability that results from associated fractures. Often, simply holding the forefoot by the toes and allowing the remainder of the extremity to act as the counter-traction results in improved alignment of the ankle dislocation without much additional effort. Gentle traction of the heel and foot may improve alignment. Immobilize the ankle with a splint. Evacuation is a necessity for this patient.

Lower Phalanges (Toes)

Reduction is achieved with the same technique that works for fingers. Splinting is seldom required. Rigid soled boots serve as splints. An evacuation is seldom, if ever, required.

Long-Term Care for Dislocations

Regular monitoring is required to ensure your patient maintains adequate circulation to the injured area. Check often for a pulse beyond the injured joint and for normal skin color and sensations beyond the joint. Give the patient the responsibility of notifying you of changes in her or his condition. This becomes a double-check, a redundant system of your checking and the patient's maintaining awareness. Once reduction has been accomplished, encourage the patient to regularly move the damaged joint through a normal range of motion to promote healing. Range-of-motion exercise after reduction should be slow and limited to motion *without* pain or discomfort.

Relocated dislocations benefit from HIRICE (hydration, ibuprofen, rest, ice, compression, elevation) and immobilization via splinting (see Chapter 12: Fractures and Chapter 14: Athletic Injuries).

Evacuation Guidelines

Patients with dislocations that successfully resist attempts at reduction should be evacuated as soon as possible. Most patients with dislocations should be evacuated for careful medical examination even if they relocate easily. There is always the chance of underlying damage that does not show up right away, especially with first-time dislocations. Exceptions to the evacuation rule include fingers and toes and chronic dislocations (repeat offenders) that the patient is able to use with reasonable comfort after the relocation. All of these guidelines may be superseded by specific protocols that differ from these guidelines.

Conclusion

It's a matter of moments before you have Deborah sitting on a foamlite pad, off the cold snow. Sitting in the snow beside her, you assess her arm on the damaged side, noting there is no numbness or tingling and no loss of normal circulation distal to the injured shoulder. You ask her to lie back, and you assist her into a supine position. Keeping her arm on the injured side flexed, you begin to pull gently in line with the way she held her arm before you took control. As the pressure on the pull is slowly increased, you encourage Deborah to keep taking deep breaths and to relax her muscles.

With the patient as relaxed as possible, you begin to slowly move the arm, under traction, up into external rotation. The final arm position looks something sort of like it would if Deborah was preparing to throw a baseball. You hold that position for several minutes.

With a soft and satisfying thump, the end of her upper arm bone rolls back into its socket. Deborah's expression changes immediately to one of relief, and she releases a long sigh.

Removing her parka, you construct a sling from the triangular bandage in your first-aid kit, a sling supporting the arm on the damaged side. Replacing the parka, you zip it up. The parka becomes a temporary swathe.

With the group sharing out her load, Deborah is able to snowshoe carefully to the van, about 2 miles away. A short visit to the hospital reveals nothing extraordinary in her shoulder. She goes home with soreness but no serious damage.

CHAPTER FOURTEEN

Athletic Injuries

You Should Be Able To:

1. Define and describe the signs and symptoms of the most common wilderness-related athletic injuries.
2. Demonstrate assessment and treatment of the most common wilderness-related athletic injuries.
3. Define and describe the treatment for tendinitis.
4. Describe ways to prevent athletic injuries.

It Could Happen to You

You're two days from the road on your third winter trip following the trackless banks of the Escalante River toward Lake Powell, Utah, but this is your first hike here since the area was proclaimed Grand Staircase–Escalante National Monument. It looks the same—startlingly beautiful, chaotic in a masterful way—but it *feels* different now. A designated national treasure: How did you get talked into guiding a half dozen 14- to 16-year-olds?

Evening at the second camp, and Tom, an impetuous teenager full of energy, climbs a rise of sand near the river, runs down headlong, leaps, and crashes. *Tweak* goes his right ankle. Tom, reluctant to admit a painful mistake, sits quietly holding the ankle when you arrive, called by another member of the group.

After removing Tom's boot and sock for a look at the ankle, it seems, almost as you watch, the injury puffs up on the outside and darkens like a rain cloud building toward an afternoon shower. Somehow, you think to yourself, you *knew* this would happen.

Muscles, the tendons that connect muscles to bones, and the ligaments that connect bones to bones work in concert to provide locomotion, movement of the human body. Under the physical stress of wilderness endeavors, muscles, tendons, and ligaments often suffer traumatic or overuse injuries ranging from mildly annoying to severely debilitating. These types of "athletic injuries" are well documented among the problems most likely to require treatment in the wild outdoors. They are called "athletic injuries" not because the patients are necessarily athletes, but because the injuries often leave the patient, after adequate management (and sometimes without proper management), able to continue the wilderness journey, as athletes are often able to return soon to the game following a similar injury. Proper intervention by the Wilderness First Responder may well promote healing and prevent these injuries from getting worse while allowing the patient to continue the wilderness adventure. If the patient must be evacuated, a properly treated athletic injury often allows the patient to leave the wilderness under his or her own power.

General Assessment of Athletic Injuries

As long as the scene is safe, *do not move the patient or ask the patient to move until your assessment has been completed.* "Walking off the pain," a common patient response to an athletic injury, is a myth that may lead to permanent damage. A patient's eagerness to prove she or he can stay with the group and/or not become a burden to the group may lead to lifelong disability. While keeping the patient in a restful position, LAF at the injury site:

Look: At skin level, is there swelling? Discoloration? How quickly is the injury swelling and/or discoloring? The degree of swelling and discoloration is indicative of the degree of damage.

Ask: How much does it hurt? Pain is a primary indicator of the severity of an athletic injury. What happened? How much force was involved? In which direction were the forces applied? Did the patient hear a popping or snapping sound at the moment of stress? Sounds may be an indication of damage. Has the patient had similar pain and discomfort before and, if so, what was the cause?

Feel: Is there pain when you press on specific points? Point tenderness is indicative of tears in connective tissues—tendons and ligaments—as well as fractures in bones.

At this point, if you see no signs of serious injury, if you have decided the injury could be "athletic," have the patient move the injured part through its normal range of motion while still sitting. Loss of range of motion may be an indicator of an athletic injury. Compare the range of motion on the injured side with the range of motion on the uninjured side to help determine the degree of the injury. When muscles, ligaments, or tendons are stressed by movement, does it cause pain? How much pain?

If the patient has tolerated self-movement of the injury without a high level of pain or a substantial loss in range of motion, you, the rescuer, may now move the injured part with your hands. Add stress at the maximum stretch points within the range of motion of the injured part. Once again, assess the level of pain and the range of motion.

If the patient has tolerated your movement of the injured part, allow the patient to attempt to use the injured part—to put weight on injured ankles and knees, or to use shoulders, elbows, and wrists as necessary to walk out and/or complete the wilderness journey. If the patient is able to use the injured part, the patient, in most wilderness cases, should be allowed to continue the journey, if he or she chooses to do so.

The single most important factor related to a wilderness athletic injury is the patient's ability to use the injured part.

If at any time in your assessment you suspect the injury is a fracture, the injury should be splinted and the patient evacuated. In other words, if the patient cannot or will not use the injured part, splint the injury (see Chapter 12: Fractures) and evacuate the patient.

General Treatment of Athletic Injuries

Athletic injuries should be managed initially and as soon as possible with **HIRICE** (hydration, ibuprofen, rest, ice, compression, and elevation). These six treatments usually limit swelling, and swelling delays the patient's return to normal activity more than any other factor.

Hydrate the patient with enough water to keep her or his urine clear. Adequate hydration is required for healing and for general well-being, and for the processing of Ibuprofen.

Ibuprofen. Doctors typically recommend a nonsteroidal anti-inflammatory drug (NSAID), such as ibuprofen, for athletic injuries. These drugs, if your protocols allow them, should be given with plenty of water and with food. The patient may follow the regimen suggested on the drug's label.

Rest the injury—stop using the injured part. Rest reduces circulation to the injury, thus it reduces swelling. Rest prevents complications that may occur when the injured area is used too soon, or when it should not be used at all.

Ice the injury—cool it with an ice pack, snow pack, chemical cold pack, soaking in a cold mountain stream, or wrapping in wet cotton (such as a T-shirt) and allowing evaporation. Ice, snow, and cold packs should not be placed directly on naked skin but separated from the skin with a layer of cloth. Cooling constricts blood vessels to limit swelling. Maintain ice for 20 to 30 minutes, then allow the injury to warm naturally—for 12 to 15 minutes—before allowing the patient to return to using the injured part.

Compress the injury with an elastic wrap from distal to proximal (from the end of the extremity toward the heart). Never wrap so tight that adequate circulation is impaired. Check and recheck CSM (circulation, sensation, and motion) often. Compression creates higher pressure on torn tissues, making it more difficult for swelling to occur.

Elevate the injury—prop it up comfortably higher than the patient's heart. Elevation reduces swelling by decreasing circulation to the area.

If adequate time is allowed for return of normal circulation between treatments, it is difficult to overdo RICE. If the wilderness situation allows, it is recommended to apply RICE to athletic injuries at least several times a day during the first couple of days following the injury, and, better still, until pain and swelling have subsided.

Note: Heat usually exacerbates swelling. Ice—or cold—is the treatment of choice, even if the injury is 48 hours old when you first see it. Heat may be applied once all pain and swelling have subsided to loosen up the damaged area for retraining.

General Management of Athletic Injuries

1. LAF (look, ask, and feel).
2. Test the injury for usability.
3. HIRICE the injury (hydrate, ibuprofen, rest, ice, compress, and elevate).

Specific Treatment for General Types of Athletic Injuries

Strains

[handwritten: no swell]

Strains are injuries to muscle fibers and tendons, resulting when those body parts are stretched too far. Only a few torn fibers, and the injury is usually called a "pull." Many torn fibers result in a "tear." There is usually little or no swelling, no matter the degree of injury, but there may be considerable discoloration, a bruised look. Pain usually subsides when the affected body part is rested, but active use can cause much pain.

[handwritten: don't stretch] Muscle pulls are probably the most common athletic injury. An injured muscle typically spasms, a protective mechanism that shortens the overstretched muscle fibers. Spasms cause pain. HIRICE, the treatment for strains, relaxes the muscle, reducing pain and encouraging healing. It is critical to remind the patient to start gently stretching the injured muscle as soon as he or she can tolerate the stretching. Gradual, nonviolent stretching of the muscle fibers reduces the chance the fibers will heal shortened, a condition that almost always leads soon to a reinjury of the muscle. Full healing is usually indicated by the patient's ability to stretch the muscle without pain and use the muscle through an active range of motion without pain.

Sprains

[handwritten: swelling]

Sprains are injuries to ligaments. A *first-degree* sprain, the one most often experienced, stretches but does not actually tear ligaments. There is pain when the affected joint is moved, stressing the damaged ligaments, and perhaps a little discoloration, but there is little swelling and little instability. With proper care, the patient will be back to normal in 1 to 2 weeks. In a *second-degree* sprain, partially torn ligaments swell and discolor, usually in a very short time. Pain may discourage the patient from stressing the damaged ligaments. Healing time can take as long as 6 weeks. *Third-degree* sprains entail serious ligament tears, typically complete tears. Discoloration may be extreme. Within 30 minutes of injury, most patients are unable to move the damaged joint due to pain and swelling. The patient could require 6 months or more of healing, and many third-degree injuries require surgical intervention. HIRICE is still the treatment of choice, and splinting is almost always required.

Note: Both strains and sprains may be associated with fractures and/or dislocations.

Tendinitis

Tendinitis is the inflammation of a tendon, a problem unlike strains and sprains in that the injury comes with overuse and not

from trauma. Tendinitis can, however, be ju[st] more debilitating than some strains and sp[rains.]

Almost all voluntary movement in the [body] tendons transmitting muscular force from [muscle] across joints. Tendons are attached to musc[le on one side and] bones at an insertion point on the other. On the way from muscle to bone, the tendon passes through a *tendon sheath* that is attached to the underlying bone. When the muscle flexes, it contracts (shortens), drawing the inflexible tendon back toward the muscle. The sheath stabilizes the tendon and acts as a pulley. Without the tendon sheath the tendon would straighten as a rope does when it is tied to a weight and pulled on by someone who wants the weight to move. The tendon sheath keeps the tendon near the bone, increasing the efficiency of the system.

The sheath secretes *synovial fluid,* the same viscous, slimy fluid that keeps joints lubricated. With repetitive motion, such as paddling or hiking, a tendon may get overworked. The sheath tries to keep things running smoothly and secretes more fluid, but the sheath cannot expand to hold the increase in fluid, so the tendon gets compressed. The tendon and the sheath swell, and inflammation begins. The tendon calls for more lubrication, and the sheath responds with more fluid, and the problem worsens each time that particular tendon is used. Overuse may also cause a tendon sheath to begin to microscopically tear loose from the bone. In both cases, inflammation—in this case tendinitis—produces insistent pain when the affected tendon is in use.

A more severe type of tendinitis can develop if the developing problem is ignored until calcium salts grow in the inflamed area. The sharp pieces of calcium irritate the bursa sac, the tough bag that surrounds all joints to the keep the lubricating synovial fluid from escaping. The irritated bursa starts to overproduce fluid. Eventually, the entire sac becomes inflamed and tense, and the whole joint becomes an agonizing burden to bear. At this point the standard treatment of tendinitis may not work.

The standard treatment is simple: Do not use the joint until it gets better. There is little chance of pain if the patient does not move the inflamed tendon. Without use, the tendon can heal. Application of cold packs 3 to 4 times a day for about 20 minutes helps speed the healing by reducing the swelling, and a regimen of a nonsteriod anti-inflammatory drug also reduces pain.

The wilderness often prevents the best treatment for tendinitis. If your assessment is tendinitis, make every effort to prevent the patient from using that joint for the next few days. When the joint is used, attempt to limit the motion of the joint to within the limits of pain. As an example, a paddler may shorten his or her stroke to reduce pain. Pain-free use stimulates healthy, healing circulation while keeping the joint from stiffening. Persistent pain, especially beyond 2 weeks, is a signal to ask for a physician's care.

Ankle Sprain

Lateral (outside) ligament damage it the primary form of ankle sprain, due to the structure of the ankle, which allows little *eversion*, or turning outward, and much *inversion*, or turning inward. Inversion sprains of the ankle are the most common wilderness sprain and, in fact, account for approximately 85 to 90 percent of all ankle sprains.

The lower leg's bigger bones, the large *tibia* and smaller *fibula*, meet the ankle at the *talus*, an upwardly rounded bone that allows the tibia and fibula to "rock" back and forth on its top. The talus, in turn, "rocks" side to side on top of the front of the *calcaneus* (heel bone) at an articulation point called the *subtalar joint*. This ability to rock allows for freedom of movement when the human body hikes, climbs, or runs. In front of the calcaneus lie two small bones, the *navicular* on the medial aspect of the foot and the *cuboid* on the lateral aspect. In front of these two bones are three even smaller bones called the *cuneiforms*. These seven bones—talus, calcaneus, navicular, cuboid, and three cuneiforms—are the true ankle bones. The bumps called "ankle bones" are actually the rounded distal ends of the tibia and fibula, bumps called the *lateral malleolus* (the end of the fibula) and the *medial malleolus* (the end of the tibia).

It takes a complex arrangement of ligaments to secure all these bone-to-bone connections, but six of these ligaments are primary targets for injury. Two of these, the anterior and posterior *tibio fibular ligaments*, hold the tibia and fibula together, preventing those bones from being wedged apart by the talus when you take a step. The *deltoid ligament*, a fan-shaped bunch of fibers, attaches the tibia to bones on the inside of the ankle. The deltoid is wide and tough, allowing little eversion of the ankle. In fact, overstressing the deltoid is more likely to pull off a fragment of bone (an avulsion fracture) than to sprain the ligament. On the outside of the ankle, the other three primary ligaments attach the fibula to the talus and the calcaneus: the anterior and posterior *talofibular* and the *calcaneofibular.* These smaller, weaker lateral ligaments allow much more inversion than the deltoid allows eversion, and they are the ones most often damaged. The ligament in the middle, the calcaneofibular, bears the brunt of inversion stress, but, as you can imagine, sprained ankles often involve more than one ligament.

Mild ankle sprains (first-degree injuries) can be taped, and the patient is usually able to move well in the wilderness. Moderate sprains (second-degree injuries) can be taped to allow the patient to limp along, self-powered, probably with most of the weight from her or his pack distributed among other group members. Severe ankle damage (third-degree injuries) require

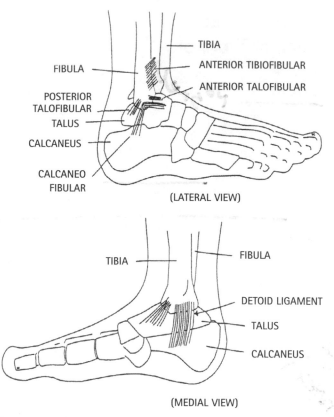

Figure 14–1: *Anatomy of the ankle.*

splinting and, most likely, carrying of the patient out to definitive medical care.

Since some fractured ankles may show less apparent damage than some badly sprained ankles, persistent pain should send the patient for an X-ray.

■ ANKLE TAPING

Taping reduces the chance of further injury or reinjury to an ankle, but the tape needs to stick well. The support from tape is reduced as the tape loosens. Apply the tape firmly, but not so tight that healthy circulation is reduced. Apply the tape with as few wrinkles as possible. Take your time.

Tape works best if applied directly to the skin. Tape feels better, especially on removal, if the hair of the ankle and lower leg is trimmed off first, but haircutting is not required. You do not need to tape from toe to calf—from mid-foot to 2 to 4 inches above the ankle bones is enough.

1. Start by marking the boundaries of the ankle taping area with tape anchors at mid-foot and above the ankle bones. The foot must be kept in position of function, at a 90-degree angle to the leg.

2. Apply stirrups that pull the foot up into the ankle. Pull firmly and equally up on both sides of the stirrup before pressing the tape into place. Remember to anchor each stirrup with a strip of tape above the ankle bones.

3. Apply a "J," a semi-stirrup that starts on the uninjured side of

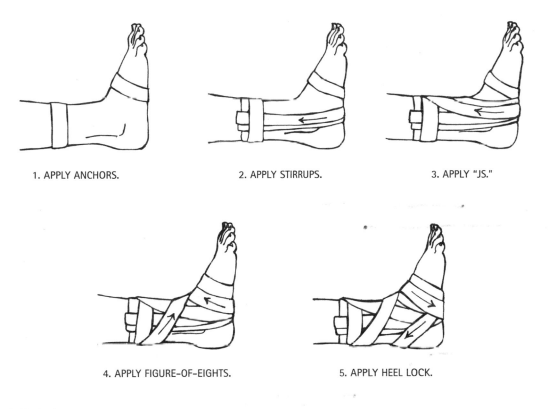

1. APPLY ANCHORS. **2. APPLY STIRRUPS.** **3. APPLY "JS."**

4. APPLY FIGURE-OF-EIGHTS. **5. APPLY HEEL LOCK.**

Figure 14-2: *Ankle taping.*

the ankle but crosses over the top of the foot to remove stress from the injured side.

4. Apply figure-of-eights to hold the ankle firmly together. Start the "eights" on the bottom of the foot and toward the injured side to reduce strain on the injured ligaments. "Eights" come over the top of the foot, around the ankle bones, and back across the top of the foot to finish where they started, forming a figure-of-eight around the ankle.

5. Apply a heel lock, a strip that starts on the bottom of the foot and goes behind the ankle, then up over the top of the foot to finish near the starting point. Apply a heel lock in both directions, lateral and medial, for the best support. Close any remaining bare spots with strips of tape.

Tape provides a flexible splint and, as a splinted extremity, the foot should be assessed before and after taping for adequate CSM (circulation, sensation, and motion). The patient may be given the responsibility of notifying you if sensations alter in the foot, such as if tingling or numbness is felt.

Athletic tape, made for taping athletic injuries, usually works best, but any tape available, including duct tape, can be used in a wilderness emergency. It is best to remove the tape at night to improve circulation and make RICE more effective.

Achilles Tendinitis

The muscles of the foot and ankle are few, movement of this area being powered by the lower leg via tendons. Too much movement, especially for an unfit hiker, may produce tendinitis. Achilles tendinitis is by far the most common such injury. The large Achilles tendon runs from the back of the heelbone on its lower end up to the muscles of the lower leg and is especially at risk for tendinitis during or after a long hike, and even more likely if the hike involved significant elevation gain. The pain of Achilles tendinitis can be startling.

Specific treatment for Achilles tendinitis includes creating a pad (about a quarter of an inch thick should do) under the heel inside the boot to relieve stress on the Achilles tendon. Two strips of padding taped in place, one on each side of the Achilles, will further reduce stress. And, once again, apply HIRICE.

Shin Splints

Beneath the skin and fat of the lower leg, the muscles lie in seven compartments walled by connective tissue called *fascia*. Beneath those muscles lies the *tibia*, which is wrapped in a tough membrane. The *shin* is the front of the lower leg, everything between the knee and ankle. Any persistent pain in that area is referred to as a *shin splint*. Shin splints are a common athletic injury, and it is often impossible to tell what part of the leg is actually damaged.

Compartment syndrome: During exercise, blood flow to muscles increases, and they grow larger. The fascia around the muscles stretches to accommodate the enlargement. With repeated exercise the muscles grow larger and pressure on the fascia increases. In most people this is not a problem, but some people have particularly small muscle compartments. The pressure on the walls of fascia causes them to stiffen, similar to the way callus forms on hands and feet when they are constantly used for rough work. The muscles are compressed against the stiffening walls of fascia, causing the pain of a shin splint. Compared with other causes of shin splints, compartment syndrome is relatively easy to diagnose. The pain comes on slowly, gradually filling the compartment. When asked where it hurts, the sufferer will sketch an outline of the muscle on the leg.

Fractures and tears: Overuse of lower legs can cause small cracks in the bone, called *stress fractures.* Repeated stress on lower legs can damage the tough membrane surrounding the tibia. Twisting or pulling motions can tear muscles, tendons, or ligaments. Just overworking a muscle that is not ready for it can inflame muscle cells and produce pain. Shin splints brought on by stress-fractured bones or torn tissues tend to cause pain most in one specific spot. These problems also tend to develop with more abruptness.

If your assessment is a shin splint, encourage the patient to take it easy on that part of the body for awhile—and use HIRICE. If the shin pain keeps getting worse, the patient should consult a doctor.

The Knee

Only the ankle has more problems coping with wilderness journeys than the knee. The demands put on the knee are great, and it is highly susceptible to trauma or overuse.

Knees are directly comprised of three bones: the *femur,* the *tibia,* and the *patella.* Another bone, the fibula, attaches behind the tibia, near the knee, but has no specific influence on the joint.

Femurs and tibias *articulate,* or rub against each other, when legs are in motion. The articulating surfaces of the femur and tibia are semi-flat and, to ensure a secure fit, each knee is padded with two C-shaped pieces of *cartilage,* one on the outer half of joint space, the other on the inner half. Cartilage also absorbs some of the shock of movement. The two pieces of cartilage are called the *medial meniscus* and the *lateral meniscus.* Placed strategically in the knee are fluid-filled sacs, called *bursae,* at points of the greatest friction.

The knee is held together by four ligaments that are attached at the points of highest stress. Connecting the femur to the tibia are the medial and lateral *collateral ligaments,* on the inside and outside of the knee. They provide stability for side-to-side motion. For back-to-front and front-to-back stability, there are the *cruciate (crossed) ligaments.* They run through the joint space, between the two menisci. Both cruciate ligaments attach on their upper end to the femur and on their lower end to the tibia, and they are named for where they attach to the tibia. The *anterior cruciate ligament* (ACL) attaches to the

femur at the back of the knee and to the tibia in front, thus preventing the knee from sliding too far forward. The *posterior cruciate ligament* (PCL) attaches to the femur at the front of the knee and the tibia at the rear, thus preventing the knee from sliding too far backwards.

When in motion, the great muscles of the leg provide additional support to the knee. The *quadriceps* (thigh) muscles are a group of four muscles. They taper down into one tendon that

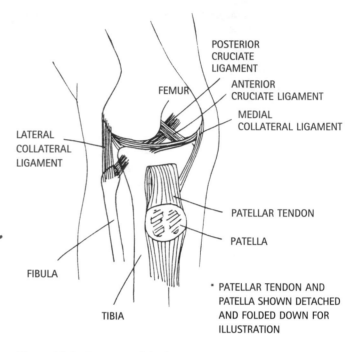

Figure 14-3: *Anatomy of the knee.*

crosses the knee and attaches to the top of the tibia. The patella is in the middle of this tendon. Three muscles in the back of the thigh, the *hamstrings,* also help support the knee; one attaches to the outside of the knee and the other two to the inside. The *gastrocnemius* (calf muscle) attaches in two places to the back of the femur and, finally, a long, thin muscle—the *gracilis,* or groin muscle—runs from the groin to the inside of the knee adding a touch more of support.

In addition, a long tough tendon, the *iliotibial band,* runs from the *gluteals,* the muscles of the hindquarters, down the thigh and across the knee, attaching to the outside of the tibia. This band, too, gives the knee a bit of support.

Any force applied to the knee can partially or totally tear a ligament. If the force is applied to the outside of the knee, the medial collateral ligament and anterior cruciate ligament may be involved, as well as the medial cartilage. If the force is applied to the inside of the knee, the lateral collateral could be torn and lateral cartilage may be ruptured. Twisting forces may significantly damage the cruciate ligaments.

During your assessment of a knee injury, have the patient move the knee through its range of motion. These tests should be done with the patient sitting down and the leg relaxed. "Relaxed" may be a relative term if the knee is causing a lot of pain. If the patient is unable to tolerate these checks, he or she needs a doctor.

The medial collateral ligament, the one on the inside of the leg, can be checked by holding the ankle, with the knee slightly bent, and pushing from the outside of the knee in. If it's loose or painful, stop pushing.

The lateral collateral ligament, on the outside of the leg, can be checked in the exact opposite way, pushing from the inside of the knee out. Again, looseness or pain is a sign to stop pushing.

The anterior cruciate ligament can be checked by bending the knee slightly and pulling out on the tibia from behind the upper calf. Watch for pain and looseness.

Posterior cruciate damage, which happens in only about 1 percent of all knee injuries, can be checked simply by lifting the relaxed leg by the ankle and letting the knee sag.

A simple active test for knee function can be performed with a high degree of accuracy, but it should not be performed until the above checks are made. If the patient can stand and walk, do deep knee bends, and jump up and down on each leg individually, the knees are almost always stable enough to remain in the wilderness.

If the knee has been traumatized to the point where it can't be used, the leg should be splinted, with the knee slightly flexed, and the patient should be carried to a doctor. Flexion of the knee can be easily maintained by padding well behind the knee, leaving the knee flexed at about 10 degrees. If the patient feels more comfort with the knee flexed more, add more padding beneath the knee (see Chapter 12: Fractures).

If the knee can be used carefully, you can build a splint that stabilizes the knee while allowing the patient to walk. The splint should leave the kneecap free. You can make this type of splint by rolling up a sleeping pad from both ends simultaneously, and securing it to the knee with the "jelly rolls," the rolled sides of the pad, against the lateral and medial aspect of the knee. A Crazy Creek chair works extremely well as a walking knee splint when placed beneath the knee and folded up around the knee. The stays within the chair give firm support to both sides of the knee while leaving the kneecap free.

Remember: Even a walking knee splint should have sufficient padding behind the knee. The patient will probably require some type of hiking staff or improvised cane.

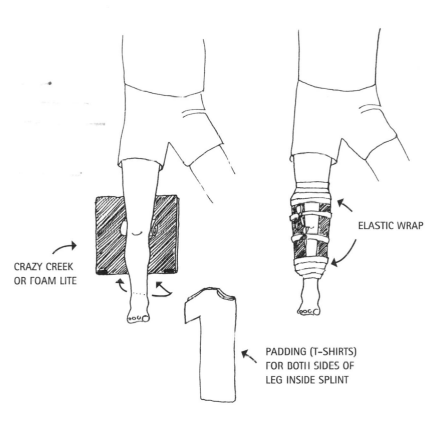

CRAZY CREEK OR FOAM LITE

ELASTIC WRAP

PADDING (T-SHIRTS) FOR BOTH SIDES OF LEG INSIDE SPLINT

Figure 14-4: *Walking knee splint.*

The most common source of knee pain is not a sprain but overuse of the muscles that support the knee. When they are stressed too much, the muscles tear—a strain—and create a great deal of discomfort. They most often strain near their attachment to the knee. Tendinitis has the same mechanism of injury. Muscle strains and tendinitis are commonly mistaken by the patient, and sometimes the rescuer, as a torn ligament or cartilage. This mistake is very common when the iliotibial band is involved. Since the band is required for uphill motion, it is often abused when someone is unused to going uphill or increases uphill activity, especially if she or he is wearing a pack. Traversing a hillside can also stress the iliotibial band, especially on the descent. The problem, called *iliotibial band syndrome*, causes pain primarily where the band attaches to the outside of the knee, simulating a torn lateral collateral ligament.

General knee pain may have other causes, including *patellar compression syndrome*, a problem created by too much pressure on the back of the kneecap, typically caused by too much walking, especially downhill. A dull ache, constant and nagging, is the common complaint. Or, perhaps, the kneecap doesn't run quite correctly in its track. The additional side-to-side motion of the kneecap puts additional stress on its inner surface that eventually causes pain for up to several hours after use. If the pain becomes chronic, the condition may be *chondromalacia* of the patella. Chondromalacia refers to a disintegration of the cartilage under the kneecap, probably caused by a chemical change stimulated by knee injury or overuse. The cartilage becomes frayed and eroded. Interestingly, the cartilage can't hurt since it has no nerve endings, so the pain must come from inflamed tissue around the cartilage.

General knee pain does not need to be specifically assessed by the WFR; leave the diagnosis to a doctor. In the meantime, general knee pain can be treated with HIRICE and a walking splint, and the patient and WFR need to decide if the wilderness trip must be cut short. A lot of general knee pain doesn't go away in the field, and the patient may eventually require a physician's care. In the meantime the patient may be encouraged to alter her or his gait and/or use a hiking staff to relieve some of the pressure on the knee.

Low Back Strain

It takes surprisingly little stress on the middle to lower back, if the angle is right, to pull muscles. The resulting agony can be surprisingly debilitating. It can take several days to weeks before the pain goes entirely away, but almost everyone can become functional in a couple of days by following a recognized routine.

Rest is initially very important for lower back strain. Have the patient rest on his or her side, or on his or her back with thick, relatively stiff padding underneath the knees. Too much rest, however, can soon lead to weakened back muscles. As soon as the patient can tolerate exercise, he or she should be exercising to tolerance. General movement, such as easy walking, is beneficial in the initial phases of the injury. The lower back can be gently exercised by having the patient lie flat on his or her back with knees raised high and feet flat on the ground followed by lifting the head and shoulders and pressing the lower back into the ground. This "abdominal crunch" stretches the lower back and, over time, strengthens the abdominal muscles, a preventive measure against lower back injury.

Ice the painful region of the back several times a day for several days. Gently massage the injured muscles several times a day. And, of course, hydration and ibuprofen are recommended.

Patients with a history of lower back pain may wish to consult a physician prior to a wilderness trip with the intention of acquiring a prescription-strength muscle relaxant.

If the pain has not appreciably decreased in 7 to 10 days, or if pain, tingling, numbness, or paralysis begins to creep down the legs, it's time to forget the wilderness for a while and find a physician.

The Shoulder

Nothing about the construction of the association of three bones at the top of the arm lends itself to long-term anatomical optimism. Shoulders are simply not very stable.

Shoulders may manifest a problem called *impingement*. The pain of impingement typically falls into one of two categories: a result of overuse by repetitive motion, or a result of an impact injury to the shoulder that disrupts normal mechanics.

The *humerus, scapula,* and *clavicle* meet at the shoulder, the only joint in the body not held together primarily by ligaments. A few shoulder ligaments prevent the bones from shifting too far in one direction or another but offer little help in keeping the joint snugly in its rightful place. Shoulders are small and shallow ball-and-socket joints. The head of the humerus has little contact with the socket, and the bones stay put because of a group of muscles known collectively as *rotator cuff* muscles: *supraspinatus, infraspinatus, subscapularis,* and *teres minor*. Rotator cuff muscles are not designed to function under the stress applied when the arm rises above a line parallel to the ground. Too much over-the-shoulder exercise, such as sea kayaking, and the muscles can stretch, with the head of the joint loosening within the socket. With the head of the shoulder loose, every time the patient reaches backward over the shoulder or lifts his or her arm above parallel to the ground, the head slips temporarily out of place.

A simple test tells if the patient is impinged. Ask the patient to lift the arm on the affected side slowly straight out to the side, at a 90-degree angle to the ground. As the patient tries to pass the 90 degree mark, pain hollers "whoa" to the motion.

Rest, ice, and anti-inflammatory drugs help the pain of impingement go away, but not the problem. As soon as the patient starts exercising again, back comes the discomfort. The cure involves strengthening the joint with exercises specific to the shoulder.

A shoulder *separation* may result from excessive forces applied to the shoulder. "Separation" is an old term describing an enlargement of the spaces between the bones. Since ligamental disruption is required for separation, the injury is a shoulder sprain and should be treated as such (see "General Treatment of Athletic Injuries," this chapter).

The Elbow

Pain in the elbow typically comes with overuse of the muscles and tendons used to bend the wrist backward and forward—a form of tendinitis referred to as *tennis elbow*. By palpation you can usually pinpoint the pain at the proximal end of either the radius or ulna. Tennis elbow feels better in some people when they wear a light constricting band about 1 inch below the elbow to help hold the tendon in place during periods of activity. If the pain remains acute, even with a band distal to the elbow, a splint may be improvised reaching from the palm of the hand to below the elbow to prevent movement.

Prevention of Athletic Injuries

Several factors can help you predict—and perhaps forestall—an athletic injury:

1. *Weak or imbalanced muscles.* The best way to prevent injury is to get in shape and stay in shape. Individuals with specific weak joints, say a shoulder, should learn and perform exercises specific to the affected joint.
2. *Excess body weight.* Shedding extra pounds reduces stress on joints, especially lower extremity joints. Individuals can carry a lighter pack if they are not in shape for backpacking.
3. *Inappropriate or damaged footwear.* Boots that provide adequate ankle support also provide better support for the knee and lower back. Worn-out footwear should be replaced before it wears out an ankle. An inward fold above the heel counter/heel cup can irritate the Achilles tendon enough to cause tendinitis in a single day of backpacking. Poor-quality boots can produce pain where other tendons meet the fibula. Boots that are too stiff or too tightly laced may cause tendinitis at the front of the ankle.
4. *Loss of balance.* Ski poles or hiking staffs can help individuals maintain balance.
5. *Failure to warm up muscles before periods of hard exercise.* Tight muscles are more likely to be strained.
6. *Previous injury.* Individuals prone to recurring injuries should follow suggestions 1 to 4 even more aggressively.
7. *Carelessness.*

Evacuation Guidelines

The ability of the patient to use the injury is the key factor in determining the need for an evacuation. Your first assessment, if done properly, will be accurate, despite predictable increases in pain and swelling as time goes by. All patients with unstable or unusable athletic injuries that do not permit continuance of the wilderness trip should be evacuated. Haste is not required or recommended.

Conclusion

On the banks of the Escalante River, you keep Tom at rest while you assess his injury. A pulse is easily palpable in his right ankle, and motion and sensation in the toes of his right foot cause you no concern. The bruising and swelling that first alarmed you don't appear quite so bad now that you've calmed down a bit.

Tom describes the accident to you. When his foot rolled in, he felt sudden pain on the outside of his ankle, but he heard no snaps, crackles, or pops. The pain, he says, has started to ease off a little.

As you probe around the injury site, you uncover mild to moderate point tenderness just below the ankle, approximately where you think one of those ankle ligament sits.

On request Tom rolls his ankle through its normal range of motion without undue pain. It hurts, sure, but he can "take it." You are relatively aggressive about manipulating his ankle, adding a bit of push to the extremes of the range of motion. Tom winces slightly when you roll his foot in, but you are satisfied the injury is not a major one. When asked to try walking, Tom does so with a slight limp.

In the shade of a cottonwood tree, Tom rests with his ankle propped up on a sleeping bag still in its stuff sack. You compress the ankle with an elastic wrap, applied from the toes toward the knee, and place a plastic bag filled with cold water from the river on the ankle. After 20 minutes, the cold bag comes off.

You will repeat the RICE treatment again before bed, and again in the morning. You tape Tom's ankle after the morning's treatment. Tom continues the hike, some of the weight of his pack distributed among the group.

Joe Costello, MS, CSCS, CMT, contributed his expertise to this chapter.

CHAPTER FIFTEEN

Wilderness Wound Management

You Should Be Able To:

1. Describe the various types of wounds, including contusions, abrasions, lacerations, avulsions, amputations, punctures, and impaled objects.
2. Demonstrate the appropriate emergency treatment and long-term care of wounds in the wilderness.
3. Describe the appropriate emergency treatment and long-term care of burns in the wilderness.
4. Describe wound infections and their management in the wilderness.

It Could Happen to You

It's a stormy night near the granite buttress of Fossil Ridge, Colorado, among the scattered and fractured rock surrounding Mill Lake. Your tent mate's vocal complaints about his need to relieve his bladder awaken you.

"I've got to go," he says, scrambling out of his bag.

"Put on your shoes," you suggest over a yawn.

"No," he says. "I'll only be a minute."

A yell of pain, a curse, and your tent mate flops back into your temporary dome-shaped home, tightly gripping his left foot with both his hands. Blood seeps through his fingers and drips onto your bag.

"Stubbed my big toe!" he grinds out from between clenched teeth.

Taking a look at the dirty digit, you see a nickel-sized avulsion. Already the blood flow has begun to ease.

Wounds are among the most common medical problems encountered in the wilderness. According to a study by the National Outdoor Leadership School (NOLS), wounds and their complications account for approximately one-third of all wilderness medical problems severe enough to prevent normal participation in course activities. Burns, including injuries from hot water and open fires, make up another large percentage of wilderness mishaps. Under the rigors of wilderness travel, even apparently trivial wounds may develop an infection. Preventing complications and attaining optimal functional and cosmetic results require careful evaluation and management of all wilderness wounds. So it is essential that Wilderness First Responders acquire a thorough understanding of the principles of wound and burn assessment and develop the skills necessary to effectively manage wounds and burns in a wilderness environment.

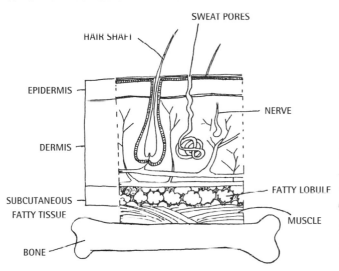

Figure 15-1: *Anatomy of the skin.*

Basic Anatomy of the Skin

The skin, or integumentary system, is composed of two major layers, the *epidermis* and the *dermis*. The thin, overlying epidermis protects the dermis from infection and desiccation (drying out). Wounds involving only the epidermis result in little or no bleeding and have a low susceptibility to infection. The dermis, the largest organ of the human body, provides most of the tensile strength of skin and contains sweat glands, sebaceous (oil) glands, hair follicles, nerves, and capillaries. Below the dermis, on most of the body, lies *subcutaneous fat,* sometimes a slim layer, sometimes not. Below the fat lies *fascia,* the membrane that covers muscle.

Types of Wounds

1. **Contusion:** A bruise.
2. **Abrasion:** A wound in which one or more layers of skin are partially or completely scraped away.
3. **Laceration:** A cut through the skin. A laceration produced by a sharp object, such as a knife, generally produces little damage to the surrounding skin. Lacerations from a blunt injury, however, typically result in a tearing or bursting of the skin, causing ragged wound edges or star-shaped patterns. Because damage to adjacent skin occurs, these wounds heal more slowly, result in larger scars, and are more prone to infection.
4. **Avulsion:** A partial amputation that leaves a "flap" of body tissue attached by skin, muscle, or tendon.
5. **Amputation:** A complete separation of a body part, such as an ear, finger, or foot, from the rest of the body.
6. **Puncture:** A wound that occurs when an object, such as a thorn, fang, or knife, penetrates the body. These wounds may introduce bacteria into deep tissues and are very difficult to clean adequately. As a result, they are particularly prone to infection.
7. **Impaled object:** A puncture wound with the puncturing object still stuck in.
8. **Bite wound:** A puncture wound caused by a bite from an animal or another human.
9. **Burn:** Tissue injury resulting from heat, electricity (lightning), radiation (sunburn), or chemicals.

General Wound Management

General wound care needs to reach three goals: (1) control of significant blood loss, (2) prevention of infection, and (3) promotion of healing.

Minor wounds may be allowed, even encouraged, to bleed to a stop, an act that may result in a cleaner wound. Significant wounds require *hemostasis* (control of bleeding). Attaining adequate hemostasis not only facilitates wound assessment and management but also may be necessary in severe wounds to prevent significant blood volume depletion. **Hemostasis can almost always be accomplished by applying direct pressure and elevating the site of bleeding** (see Chapter 6: Bleeding).

If the patient has known allergies—such as allergic reactions to latex (perhaps your gloves) or to adhesives that may be on your wound dressing materials—he or she should tell you this before you continue wound treatment.

It is important to learn, if you do not already know, when the injury occurred, because bacteria begin to proliferate even in relatively clean wounds after approximately 6 hours. Knowing how and where the injury occurred can give you additional clues about possible contaminants (such as the presence of feces, saliva, or soil) that increase the risk of infection. All wounds should be visually assessed for the presence of foreign bodies. Involvement of important underlying structures such as nerves, major blood vessels, tendons, and joints should be determined, if possible. Motor and sensory function and pulses distal to deep injuries should be checked before and after wound management.

Determining the exact mechanism of injury may aid you in assessing the extent of the wound and may indicate the need to assess for additional, less obvious injuries.

The *tetanus immunization* status of the patient should be determined. The bacteria that cause tetanus are ubiquitous. Every patient treated for an open wound should receive immunization against tetanus as soon as possible if (1) the patient has never been immunized, (2) the patient has not received a tetanus booster within the last 10 years, or (3) the patient has a large and/or highly contaminated wound and has not received a booster within the last 5 years. Tetanus immunization must be received within approximately 72 hours of a wound to be effective. If you serve as an outdoor leader, you would do well to be sure all group members are immunized against tetanus prior to the start of a wilderness trip.

General Wound Cleaning

Note: A rescuer should wash his or her hands and put on protective gloves and protective eyewear before cleaning an open wound.

All wounds acquired in a wilderness environment should be regarded as contaminated and, therefore, require cleansing to prevent infection and promote healing. There are three effective methods of wound cleaning available to the WFR: You can *scrub, irrigate,* and *debride.*

Scrubbing: Disinfectants (such as isopropyl alcohol, povidone-iodine, and hydrogen peroxide) and soaps and detergents should *not* be put directly into wounds because they can damage viable tissue and may actually increase the incidence of wound infection. These substances may be used to scrub *around* a wound prior to wound cleaning, with soap and water working as well as anything else.

Irrigating: The most effective and practical method of removing bacteria and debris from a wound involves using a high-pressure irrigation syringe. Irrigation syringes that supply adequate pressure are available commercially in quality first-aid kits. Without an irrigation syringe, you can put water in a plastic bag, punch a pinhole in the bag, and squeeze the water out forcefully, or you can melt a pinhole in the center of the lid of a water bottle with a hot needle and squeeze the water out forcefully. These and other improvised methods are not nearly as effective as an irrigation syringe, but they may be the best you can do. Simply rinsing or soaking a wound is inadequate to remove bacteria. The cleanest water available, most preferably water disinfected for drinking, should be used for irrigating. The tip of the irrigating device should be held 1 to 2 inches above the wound surface, and the plunger of the syringe forcefully depressed. Be sure to tilt the wound to irrigate contaminants out and away from the wound. The volume of irrigation fluid required varies with the size of the wound and the degree of contamination, but plan on using at least a half liter of water.

Note: Wound irrigation is the single most important factor in preventing infection.

Debriding: Deeply imbedded, visible debris not removed by irrigation may be removed carefully with forceps (tweezers) sterilized by either boiling or with an open flame, such as a match or lighter. Carbon (the black stuff) left on forceps after holding them in an open flame is sterile. Removal of visible debris from a wound and/or dead skin from around a wound is called *debridement*.

When the protective layers of the epidermis are opened, the superficial dermal cells dry out and die. These dead cells, together with *serum*, the watery portion of blood that seeps from the wound, form the familiar *eschar* (scab). Although wounds heal beneath scabs, the application of occlusive wound dressings, after thorough cleaning, prevents the formation of an eschar by keeping the dermis moist with fluids from the patient's body, speeding the growth of new skin and wound healing.

After closing and/or dressing a deep wound on an extremity, immobilization by splinting reduces lymphatic flow and the spread of microorganisms. Elevation of the extremity decreases

swelling. Both measures reduce the likelihood of wound complications and should be employed whenever possible.

Prophylactic antibiotics are not indicated for most wounds. Many authorities would recommend antibiotics for wounds involving tendons, particularly of the hand, bones, or joint spaces, as well as for wounds heavily contaminated with saliva, feces, or soil containing large amounts of organic material. If antibiotics are used, they should be started as soon as possible after the injury, and a broad-spectrum agent should be chosen. Antibiotics require a prescription, and a physician should be consulted well before you start a wilderness trip. Follow the physician's instructions precisely when using antibiotics.

Note: Antibiotics should never be considered a substitute for a vigorous wound cleaning.

Wound Healing

In general, wounds in highly vascular areas such as the face or scalp heal faster and with a lower risk of infection than wounds in less vascular regions, such as the distal extremities.

Management of Specific Wounds

Once irrigation has been accomplished, eyewear is typically not necessary, but the rescuer should still be wearing protective gloves. Remember to wash your hands before and after donning the gloves. Soap, water, and gloves are a tough trio on germs. In the absence of protective gloves, the rescuer may improvise with clean plastic bags over her or his hands. With relatively minor wounds, to prevent sharing germs, the patient may be directed in the management of his or her own wound (including control of blood loss).

Contusions

Bruises seldom require emergency care, but large bruises benefit from cold, compression, and/or elevation. Substantial bruises should cause you to assess the patient for damage to underlying structures, such as bones and organs. Large bruises should be protected from freezing in extremes of cold because a bruised area will freeze sooner than normal skin.

Abrasions

Abrasions are the exception to the rule of wound cleaning: You need to scrub within the wound to

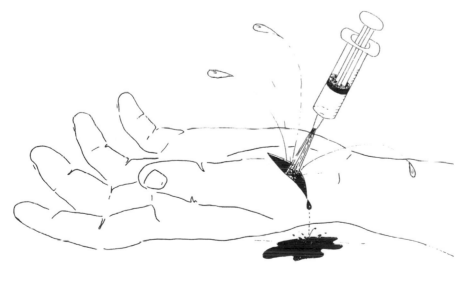

Figure 15-2: *Irrigating a wound.*

achieve adequate cleaning. A sterile gauze pad is adequate for scrubbing. Scrubbing may be enhanced by using any soap, but all soap should be carefully rinsed and then irrigated from the wound after scrubbing. Green Soap Sponges are packaged with soap and water already in the sponge, making them useful additions to first-aid kits. It is important to remove all embedded debris not only to reduce the risk of infection but also to prevent subsequent "tattooing" (scarring) of the skin. With a deep abrasion, self-scrubbing is seldom successful due to the high level of pain associated with the exposed nerves.

After cleansing, abrasions can be kept moist to avoid desiccation and speed healing with microthin film dressings that can be left in place until healing occurs. Without microthin film dressings, a topical agent, such as an antibiotic ointment, can be applied, followed by a dressing of a sterile gauze pad or a roll of sterile gauze to keep the ointment in place. Tape, an elastic wrap, or some other holder may be used to hold a sterile gauze pad in place. Ideally, gauze dressings should be changed twice a day, or at least once a day, as well as any time the gauze gets wet.

Dressings and Bandages

Historically speaking, the dressing goes directly on the wound and the bandage holds the dressing in place. But with the advent of a plethora of modern dressing and bandaging material, the lines have become somewhat blurred. Microthin film dressings, for instance, require no bandage. And the Band-aid, the progenitor of modern dressings, is actually a dressing and bandage combined. What you can find today and what you may choose to carry in your first-aid kit include, but are not limited to, dressings that absorb exudate from the wound, antibacterial dressings, dressings that help stop bleeding, and dressings that pad, occlude, stretch, transpire, and/or cover the wound with colorful cartoon characters.

Lacerations

When cuts are in hairy areas, closely clipping the hair adjacent to the wound with a pair of scissors facilitates wound management. Shaving hair should be avoided because it might increase the risk of infection. Eyebrow hair, however, should never be removed. An intact eyebrow is necessary for proper alignment of wound edges, and eyebrow hair grows back slowly, creating cosmetic problems if removed.

Because lacerations may at least partially close after the cut, be sure to hold the wound open during irrigation to make sure you clean to the bottom of the wound.

Minor lacerations that do not gape open after cleaning or that gape open no more than a half inch, can be closed with "butterfly" adhesive strips or, even better, with skin closure strips. A thin coat of tincture of benzoin compound applied to the skin around the wound prior to applying the strips aids in adhesion of the strips. Benzoin is an irritant, so take care to keep it out of the wound. Let the benzoin dry for at least 30 seconds before applying the strips. In the absence of benzoin, ethyl-2-cyanoacrylate glue ("superglue") may be used. Use of superglue to glue a wound closed is not recommended, although putting glue in small cracks in, say, fingertips, is OK. The closure strips should be applied perpendicular to the wound. Apply one to one side of the wound and another to the opposite side. By using the opposing strips as handles, you can pull the wound edges together, pulling the skin as close as possible to where it should lie naturally, but without pulling the wound tightly shut. In the absence of butterfly strips or closure strips, wound closure may be accomplished with whatever tape happens to be available. Adhesive wound closing strips may be left in place until healing occurs. Small wounds cleaned and closed properly with adhesive strips have a very low incidence of infection.

Because optimal conditions for wound care rarely exist in remote settings, suturing or stapling lacerations in the wilderness is controversial. Individuals with advanced medical skills may elect to suture or staple lacerations in low-risk, cosmetically important locations like the face after thorough cleansing. However, subcutaneous or deep sutures should be avoided. Heavily contaminated wounds should be treated without suturing.

After closing, lacerations may be dressed with antibiotic ointment and sterile gauze. Alternatively, occlusive microthin film dressings may be used. These dressings prevent eschar formation and provide a humid environment for new cell growth. Because microthin film dressings are transparent, they allow monitoring of the wound for signs of infection. These dressings also create a barrier to bacteria and debris that may be particularly important in wilderness settings. If a microthin film dressing is used, do not apply an ointment under the dressing, a procedure that is unnecessary and usually causes the microthin dressing to slip off.

Heavily contaminated lacerations that are left open acquire increased resistance to infection over a 3 to 4 day period. After thorough cleansing, these wounds should be loosely packed with moistened, sterile gauze and covered with a dry, absorbent dressing. The best solution to moisten the gauze is probably sterile saline; you can use contact lens solution if you have any available. Sterile water is the next best option. Water safe to drink is third best. The inner moist gauze eventually dries and, when removed, takes some wound debris with it, which helps keep the wound clean and open. Unless a wound infection is suspected, the dressing can be left in place until definitive care

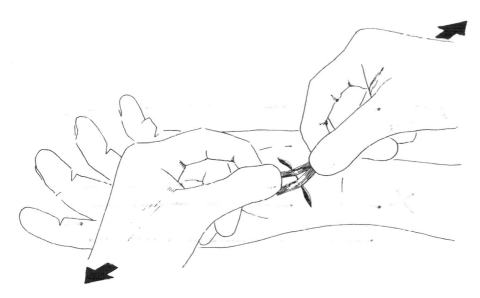

Figure 15-3: *Closing a wound.*

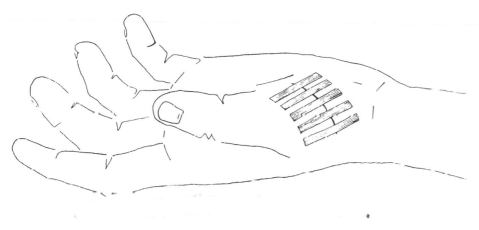

Figure 15-4: *Closed wound.*

is available, but, if materials allow, changing the dressings approximately every 12 hours provides better care. These wounds can be sutured (delayed primary closure) in 4 to 5 days without impairment in wound healing. This often allows time for an evacuation to a hospital.

Avulsions

In the case of an avulsion, treat the wound as you would a laceration. Do not cut off the "flap." Adequate cleaning often requires that you hold the avulsion open during irrigation.

Amputations

In the case of an amputation, hemostasis is the first priority. The wound should then be irrigated and bandaged as previously described. **Do not attempt to reattach the body part.** If a detached body part can be easily recovered and the patient can be brought to a hospital within 6 hours, even longer for fingers

and toes, reattachment may be possible. The detached body part should be quickly irrigated to remove contaminants. The part should then quickly be wrapped in slightly moistened sterile gauze and placed in a plastic bag, if available. Keep the part cold with an ice or cold pack, if available, or by placing the plastic bag containing the body part in a water bottle filled with the coldest water possible. Do not apply ice directly to the body part or frostbite may ensue, and do not place the part directly in water.

Punctures

Puncture wounds carry a particularly high risk of infection despite your best cleaning efforts, specifically because you cannot get to the bottom of the wound to clean it. Irrigation must be limited to the surface of the wound because forcing water under pressure into the puncture may drive contamination deeper into tissue inside the wound. Evaluation of the wound, in terms of the need for an evacuation, should take into account structures, such as organs, that might lie beneath the puncture and that could have been damaged by the injury.

Impaled Objects

Large objects found impaled in a wound should be left in place if you can get to a medical facility with relative ease. Yanking on an object can stimulate serious bleeding and damage underlying structures, especially if the object is impaled in a body cavity, such as the chest, abdomen, or head (see Chapter 10: Chest Injuries and Chapter 11: Abdominal Injuries). The object should be stabilized with padding to prevent movement. The padding should be as high as the object to protect it from being bumped during transport, and the patient should be carried out.

In many cases there is nothing easy about getting to help, and impaled objects sometimes make evacuation very difficult. Removal of impaled objects in the wilderness is an oft-debated subject. They can be impossible to stabilize over rough terrain. Metal objects, such as an ice ax, can suck a significant amount of heat out of a cold patient.

If the object is through a cheek, or in any other way endangering the patient's airway, there is ample agreement on removing it carefully. If the object is loosely embedded, removal is simple and beneficial to the patient and you, since evacuation is simplified. If the object is in an extremity, removal is often safe. Note the object's shape and angle of entry, and remove accordingly.

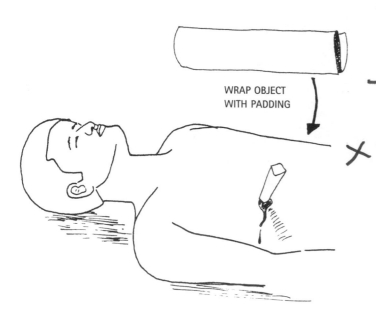

WRAP OBJECT
WITH PADDING

USE TAPE TO
SECURE PADDING

Figure 15-5: *Stabilization of impaled object.*

Note: Objects impaled in the eye should never be removed. A patient with an impaled object in the eye should have the object well protected against movement, both eyes should be covered, and the patient should be carried out, preferably propped up at approximately 45 degrees to prevent an increase in pressure in the damaged eye (see Chapter 31: Common Simple Medical Problems).

Bite Wounds

Saliva from human and other animal mouths contains large amounts of bacteria. All bite wounds should be considered heavily contaminated. The wounds should be copiously irrigated. Following irrigation—another rule exception—1.0 percent povidone-iodine solution, if available, should be used to scrub directly in to the open wound. Although many bite wounds, particularly dog bites, can be sutured after thorough irrigation and debridement in an urban emergency setting, in the wilderness it is safer to leave all bite wounds open. The wound should be loosely packed with moistened gauze and covered with an absorbent dressing, as for lacerations. If possible, the bite site should be splinted and elevated. Delayed primary closure can be attempted after visiting a hospital.

Animal bites on the hand, foot, wrist, or over a joint are at a particularly high risk for serious infection and should be managed with extreme care. Human bites to the hand, particularly those that occur when a clenched fist strikes a human mouth, are also at high risk for infection. These wounds should be cleaned with the hand as close as possible to the position it was in when it received the wound because the shifting of underlying muscles in hands may trap contamination. Many physicians recommend prophylactic antibiotics for these wounds. If available, the antibiotics should be started as soon as possible after the bite. Consult a physician concerning antibiotics prior to a wilderness trip, and follow his or her instructions precisely.

Rabies virus infection is more common in wild animals than in domestic animals. Washing a bite wound immediately with soap and water or povidone-iodine solution prior to irrigation deactivates the rabies virus on the surface of the wound. All victims of animal bites should be referred as soon as possible to a physician to determine the need for antirabies serum and rabies vaccination (see Chapter 21: North American Bites and Stings).

Evacuation Guidelines for Wounds

While not a matter of urgency, patients with wounds that hinder the ability to participate in and/or enjoy the wilderness experience should be evacuated. Patients with wounds that require careful closure for cosmetic reasons, such as facial wounds, should be evacuated, but delayed primary closure for face wounds can be accomplished in 3 to 5 days if the wound has been kept clean and packed open. Patients suffering amputations and impalements (except small splinters) should be evacuated. Rapid evacuation is advised, due to the high risk of infection, for (1) deep and/or highly contaminated wounds, including large imbedded objects and deep puncture wounds; (2) wounds that open a joint space; (3) wounds that reveal underlying structures, such as ligaments and tendons; (4) wounds caused by a crushing mechanism; and (5) severe wounds from bites and/or from animals that might be rabid.

Wilderness Burn Management

Burns are among the most painful and emotionally distressing of injuries. Even relatively minor burns may disrupt the wilderness experience for an individual or an expedition. Although burns may result from electricity, radiation, and chemicals, serious wilderness burns most often result from high heat sources, such as scalding hot water, open flames, (campfires or camp stoves), or hot objects (such as pots). Burns may also occur from lightning strikes (see Chapter 20: Lightning). The most common wilderness burn is a radiation burn from solar rays (sunburn), seldom a serious injury (see Chapter 31: Common Simple Medical Problems).

Initial Burn Management

The critical first step is to **stop the burning process**—the faster the better, and within 30 seconds, if possible. Burns can continue to injure tissue for a surprisingly long time. No first aid can be effective until the burning process has stopped. Smother flames, if appropriate, then *cool the burn with cool water.* There are also several burn treatment materials, such as Emergency Burncare, appropriate for application immediately to a burn site. Remove clothing and jewelry from the burn area. *Do not try to remove tar or melted plastic that has stuck to the wound.*

Burn Assessment

Every aspect of burn treatment depends on your assessment of the depth, extent, and location of the injury, and the amount of pain the patient suffers. Even though this assessment may be an estimate, it will be your basis for deciding how the patient will be managed, whether evacuation is required, and how urgently evacuation is needed.

■ EXTENT

To assess the extents of a burn injury, use the *Rule of Nines:* Each arm represents approximately 9 percent of a person's *total body surface area* (TBSA). Each leg represents 18 percent (the front of the leg 9 percent, and the back of the leg 9 percent). The front of the body's trunk represents 18 percent TBSA, and the back of

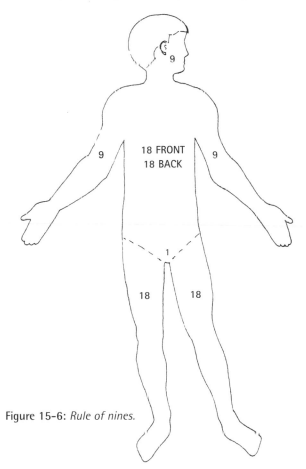

Figure 15-6: *Rule of nines.*

Depth-of-Burn Assessment Guide

Skin Characteristics ╲ Depth	Superficial Burn	Partial-Thickness Burn	Full-Thickness Burn
Skin Layer	Epidermis	Epidermis/Dermis	All layers
Color	Bright red	Red to pale	Pale (for scalds), charred (for open flames)
Blisters	None	Large, fluid-filled	Dry
Pain	Mild to moderate	Severe	Dull to severe
Healing	Spontaneous (3 to 5 days)	Spontaneous (1 to 3 days)	Very slow or never
Scarring	None	Moderate	Severe

the trunk 18 percent. The head represents 9 percent, and the groin 1 percent. For infants and small children, the head represents a larger percentage and the legs a smaller percentage.

For smaller areas use the *Rule of Palmar Surface:* The *patient's* palmar surface equals about 1 percent TBSA. A palmar surface is the surface of the palm and the fingers when the fingers are held together.

Partial-thickness burns and full-thickness burns that cover more than 15 percent TBSA constitute an immediate threat to life due to fluid loss (see Chapter 7: Shock).

■ LOCATION

Burns to the face—indicated by burned tissue, singed facial hair, soot around and/or in the nose and mouth, or coughing—may compromise the airway and demand immediate evacuation. Deep burns to areas of special function—hands, feet, armpits, and groin—often disable the patient permanently, and, therefore, demand immediate evacuation. Deep circumferential burns, burns entirely around an arm or leg, may swell to the point that they cut off circulation distal to the burn, so they demand immediate evacuation.

■ PAIN

In addition to depth and extent, do not underestimate the value of pain as a burn assessment tool. If the patient is in a lot of pain, that is an indication of the need for a physician's care.

Specific Burn Management

Specific burn management is aimed primarily at keeping the wound clean and reducing the pain.

1. Gently wash the burn with slightly warm water and mild soap. It is highly preferable to use an antibacterial soap. Pat dry. Remove the skin from blisters that have popped open, but do not open blisters. Gently wipe away serum and obvious dirt.
2. Dress burns with a thin layer of antibiotic ointment. Silvadene cream, a prescription medication, seems to work especially well.
3. Cover the burn with 2nd Skin, a commercially available product, instead of ointment if the burn is small enough, or cover with a thin layer of gauze coated with ointment. You can apply dry gauze or clean dry clothing (if nothing else is available). Covering wounds reduces pain and evaporative losses, but do *not* use an occlusive dressing.
4. When evacuation is imminent, say, within 12 hours or so, do not re-dress or reexamine the injury. But if evacuation is distant, re-dress twice a day by removing old dressings, recleaning (and removing the old ointment), and putting on fresh ointment and a clean, dry covering. You may have to soak off old dressings with clean, tepid water.

5. Do *not* pack wounds or patients in ice. Do *not* leave coverings wet with water on burns for very long. Coverings "wet" with ointment, 2nd Skin, burn gels, and other such substances, may be left in place for extended periods of time.
6. Elevate burned extremities to minimize swelling. Swelling retards healing and encourages infection. Get the patient, as much as possible, to gently and regularly move burned areas. Continue regular range-of-motion movement until healing is complete.
7. Ibuprofen is probably the best over-the-counter painkiller for burn pain, including sunburn.
8. If you have no ointment, no dressings, and/or no skill, leave the burn alone. The burn's surface will dry into a scab-like covering that provides a significant amount of protection.

General Management of the Burn Patient

1. Keep the patient warm. When enough skin is lost, so is the patient's ability to thermoregulate. Hypothermia may become a problem, and it will increase the chance of shock.
2. Elevate injured parts.
3. Get the patient to drink as much fluid as he or she will tolerate, unless the patient complains of nausea. Vomiting should be avoided, if possible.
4. **Remember:** An unconscious patient is unconscious from something other than the burn.

Evacuation Guidelines for Burns

Partial- and full-thickness burns covering 10 to 15 percent TBSA call for a rapid evacuation. Partial- and full-thickness burns covering more than 15 percent TBSA are a threat to life, requiring an evacuation as rapid as possible. Any serious burn to the face should be considered for rapid evacuation, as well as deep burns to genitals, armpits, hands, and/or feet. Circumferential burns warrant immediate evacuation.

While not a matter of urgency, patients with partial-thickness burns that hinder the ability to participate in and/or enjoy the wilderness experience should be evacuated. Superficial burns, even extensive ones, rarely require evacuation. Partial-thickness burns covering less than 10 percent TBSA should receive definitive care but seldom warrant rapid evacuation. Full-thickness burns need definitive medical care to heal best but do not usually require rapid evacuation unless they are extensive.

General Wound Infection Management

Infection, the multiplication of harmful microorganisms in tissues of the body, is a major cause of death worldwide. In the United States death from wound infection has been virtually eliminated by the administration of antibiotics. In a wilderness environment, however, organisms that invade open wounds, coupled with less than ideal wound-cleaning techniques, may create a serious and potentially life-threatening infection.

Wound Inflammation

As a wound heals, an unseen barrier develops on the inside of the body, sealing off the wounded area. *White blood cells* attack entrenched germs that have worked their way inside. Much of the contamination that migrates to the surface and drains out as white or faintly yellow pus is composed of white blood cells that died in the line of duty.

Some of the contamination is dumped internally into the *lymphatic system* and carried to lymph nodes, where it is picked up by the patient's circulatory system and eventually cleaned out of the blood and excreted from the body. The wound wonderfully maintains itself. This complex body process is called *inflammation*. The wound looks a little red and swollen and feels a little warm and tender. The wound may itch.

Wound Infection

If the bacteria trapped in a wound are too abundant for the inflammation process to handle, *infection* will occur. The redness becomes redder, the warmth becomes heat, the swelling continues, and tissue begins to harden past the borders of the wound. Pus will increase in amount and darken in color, typically appearing white, then yellow, then green, then brown. Pain may become great, and mobility of an extremity may be limited. These signs and symptoms usually show up in 24 to 48 hours, but they can develop much more rapidly with some strains of bacteria. Infection can also develop rapidly in deep tissue, creating gross swelling and pain, without the evidence of pus or red streaks. **Increasing swelling and persistent and/or increasing pain are indicators of serious infection.** Gas gangrene, the result of an anaerobic bacterium that produces rapidly and spreads tissue destruction, can cause death in as little as 30 hours. Gangrenous wounds stir up an abundance of dark, foul-smelling pus.

As infection spreads, lymph nodes can grow large and painful as they attempt to catch and kill the contamination, a condition known as *lymphadenitis*. The principal lymph nodes are located in the elbows and knees, neck, groin, and along the mid-to-lower backbone. Less common in infected wounds, *lymphangitis* produces red streaks that sometimes appear just under the skin. The streaks move from the wound toward the nearest lymph node as lymphatic vessels become inflamed. Should the infection reach the bloodstream, the likely result is *septicemia*, sometimes called *blood poisoning*, and the possibility of life-threatening shock (see Chapter 7: Shock).

Some wounds are more susceptible to infection than others. Wounds on the hand commonly become infected, largely due to less than ideal circulation normally flowing to hands. Flesh torn open by the teeth of animals has a high rate of infection, the bites of cats and humans being especially infectious. Infection is likely to result from crushing injuries and burns that open the skin. Heavy foreign-body contamination is responsible for many infections, as with an unnoticed splinter of wood buried in a foot or hand.

Any infection, whatever the source, produces similar signs and symptoms once it spreads throughout a human body. The patient develops a fever with accompanying chills and malaise, a fever that typically maintains a temperature of 102° F or higher. Serious infections often cause headache, nausea, vomiting, and/or back pain.

General Treatment for Wound Infection

1. Open the wound and clean it thoroughly. Soaking the wound in hot water—water as hot as the injured person can tolerate—may be very helpful in the cleaning process. Hot compresses may also stimulate the wound to drain. If scabs have formed, they will have to be removed.
2. Pack the wound open with sterile gauze or clean cloth boiled to make it as sterile as possible. The inner dressings should be wet with water that is as sterile as possible; the outer dressings should be dry.
3. Provide rest, warmth, and a high fluid intake for the patient.
4. Give nonsteroidal anti-inflammatory drugs, such as ibuprofen, or acetaminophen for fever and pain.
5. Stronger painkillers, if available, are usually indicated.
6. Antibiotics may be indicated if they are available and if the evacuation will be prolonged. A physician should be consulted prior to a wilderness trip for advice on antibiotics.
7. Infected wounds that do not respond to treatment in 12 to 24 hours should be evaluated by a physician.

Evacuation Guidelines for Infected Wounds

Rapid evacuation is advised for any patient with the signs and symptoms of serious infection: gross swelling, persistent and/or increasing pain, darkening pus, hardening edges of the wound, swollen lymph nodes, red streaks, or fever.

Conclusion

Assuring yourself your tent mate's blood loss is inconsequential, you gather supplies. After putting on protective gloves and eyewear, you gently but aggressively wash the foot, especially the big toe, clean with soap and water. With an irrigation syringe and plenty of safe-to-drink water, you irrigate the wound clean, lifting the avulsed flap of skin to flush as much contamination from the wound as possible. You pat the toe dry with sterile gauze. Smearing tincture of benzoin compound along both edges of the avulsion, you use wound closure strips to secure the avulsed flap as close to its normal anatomical position as possible—not too tightly closed, not too loosely closed. After covering the wound with antibiotic ointment, you apply sterile gauze to the site, taping it in place.

Following a brief condemnation of your tent mate for not listening to your sage advice about slipping on his boots before slipping outside, you settle down and try to get back to sleep.

Section Four

Environmental Emergencies

CHAPTER SIXTEEN

Cold-Induced Emergencies

You Should Be Able To:

1. Describe thermoregulation and its relationship to cold-induced emergencies.
2. Define and describe the signs and symptoms of hypothermia, frostbite, and nonfreezing cold injury.
3. Demonstrate the treatment for hypothermia, frostbite, and nonfreezing cold injury.
4. Describe the prevention of hypothermia, frostbite, and nonfreezing cold injury.

It Could Happen to You

As you head toward a lake high in North Cascades National Park of Washington, a June drizzle begins to fall. You've let yourself grow a little out of shape, and you fall behind on the trail, behind Mark, who forges ahead. The rain's chill begins to penetrate your clothing, but you can't stop to put on rain gear—you'll fall even farther behind. Besides, it's not *that* cold.

You stumble, almost losing enough balance to hit the ground. For a moment, the rush of adrenaline warms you, but that passes quickly as you move on. You'd like to take a break and dig out a candy bar. Mark will be worrying about you. You press on. Your water bottle lies forgotten in an outside pocket of your pack.

At last, the surface of the lake spreads out before you, dimpled by rainfall. Mark is already setting up the tent. You drop pack and try to help, but your hands don't seem to work right.

"Never mind," Mark says. "Put on your rain parka."

You fumble your rain gear out of your pack and struggle into it, but you can't get the parka zipped up. Shivers have begun to shake you. Mark zips up your parka and returns to setting up the tent, a look of concern wrinkling his forehead.

By the time the tent is set, you have begun to tremble violently and uncontrollably. You hear Mark suggest you crawl into your sleeping bag, but his words have lost importance. You sit and stare through your shivers at the yellow dome-shaped domicile.

As a warm-blooded species, humans have the ability to function well in a great variety of environmental conditions, maintaining a core temperature of approximately 98.6° F (37° C). *Thermoregulation*—heat regulation—is the key to this successful environmental adaptability. Human thermoregulatory centers are located at the base of the brain in the *hypothalamus;* those centers regulate heat production and heat loss, especially heat loss. The hypothalamus is strongly influenced by nerve impulses from receptors in the skin and by the temperature of the blood flowing through the receptors. Thus, skin is intimately involved in the maintenance of core temperature. Under the direction of the brain, skin has the ability to lose excess heat if the core temperature begins to rise or to conserve heat if the core temperature begins to fall.

The thermoregulatory centers do their best to balance heat production against heat loss. Heat conservation mechanisms, however, are not as well developed as heat shedding mechanisms. Humans work well in warm climates but survive in cold climates only by using their brains to plan and perform ways to conserve body heat. When the thermoregulatory system can no longer maintain an adequate core temperature, life-threatening situations may arise. Human biochemistry works best at around 98° F, and while adaptation to living in and visiting different environments is possible, from the Mojave to the moon, adaptation to altered body core temperatures is not possible.

Heat Production

There are two major internal sources of heat. Even as you sit quietly reading this book, you are making heat via *basal metabolism,* the energy necessary to sustain your life at complete rest. The *basal metabolic rate* (BMR) is the constant rate at which a human body consumes energy to drive chemical reactions and produce heat. Food and water are the fuels that are "burned," in the presence of oxygen, to maintain the BMR.

When you start doing anything powered by your voluntary muscles, a second heat source kicks in, *exercise metabolism.* Strenuous exercise metabolism may produce 15 to 18 times the amount of heat of basal metabolism, depending on your level of fitness. Nothing increases internal heat faster and more efficiently than exercise.

In addition to internal heat production, the human body can absorb a small amount of heat from external sources, such as the sun, a fire, another warm body, the ingestion of hot drinks, and the inhalation of warm air.

You end up with far more heat than you need on an average day and, if you couldn't shed the excess, you would literally cook in your own juices.

Heat Loss

Four mechanisms allow heat to be lost from the skin. *Conduction* is heat lost from a warmer object when it comes in contact with a colder object. The most extreme form of conductive heat loss occurs when your warm body falls into icy cold water. On a very hot day, you may not be able to shed heat via conduction, and you may even gain heat from the environment. You could, for example, gain heat from lying on a hot, sandy beach. *Convection* describes heat lost directly into air or water by movement of the air or water. Convection is a facilitated form of conduction, the result of warm molecules leaving a surface by the movement of air or water. An extreme form of convective heat loss on a cold day is defined by the wind-chill factor, which means, essentially, that the faster the wind blows, the faster you lose heat into the air. In this case your warm body has heated the air surrounding it, but that warm air is replaced by cold air blowing across it. *Radiation* describes heat given off as electromagnetic radiation constantly by a warm object. The warmer the object, the more it radiates; at around 98° F, you are often the warmest thing in the environment. You persistently lose some heat into the environment, but you lose it fastest into a black body, such as a cloudless night sky. On a cold night you can shed a massive amount of heat via radiation. *Evaporation* is the process of a liquid changing into a vapor. As water evaporates it requires energy to convert from a liquid to a gas. In this case the energy is heat, and the evaporation takes place on your skin. This is your most important cooling mechanism.

Hypothermia

When a body loses heat to the environment faster than it can produce heat, the body's core temperature starts to drop—a condition called *hypothermia*. Hypothermia is encouraged by inadequate hydration, insufficient nutritional intake, and fatigue. Untreated, hypothermia is progressive, a continuum that begins with a cold and unhappy person and ends with a cold and dead person.

Immersion Hypothermia

Immersion (or submersion) in cold water may cause the sudden onset of a form of hypothermia, sometimes called *immersion hypothermia* (see Chapter 19: Immersion and Submersion Incidents). The rapid onset of muscle rigidity often causes the

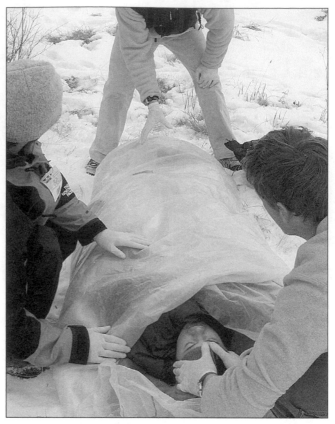

Figure 16-1: *Hypothermia occurs when the body loses heat to the environment faster than it can produce heat.*

rapid onset of drowning. In contrast a lowering of the core temperature is not sudden with immersion hypothermia, although it will be quite a bit faster than the onset of hypothermia in, say, the snowy mountains.

Urban Hypothermia

The slow onset of "urban" hypothermia, a condition that takes several days to months to show obvious signs and symptoms, is typical of the elderly and often poor and/or homeless who are exposed to less than adequate environmental temperatures. Almost never a wilderness medical problem, the loss of core temperature is so slow some individuals reach the stage of severe hypothermia without demonstrative shivering.

Generalized Hypothermia

Called "exposure" in the old days, the common wilderness form of hypothermia—generalized hypothermia, mountain hypothermia, or exposure hypothermia—usually requires an exposure time of at least several hours. Although you may think cold is the dominant environmental factor in hypothermia, it is the combination of cold and wet that poses the greatest threat. Well documented are serious cases of hypothermia in Florida and Baja, California, usually when rain falls under a high wind. One of the leading causes of wilderness emergencies, hypothermia impairs the level of judgment and blunts natural protective instincts.

Degrees of Hypothermia

■ MILD HYPOTHERMIA

Mild hypothermia has been termed by some experts as "a case of the umbles": the patient typically first stumbles, then fumbles, grumbles, and, later, mumbles. As gross motor skills are affected, a stumbling gait begins. Fine motor skills decrease and give rise to fumbling. The patient begins to draw inward, becoming less and less sociable. Designed to function optimally at approximately 98.6° F (37° C), the brain will begin to malfunction when its temperature drops below the ideal. In the case of hypothermia, normal thought processes become impaired. Mild hypothermia could be termed "mild stupidity." Patients begin to make poor decisions, such as not putting on rain gear when rain begins to fall. Patients typically show increasing confusion and apathy. Fine shivering, relatively controllable by the patient, begins. A healthy sign, shivering is the body's involuntary form of exercise to increase core heat. But mild hypothermia is insidious, affecting the ability of the patient to think, to be aware of its onset, to take care of self.

When the brain first senses heat loss is gaining on heat production, it stimulates the primary defense mechanism against further heat loss—vasoconstriction of the peripheral circulation (shrinking of the blood vessels in the skin). This vasoconstriction dramatically slows blood flow to the surface of the skin, where it will lose heat into the surrounding environment. The lack of blood causes the skin to become pale and cool. BMR will increase in response to the threat of cold, with an accompanying increase in heart rate and respiratory rate.

Signs and Symptoms of Mild Hypothermia

1. The "umbles."
2. Lack of sound judgment, confusion, apathy, "mild stupidity."
3. Increased heart rate and respiratory rate.
4. Pale, cool skin.

■ MODERATE HYPOTHERMIA

If the core temperature continues to fall, the brain stimulates an increase in shivering to the point where it is violent and uncontrollable. The patient has now entered the realm of *moderate hypothermia*. Shivering requires an immense amount of energy. If the moderately hypothermic patient is not properly treated, heat rushes from the patient into the environment. Radiative heat soars into the sky from an uncovered head. Heat conductively floods into the ground from a patient poorly insulated from the ground. A breeze rips heat away via convection. A drop in core temperature is rapid for an unprotected, shivering patient. The "umbles" worsen. A patient may find it impossible to walk, and he or she finds it increasingly difficult to speak and to think. Staring dully with a faraway gaze is not uncommon. Lack of circulation to the surface of the body causes the skin to turn very pale, perhaps a dusky color. Heart and respiratory rates increase further.

Signs and Symptoms of Moderate Hypothermia

1. Uncontrollable shivering.
2. Worsening of the "umbles."
3. Increased confusion.
4. Increased heart and respiratory rates.
5. Cold and pale (perhaps dusky) skin.

■ SEVERE HYPOTHERMIA

When energy stores are depleted and/or when the patient cools to the point where nerve messages from the brain saying "shiver" are blunted, the patient will stop shivering. When shivering begins to weaken, and then stops, the patient has either warmed or entered *severe hypothermia*, a condition characterized by increasing muscular rigidity and stupor progressing to unconsciousness. The heart rate slows and weakens to the point where it may no longer be palpable in a patient with severe hypothermia. The patient's respiratory rate slows down and grows so shallow that it may not be detectable. Skin grows deeply cold, and turns cyanotic. **A severely hypothermic patient may appear dead and yet be alive.**

Signs and Symptoms of Severe Hypothermia

1. Cessation of shivering.
2. Muscle rigidity.
3. Stupor progressing to unconsciousness.
4. Slow and/or nonpalpable pulse and respirations.
5. Cold and cyanotic skin.

Note: Core temperatures measured in the field are approximate and must be taken rectally to be considered of value. Taking rectal temperatures in the wilderness requires a hypothermia thermometer, one that registers lower than a usual thermometer, and is impractical at best. At worst, the process may dangerously expose an already cold patient. It is recommended to use obvious signs and symptoms in determining the treatment for hypothermia instead of temperature measurements.

Treatment of the Hypothermic Patient

■ MILD AND MODERATE HYPOTHERMIA

For the mild and moderate hypothermia patient, the most important treatment is to change the environment from cold and wet to warm and dry.

1. Immediately replace damp clothing with dry clothing. Dampness in clothing may be subtle, so check carefully. Pay attention to details: Snug up the drawstring on a parka's bottom, snug up a hood, change socks, and so on.
2. Add a windproof/waterproof layer and/or place the patient within shelter from wind and water.
3. Add extra insulation under and around the patient.
4. To stoke the inner fire of mild and moderate hypothermic patients, give water and food. Water is probably more important to the patient in the initial stage of treatment. *Caffeine and alcohol should be avoided.* Food intake should be encouraged, especially simple carbohydrates—sugars—which the body assimilates most quickly. Warm, sweet liquids, such as a cup of warm Jell-O, serve with excellence, adding fluid and simple carbohydrates to the patient.

Once the patient is thickly bundled in dry insulation, he or she will warm up. A patient violently shivering will warm fastest to a normal core temperature when he or she is adequately protected against losing the heat being manufactured by shivering. The addition of heat, such as an open fire (when environmentally acceptable and possible), may increase patient comfort psychologically, but it does not significantly increase the rate at which the bundled patient warms.

In the wilderness, cold patients may *seem* to rewarm faster when they snuggle inside a couple of sleeping bags with one or two warm bodies. The addition of heat via snuggling with a warm rescuer did *not* increase the rate of warming in laboratory tests. Whether or not the snuggled patient and rescuer or rescuers should be naked versus lightly clothed remains somewhat controversial. Skin-to-skin contact transfers heat faster, but rescuers and patients alike often report feeling more comfortable in light clothing, such as long underwear. Light clothing limits moisture transfer, probably the reason for greater comfort. Another disadvantage of person-to-person warming, especially in small groups, is the loss of personnel who could otherwise be heating water, cooking, making camp, and, in general, protecting and caring for the patient.

The addition of heat via heat packs or hot water bottles does increase the rate at which hypothermia patients warm up, at least a little bit. The use of heat packs should be focused on

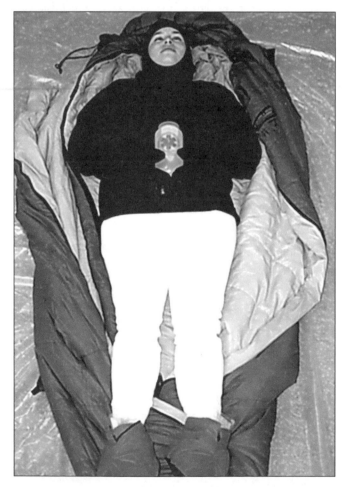

Figure 16-2: *Preparing a hypothermia wrap.*

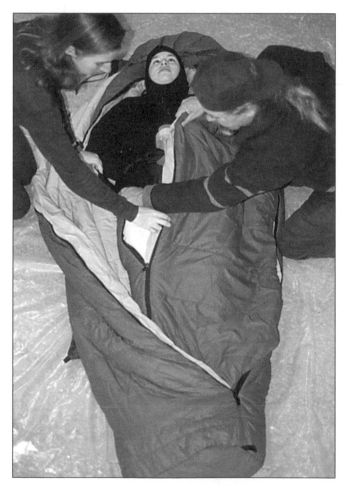

the chest, and the patient should be encouraged to hold a heat pack in his or her hands. Heat packs at the patient's feet will additionally provide comfort and protection against frostbite. This heat should not be applied directly to the skin because heat damage can occur on cold skin. Some type of dry material, such as a bandanna, should separate the heat pack from direct contact with the skin.

A mildly hypothermic patient may be encouraged to exercise once he or she has donned dry clothing. A moderately hypothermic patient typically has too little coordination to be exercised safely, although he or she may be able to perform simple exercises, such as sit-ups within a sleeping bag. The completely exhausted moderate hypothermic patient—exhaustion is a common component of hypothermia would do well in a hypothermia wrap (see "Severe Hypothermia").

As long as the patient returns to normal, an evacuation is not required. The patient, however, is often exhausted, requiring a rest day to rehydrate and refuel. Do not risk a return to a hypothermic condition by pushing a recovered patient to expend too much energy too soon.

■ SEVERE HYPOTHERMIA

Do not rush. Of the utmost importance is extra gentle handling of the patient—even if the patient appears dead. Rough handling is likely to cause a weak, cold, fragile—perhaps even undetectable—heartbeat to stop.

A patient who is cold, stiff, and blue needs oxygen. Supplemental oxygen would be of great benefit to the patient with severe hypothermia. Slow respirations and slow heart rate lead to an insufficiency of oxygen, the eventual cause of brain death. If supplemental oxygen is not available, mouth-to-mouth (or mouth-to-rescue-mask) breathing, started when breathing is undetectable or barely detectable, may help keep a severely hypothermic patient alive. Supplemental oxygen or artificial respirations should be started as soon as possible—before you move the patient—and even before removing the clothing. Oxygen will give a cold heart a better chance of survival during movement. Breathe for the patient for a minimum of 3 to 15 minutes before moving the patient. (For guidelines on managing a cold patient who appears breathless and pulseless, see Chapter 5: Cardiopulmonary Resuscitation.)

You want the severely hypothermic patient to end up in a *hypothermia wrap,* a cocoon of protection against the mechanisms of heat loss. It is best to prepare the hypothermia wrap prior to any movement of the patient. The wrap is like a burrito: The "tortilla" is an occlusive layer, windproof and waterproof (such as a sheet of plastic, a tent fly, or a tarp), and the "filling" is insulation, with the patient at the core.

Gently remove clothing, cutting it away if at all possible. You may, in fact, need to cut it to remove it, and the cutting process itself needs to be gentle. Lift the patient gently onto the dry "bed" of insulation. The thicker the insulation, the better for the patient. If snow or rain is falling, take great care to keep the insulation as dry as possible. Insulation such as sleeping pads must adequately protect the patient from the ground. The resulting cocoon leaves only the patient's face exposed, and that, too,

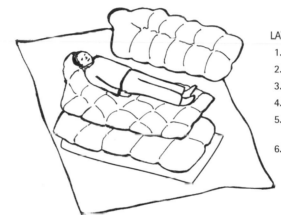

LAYERS (FROM GROUND UP):
1. TARP.
2. SLEEPING PAD(S).
3. CLOSED SLEEPING BAG.
4. OPEN BAG (PATIENT ON TOP).
5. PATIENT IN DRY CLOTHING OR NAKED WITH HEAT PACKS.
6. THIRD SLEEPING BAG OPEN AND UPSIDE DOWN.

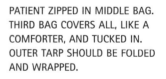

PATIENT ZIPPED IN MIDDLE BAG. THIRD BAG COVERS ALL, LIKE A COMFORTER, AND TUCKED IN. OUTER TARP SHOULD BE FOLDED AND WRAPPED.

PATIENT WRAPPED LIKE A BURRITO.

Figure 16-3: *Layering and wrapping to form the burrito-style hypothermia wrap.*

should be protected from wind and wet by an improvised roof. Even though it is extremely unusual for a severely hypothermic patient to return to a normal core temperature in the wilderness, he or she has the possibility of surviving a relatively long time, perhaps a couple of days, within an adequate hypothermia wrap. Once again, heat packs may be added. Initiate an evacuation by the most gentle means available—in other words, don't try to haul the patient out. In the meantime, keep up periodic rescue breathing.

Prevention of Hypothermia

1. Know the environment in which you intend to travel, and travel prepared to prevent hypothermia with adequate clothing, gear, food, and water.
2. Stay well hydrated. Avoid thirst. Keep your urine clear.
3. Stay well fed. Avoid hunger. Internal "fires" need fuel.
4. Plan to stay dry. Avoid cotton. Wear layers of clothing, taking off layers *prior to* sweating, adding layers back on *prior* to losing heat.
5. Pace yourself and your group to avoid overexertion with resulting sweat, fatigue, and loss of stored energy.
6. Make sure all members of your party understand hypothermia, and watch one another for early signs and symptoms.

Frostbite

Frostbite is localized tissue damage caused by freezing, a problem most likely to occur at the extremities of the body: the ears, nose, fingers, and toes. Frostbite creates a spectrum of injuries depending on how cold the tissue becomes. Frostbite is progressive, moving from mild to severe if untreated. Ranging from little or no damage to extensive damage resulting in substantial tissue loss, this spectrum of injuries can be classified in three basic categories: superficial, partial-thickness, and full-thickness.

Predisposing factors related to frostbite include moisture, low ambient temperatures (which must be below freezing for frostbite to occur), high winds (which speed heat loss), dehydration, and poor nutrition.

Damage from frostbite occurs in two phases, the freezing phase and the thawing phase. Blood flow decreases during the freezing phase, and ice crystals form in the fluid between cells, drawing fluid out of the cells and dehydrating them. More damage can occur mechanically if the ice crystals rub against one another. As blood clots form in blood vessels, circulation diminishes, causing more destruction. During the thawing phase, substances are released by the damaged cells that promote clot formation and vasoconstriction. More damage occurs during the thawing phase than during the freezing phase.

Superficial Frostbite

The initial stage of injury can be termed *frostnip*, considered by many experts to fall short of a true frostbite injury. Skin is white, perhaps waxy, numb, cold to the touch, but still relatively soft and pliable. On warming, no damage may be apparent. The outer

Facts about Wind Chill

1. The wind chill factor refers to the rate of cooling and not the temperature of the air.
2. Wind cannot lower skin temperature to less than the ambient air temperature under normal conditions.
3. A high wind blowing over exposed skin wet with gasoline or alcohol could drop the skin temperature measurably below ambient air temperature, but the drop would not be extreme.
4. Wind chill is definitely a factor in how fast exposed skin can be damaged by frostbite when the air temperature is below freezing.

layer of skin may have been frozen, and may turn reddish and later peel as sunburn peels. Superficial frostbite is, in fact, similar in physiology to a superficial burn (see Chapter 15: Wilderness Wound Management).

Frostnip is warmed optimally by submerging the affected body part in warm water, water heated to 104 to 108° F (40 to 42° C). It's better if the water circulates around the affected part. In the wilderness, however, warm water is often impractical, at best, but skin-to-skin warming is safe and effective. The cold skin should be in contact with warm skin. For example, a cold finger can be cradled in a warm armpit, or cold feet placed against a warm belly. *Do not massage cold tissue or place it near a strong radiant heat source,* both of which may increase damage. Every attempt should be make to prevent refreezing the area. Ibuprofen will ease pain and, more importantly, decrease the damage done on warming, and should be considered a standard operating procedure for all warming of frostbite. Aloe, applied topically, may aid in healing.

Partial-Thickness Frostbite

The first true frostbite is partial-thickness frostbite, an injury affecting a partial-thickness of the skin, similar to a partial-thickness burn. Some signs and symptoms of a partial-thickness injury mimic a superficial injury: Skin is white, waxy, numb, and cold to the touch. The affected skin, however, will often feel harder than frostnipped skin, and it will dent when you push gently on it. The dent may linger. It is, however, often impossible to tell the depth of injury prior to warming.

As with superficial frostbite, warming should be accomplished by submerging the affected part in warm water (104 to 108° F/40 to 42° C) to minimize damage. Skin-to-skin thawing is acceptable in the wilderness. Once again, no massage, no strong radiant heat, give ibuprofen, and use aloe when it is available.

When partial-thickness frostbite is thawed, a fluid-filled blister, sometimes called a *bleb*, forms. Blebs can form in minutes to hours, perhaps as long as 48 hours, after warming. Bleb formation is typically the deciding factor in whether you are treating superficial or partial-thickness frostbite. If the fluid is relatively clear, the damage to tissue did not go very deep, and hope for the patient's tissue is generally high. Only a little skin, perhaps, will be lost. If the fluid filling the bleb is reddish, the damage penetrated deep, into the vasculature, and the loss of tissue will probably be higher. If the fluid is deep red or reddish-blue, the damage is significant, and substantial loss of tissue is likely. If the blebs extend fully to the end of digits, the outcome for the patient tends to be better than when the blebs form short of the end of digits. Blebs forming short of the end of digits cut off circulation to the distal extremes.

Blebs should be protected with dry dressings. Care should be taken to prevent rupture of the blebs, which turns closed wounds into open wounds. Ruptured blebs should be covered with antibiotic ointment and sterile dressings. Ointment and gauze should also be placed between digits when fingers and toes are involved. All care should be taken to prevent refreezing of blebs. Since circulation is impaired, tissue that has been thawed suffers more extensive damage when it refreezes. If you can keep the affected part elevated above the level of the patient's heart, you may ease some of the pain that accompanies swelling. Keeping the patient well hydrated remains important.

Full-Thickness Frostbite

Full-thickness frostbite can be recognized because the affected area is pale white and frozen solid. It may be described as "wooden" by the patient, and it may feel wooden to the rescuer. Full-thickness damage feels icy cold to the rescuer and completely numb to the patient.

It is impossible in the wilderness to detect how far freezing has extended into the tissue. In other words, a toe might be only frozen solid on the surface, or it might be frozen through the bone. The frozen part is dead, but proper treatment might save part of an extremity.

Full-thickness frostbite is extremely painful when thawed—too painful to use. Once thawed, severely frostbitten feet typically cannot be used by the patient for walking, a consideration when the patient must then be carried from a remote and frigid

environment. (You may also find patients with blebs on their feet in too much pain to walk.) Although the outcome for the patient is better if proper thawing occurs immediately, it is usually better to have the patient keep walking on a painless, frozen foot than to stop and thaw when the situation is inadequate to manage thawing.

When the situation is manageable, such as in a base camp, full-thickness frostbite should be thawed. Warming should be rapid without being damaging. The best method, as with less serious frostbite, involves soaking the frozen tissue suspended in water preheated to 104 to 108° F (40 to 42° C). Too much heat should be carefully avoided. You can burn the patient. Too much time in appropriately warm water is not a problem. You cannot "overthaw," but you can "underthaw." Monitor the temperature of the water: It cools off quickly, necessitating the addition of more warm water fairly often. Without a thermometer, you can guess the proper temperature by dipping your naked elbow in the water. If it feels comfortably warm but not hot, it is probably close to the right temperature.

Watch for indications of adequate warming: the flush of pink to the affected area, and the flush of severe pain on the patient's face. The skin will later turn deep red, perhaps cyanotic, or mottled. This looks serious, but it's not a final indication of the severity of the injury. Only time will tell. The black, dried, dead look of mummification may develop rapidly or slowly. Once warming is accomplished, the affected parts should be air-dried, covered with an antibiotic ointment, and kept elevated. Once again, place ointment and gauze between damaged digits. Painkilling drugs would benefit the patient. Ibuprofen, once again, is strongly recommended. Refreezing must be avoided. Trauma, even mild trauma such as walking on thawed feet, must be avoided.

Prevention of Frostbite

1. Wear adequate cold weather clothing, including boots, and keep it all as dry as possible.
2. Avoid constricting clothing, especially tight boots.
3. Wear mittens instead of gloves.
4. Avoid skin contact with cold metal and cold gasoline, both of which can cause frostbite on contact.
5. Stay well hydrated.
6. Maintain a high-calorie diet.
7. Watch other party members for the early signs of frostbite, including white patches on the face.
8. Pay attention to your fingers and toes, and stop and treat frostbite early, while it still causes tingling and pain and before it goes numb.

Nonfreezing Cold Injury

Nonfreezing cold injury (such as immersion foot, also known as trenchfoot) is a cold-weather emergency resulting from prolonged contact with cold—and usually also with moisture—that causes inadequate circulation with resulting tissue damage. Nonfreezing cold injury (NFCI) may be divided into three phases. Phase one, the *prehyperemic phase,* involves decreased blood flow, the period of time when blood vessels are constricted by the cold and dampness inside a shoe or boot, and too little oxygen is carried to the cells of the foot. When you check the foot, it is cold to the touch, there may be a bit of swelling, and discoloration is typically evident (usually white or bluish). The patient may complain of numbness. The problem can occur in less than 12 hours of exposure.

Phase two, called the *hyperemic or increased-blood-flow phase,* is the period when the foot warms. The cells of the foot have become deeply damaged by the lack of adequate circulation. On warming, the blood vessels open back up, and the tissue swells with excess fluid. The damaged tissue typically looks red. Patients complain of tingling pain, often severe, that never lets up. On warming, blisters sometimes form, and, later, ulcers where the blisters have fallen off to reveal dead tissue underneath. In severe cases, gangrene can result. This phase typically takes at least 12 hours of exposure to occur.

Phase three, the *posthyperemic or recovery phase,* may last weeks to months. During this phase, the patient may complain of increased perspiration in the foot, increased sensitivity to cold, and varying degrees of pain, itching, and paresthesia (a creeping, tingling, prickly feeling). Pain medications are often prescribed. The damaged foot may be more susceptible to cold injury in the future.

Here's what should be done if you think a member of your group is developing NFCI. You can warm NFCI in warm water, as prescribed for frostbite. You may also choose to keep it elevated above the level of the patient's heart while you gently warm the foot with passive skin-to-skin contact. Then carefully dry the foot. If the foot has not been soaked and/or looks dirty, carefully wash it before drying it. Do not rub the foot or place it near a strong heat source such as a fire or stove, both of which can damage the tissue of the cold foot. Start the patient on a regimen of over-the-counter anti-inflammatory drugs (aspirin or, even better, ibuprofen), following the directions on the label. Remember it will probably take 24 to 48 hours before the severity of the damage is fully apparent. If you end up with a painful, obviously swollen foot and/or one that develops blisters, the patient needs the attention of a physician. Whether or not the patient can walk out to a physician will be determined by the patient.

Evacuation Guidelines

Any patient treated for hypothermia who recovers may safely stay in the wilderness. Any patient treated for severe hypothermia—one in whom shivering has stopped and/or the level of consciousness stays altered—requires rapid yet gentle evacuation. A patient with frostbite who develops blebs and/or dusky or blue-gray skin should be evacuated. A patient who is assessed with full-thickness frostbite should be evacuated. Patients with nonfreezing cold injury should be evacuated, unless the problem is mild and the injury usable.

Prevention of Nonfreezing Cold Injury

1. NFCI is encouraged by poor nutrition, dehydration, wet socks, inadequate clothing, and the constriction of healthy blood flow in the feet by shoes and socks that are too tight. Make sure your boots fit with plenty of room for the socks you choose to wear.
2. Keep a dry pair of socks on hand at all times, preferably packed in a plastic bag to make sure they stay dry.
3. Don't add more socks if your feet get cold—get bigger boots, or boots with more insulation.
4. People who sweat heavily are more susceptible; an antiperspirant spray can reduce sweating and thus reduce the risk of NFCI.
5. Periodically, preferably twice a day, dry your feet and gently massage them before stuffing them back into boots.
6. Do not sleep in wet socks.

Conclusion

In North Cascades National Park, Mark helps you into the tent. Unrolling a sleeping pad, he fluffs out your sleeping bag while you fumble ineffectively with your clothing. Mark has to help you undress and wiggle into dry pile pants and a pile sweater. Once you've crawled into your sleeping bag, your friend spreads his bag over you for additional insulation. He leaves the tent, returning soon with your wool stocking cap, the one you left lying unused in your pack.

Firing up the stove, Mark brings water to a boil, pours a cup, drops in a bag of peppermint tea and two teaspoons of sugar. By the time the tea is ready, your shivers have diminished to an occasional tremor. You sit up and gratefully sip the tea, feeling assured you'll see tomorrow's dawn.

Murray Hamlet, DVM, contributed his expertise to this chapter.

CHAPTER SEVENTEEN

Heat-Induced Emergencies

You Should Be Able To:

1. Describe the signs, symptoms, treatment, and prevention of dehydration.
2. Describe the signs and symptoms of heat cramps, heat exhaustion, heatstroke, and hyponatremia.
3. Demonstrate the treatment for heat cramps, heat exhaustion, heatstroke, and hyponatremia.
4. Describe the prevention of heat cramps, heat exhaustion, heatstroke, and hyponatremia.

It Could Happen to You

The group you're guiding on this August trip down the Wild and Scenic Chattooga River looks a little older than your average clients, but their enthusiasm is undiminished by age. You raft past the point in the Ellicott Rock Wilderness where Georgia, North Carolina, and South Carolina meet. Although the movement of the raft you're oaring, one of two rafts traveling together, creates a breeze that makes you feel relatively cool, you know the high temperature, somewhere in the 90s, and high humidity, not far behind the temperature, are sending your body's need for water skyrocketing. You lean on an oar, unclip your water bottle from where it rides near at hand, and take a big swig. You told your clients to drink up throughout the day during your pre-put-in talk, and you hope your example is noticed.

Toward mid-afternoon, however, jovial David, an overweight man in his late fifties, quick to laugh, doesn't seem as jovial anymore. He begins to act strangely—slightly disoriented and very argumentative. He states rather firmly that he is not hot, tired, or thirsty. By the takeout, David becomes more disoriented and combative, insinuating, in a nasty way, that you've ruined his day. His skin looks unusually red. On shore he begins to hallucinate, swinging at invisible large black birds. He is convinced the birds are "going for my eyes."

Of the two environmental temperature extremes, heat and cold, the human body is better adapted to deal with heat. With virtually hairless skin filled with abundant sweat glands, powered by a cardiovascular system of marvelous endurance, humans function well when the mercury rises. You are not, however, a foolproof design. Overheating can ruin your day—and your life.

To understand heat illness requires an appreciation of the fundamentals of human *thermoregulation*—body heat production and body heat loss (see Chapter 16: Cold-Induced Emergencies). As the body's core temperature begins to rise, the excess heat is absorbed by the blood. As the thermoregulatory centers in the brain detect the increase in blood temperature, the brain causes the peripheral circulation, the vasculature (blood vessels) of the skin, to dilate, or open up. This dilation of peripheral vasculature increases blood flow to the skin, where the blood can be cooled. At the same time, to further increase the rate of heat loss from the skin, the brain stimulates the sweat glands to produce sweat. The evaporation of sweat from the skin increases the rate of cooling by many times. As long as you can sweat and the sweat can evaporate, you can, in most cases, continue to cool efficiently. But if for some reason either the sweating mechanism begins to fail or the sweat cannot evaporate, then the cooling mechanism will fail. On hot, very humid days, for instance, cooling becomes extremely inefficient because sweat cannot evaporate, and it is relatively easy to overheat.

Sweat consists primarily of water with some electrolytes, specifically sodium and chloride ions that are necessary for normal body function. It is this combined water depletion—*dehydration*—and electrolyte depletion that forms the basis of a spectrum of problems with one general name: heat illness.

Dehydration

Without water, there could be no life—at least no life as you know it. As a developing embryo, you nestled in a watery bed and your body weight averaged around 80 percent water, an average that dropped to approximately 74 percent by birth. As an adult, you gurgle along at somewhere around 62 percent water overall, and healthy blood, the red tide of life, surges at between 85 to 90 percent water.

Water puddles inside every one of your cells and flows through the microscopic spaces between cells. Oxygen and nutrients in that water float to all parts of your body, and waste products are carried away. When your kidneys remove wastes from your body, those wastes have to be dissolved in water. Digestion and metabolism are water-based processes, and water is the primary lubricating element in your joints. You even need water to breathe, your lungs requiring moisture to expedite the transfer of oxygen into blood and carbon dioxide out of blood. Sweat, as mentioned, is mostly water. The water in your blood carries heat from warmer body parts to cooler areas of your anatomy when you are exposed to cold. In short, if you aren't well hydrated, you won't be able to stay healthy, maximize your performance, or even maintain joy in being outdoors.

The water in your body, the fluid that keeps you alive and active, leaves you at an alarming rate. Estimates vary widely, but an average person at rest on a normal day loses 2 to 3 liters of water. About 1 to 1.5 liters rushes out as urine and another 0.1 liter in defecation. Moisture is lost from the act of breathing, more than 0.5 liter per day, a rate that increases in dry, winter air.

Then there's sweat. The fluid lost in perspiration can climb to 1 to 2 liters per hour during periods of strenuous exercise. Compared with watching TV all day, 1 hour of exercise may lead to an approximately 50 percent increase in the amount of water your body uses.

How can you tell you're running low on water? At first the signs are subtle. You'll get a mild headache and then start feeling tired. Down merely 1.5 liters, your endurance may be reduced 22 percent and your maximum oxygen uptake (a measure of heart and lung efficiency) can be lowered 10 percent. Down 3 to 4 liters can leave your endurance decreased to 50 percent and your oxygen uptake reduced close to 25 percent. By now, if you're observant, you'll have noticed your urine has turned a dark yellow. You may suddenly find yourself seriously dehydrated, with a rapid pulse and respirations and utter exhaustion.

Dehydration can be classified into three levels:

1. **Mild,** characterized by dry mucous membranes (lips and mouth), normal pulse, darkened urine, and mild thirst.
2. **Moderate,** characterized by very dry mucous membranes, rapid and weak pulse, darker urine, and thirst.
3. **Severe,** characterized by very, very dry mucous membranes, an altered level of consciousness (drowsy, lethargic, disoriented, irritable), no urine, no tears, and shock (indicated by rapid and weak pulse, rapid breathing, and pale skin).

Heat Cramps

On the minor end of the spectrum of heat-induced emergencies are *heat cramps,* a painful spasm of major muscles that are being exercised. Heat cramps are often associated with heat exhaustion. Those most often cramped are people unacclimatized to heat who are sweating profusely. Heat cramps are poorly understood; they probably result not only from the water lost in sweat but also the salt lost in sweat. Rest, gentle massage, and gentle stretching of the affected muscles usually provide relief. Drinking water, preferably with a pinch of salt per liter added, is advisable. Heat cramps do not often occur in someone who is adequately hydrated. Once the pain is gone, exercise may be continued. If the cramps return and/or worsen, a day of rest with adequate water and food is the recommended treatment.

Heat Exhaustion

Heat exhaustion is a volume problem characterized by headache, nausea, rapid pulse, rapid breathing, and, of course, exhaustion. Light-headedness or dizziness is common, especially if the patient changes position from lying to sitting or sitting to standing (orthostatic changes). Dehydration is the major threat, and the patient is experiencing the early stage of shock—compensatory shock (see Chapter 7: Shock). Thirst is a common complaint. Sufferers are so sweaty they often feel cool, grow goose bumps, and complain of chills. Core temperature may or may not be elevated.

Treatment for Heat Exhaustion

Treatment should include (1) changing the patient's environment from hot to cool, such as moving the exhausted person to a shady spot, pouring water on his or her head, and fanning; and (2) orally rehydrating the patient with water, preferably with a pinch of salt and, perhaps, a couple of pinches of sugar. It will take about 1 hour to get 1 liter of fluid back into circulation, and the patient needs the hour to rest. Heat exhaustion is not physiologically damaging, but it should be treated aggressively before it progresses to a more serious condition. Heat exhaustion can lead to serious dehydration and/or heatstroke.

> ### *Signs and Symptoms of Heat Exhaustion*
>
> 1. Increased heart rate.
> 2. Increased respiratory rate.
> 3. Headache.
> 4. Dizziness.
> 5. Nausea.
> 6. Thirst.
> 7. Fatigue.

Heatstroke

On the serious end of the spectrum lies *heatstroke,* a true life-threatening emergency. Death has been known to occur within 30 minutes of assessment.

There are two varieties of heatstroke. In *classic heatstroke*, the patient is usually elderly or sick, or both—and terribly dehydrated. Temperature and humidity may have been high for several days, and the patient has dehydrated to the point where his or her heat-loss mechanisms are overwhelmed. You might say the patient simply ran out of sweat. Skin gets hot, red, and dry. She or he lapses into a coma and, if untreated, dies.

In the wilderness you are far more likely to encounter the second variety, *exertional heatstroke*—a brain problem. The patient is usually young, fit, and unaccustomed to heat, sweating but producing heat faster than it can be shed. Signs include, primarily, a sudden and very noticeable alteration in normal mental function: disorientation, irritability, combativeness, hallucinations, bizarre delusions, or incoherent speech. Skin is hot and red but wet with sweat. Rapid breathing and rapid heart rates are almost universal. Loss of coordination and, later, seizures are common. These signs and symptoms may show up without earlier indications of heat exhaustion. Clinically speaking, the patient's core temperature has risen to at least 105° F (40.5° C), and this rise in core temperature is the primary threat. The brain is beginning to cook. Cardiovascular and neurological collapse, if they haven't happened already, are imminent.

Treatment of Heatstroke

Rapid cooling is required to save the patient's life. Change the patient's environment as much as possible, moving her or him to shade. Remove clothing that retains heat. Cotton clothing is OK. Keep the patient wet while you vigorously fan the body. These procedures maximize evaporative heat loss. Massaging of arms and legs and application of ice packs at the neck, groin, and armpits increase heat loss. Easing the patient into cold water is less effective—

cold water may cause the patient to vasoconstrict, reducing the return of cooled blood to the core. And immersion can be dangerous because the patient is difficult to manage and may drown. In the absence of an abundance of water for external cooling, concentrate on cooling the head and neck of the patient. The process of cooling is often lengthy.

Because heatstroke patients are dehydrated, rehydration is critical. Unfortunately, getting the patient to drink is typically impossible due to the decreased level of consciousness. With impaired mental function, it is inappropriate to force fluids. Continue cooling externally in the hope that the patient will recover enough to begin oral rehydration.

Give no drugs to the heatstroke patient. You do not want the hypothalamus, the body's "thermostat," altered in its ability to function by antifever medications.

Heatstroke patients should be seen by a physician as soon as possible, which means a rapid evacuation is appropriate. During the evacuation, the cooling process may have to be maintained. With a patient who seems to have recovered, too much internal heat can cause breakdowns in some body systems that show up later, a cause for evacuation even if your treatment appears successful. Since relapses into heatstroke are not uncommon for recovered patients, all patients should be closely monitored until turned over to a physician.

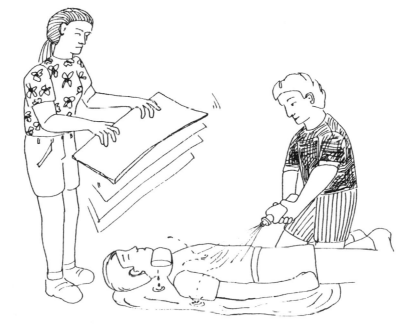

Figure 17-1:
Patient being cooled.

Hyponatremia

When someone drinks more than enough water and fails to take in adequate salt (sodium), the salt lost in sweat, the sodium level in the blood starts to drop. When blood sodium gets too low, a case of *hyponatremia* develops.

Signs and symptoms of hyponatremia may vary among individuals and depending on how low the sodium level has sunk. Common complaints include headache, weakness, fatigue, light-headedness, muscle cramps, nausea with or without vomiting, sweaty skin, normal core temperature, normal or slightly elevated pulse and respirations, and a bit of anxiety. Sound familiar? Yes, it sounds like heat exhaustion. But if you treat hyponatremia like heat exhaustion—just add water—you are harming the hyponatremia patient. More severe symptoms of hyponatremia include a patient who is disoriented, irritable, and combative—which gives the problem a more common name: water intoxication. Untreated, the ultimate result will be seizures, coma, and death.

It is of critical importance to get an accurate history for the patient. Little or no food intake combined with high fluid intake should make you highly suspicious. If the patient's urine output is clear and copious (urination occurring every few hours to several times per hour) combined with a lack of thirst, you'll draw closer to assessing hyponatremia. Heat exhaustion patients typically have a low output of yellowish urine (urinating every 6 to 8 hours) combined with thirst.

Hyponatremia patients with mild to moderate symptoms and a normal mental status may be treated in the field. Move the patient to shade for rest and allow *no* fluid intake and a gradual intake of salty foods to help the kidneys reestablish a sodium balance. Once a patient develops hunger and thirst combined with normal urine output, the problem is solved. Restriction of fluids for someone who is well hydrated, fortunately, is harmless. Giving oral electrolyte replacement drinks alone might damage the patient. These drinks are so low in sodium and so high in water the imbalance in sodium may be increased.

Only a blood test confirms hyponatremia beyond doubt. If you just can't make up your mind—is it heat exhaustion or hyponatremia?—give the patient electrolyte replacement drinks and salty food and monitor closely for improvement. Concerning patients with an altered mental status there is no question: They demand rapid evacuation to a medical facility.

Prevention of hyponatremia is a matter of paying attention. Many people heard the message—stay hydrated—but failed to hear the rest of message—keep eating, too. Stay hydrated, yes, but eat salty foods regularly while exercising in heat.

Prevention of Heat-Induced Emergencies

Hydration

Internal water is sometimes used at a faster rate than the need for replenishment is felt. There are individual differences, but thirst is not always an accurate indicator of hydration status during strenuous exercise. To be safe, drink before you get thirsty, and if you become thirsty, drink until after your thirst is quenched. Remember also that dehydration can be cumulative, carrying over from previous days.

To remain hydrated, water should be consumed at a disciplined rate. There is considerable benefit from starting each day with an ingestion of a large volume of water, about 0.5 liter. Following that, the International Sportsmedicine Institute recommends 0.5 to 0.67 ounce of water per pound of body weight per day, ingested periodically throughout the day. Figured in liters, that's about 3 to 4 liters per day for the average-sized person. Drink water with meals and snacks, and suck down a few swallows before bedtime.

In extremes of heat, such as summer hiking in the Grand Canyon, a body's need for water can also be extreme, more than the average. It's best to check with local experts and/or land managers about fluid intake and follow their directions.

You should also be drinking water on a disciplined schedule when you are traveling hard outdoors. Because the human body can only absorb so much water at one time, the rate of ingestion should be matched, as closely as possible, to the rate of absorption. Most people fall into a rate-of-absorption range of 250 to 300 milliliters per 0.25 hour. Drinking at least 0.25 liter of water (about 8 ounces) every 15 to 20 minutes during periods of exercise should meet the needs of most bodies. There are, however, individual differences, as noted above. The most important factor in hydration may be self-monitoring: Do you feel OK? Are you thirsty? Is your urine relatively clear?

The bottom line is sensible hydration. The wise outdoors enthusiast drinks enough water and eats enough food not only to prevent dehydration and hyponatremia but also to function at her or his best.

Temperature, Sugar, Salt, and Water

Some research indicates that cold water, water at a temperature in the 40 to 50° F range, is emptied from the stomach at a higher rate than water at other temperatures and is therefore absorbed faster by the body. The data remains inconclusive, and the most important impact of fluid temperature may be palatability. Cool water usually tastes better.

There is a lot of information about the need to add sugar to water. This is primarily of interest to the competitive athlete. On wilderness expeditions most people have their needs met by well-planned meals and on-trail snacks.

Most people obtain ample salt with a balanced diet. Salty snacks when exercising hard are probably of benefit. There may be small advantages to having a tiny bit of salt—a "pinch"—in a liter of water when you're active. Salt not only helps retain water during exercise but also stimulates the need to drink. Too much salt, of course, is counterproductive, causing you to need more water than normal.

Sports Drinks

During exercise lasting less than 1 hour, there is little evidence of physiological or physical performance differences between consuming a carbohydrate-electrolyte drink and plain water. Participants in longer events should consider solutions containing 4 to 8 percent carbohydrates. The carbohydrates can be sugars (such as glucose or sucrose) or starch (such as maltodextrin). For most wilderness travelers the primary value in sports drinks probably is palatability. Most sports drinks have acceptable concentrations of sugar, but on an individual basis people may like to dilute the drink, certainly an acceptable act. Sport drinks may not provide enough sodium.

Hydration Guidelines

You cannot train yourself to need less water. In fact, the harder you work, the more water you need. To prevent dehydration, follow these recommendations:

1. Drink 0.5 liter of water first thing each morning.
2. Drink 0.25 liter of water every 15 to 20 minutes during periods of strenuous exercise.
3. Drink water with meals and snacks.
4. Keep track of your fluid consumption and drink at least 3 to 4 liters of water every day you're "on the trail."
5. Avoid alcohol.
6. Monitor your urine: Keep it clear and copious. Clear to light yellow is fine; dark yellow is an indicator of dehydration. Copious describes normal, healthy urine output.

As a group leader, maintain a pace that allows everyone to adapt to the heat. If anyone feels the symptoms of heat exhaustion coming on, the group is going too fast. It is especially important to maintain a gradual pace early in the hot, humid season. The thermoregulatory system will become more efficient as it gets used to hot weather. It takes approximately 10 days to 2 weeks for a human body to adapt to heat and humidity. Remember that the elderly, the very young, the very muscular, and the very overweight take longer to acclimatize to heat.

Take a break during the hottest part of the day, the mid-afternoon hours. Wear loose-fitting, cotton clothing that lets air pass through and sweat evaporate. Wear a brimmed hat or cap that provides shade for the face and head.

Avoid drugs known to contribute to heat-induced emergencies, including alcohol, antidepressants, antihistamines, some anesthetics, cocaine, and amphetamines.

Evacuation Guidelines

Evacuate all patients with heat-induced problems who have or have had an altered level of consciousness. Monitor the patient carefully during the evacuation even if he or she appears to have recovered. Relapses are not uncommon.

Conclusion

With the help of other clients on the Chattooga River, you take control of David, lowering him to the ground in the shade of a tree near the bank. He is wearing cotton shorts, a cotton T-shirt, and a personal flotation device (PFD). Removing the PFD, you soak David with water from the river. Using the PFD as a fan, you aggressively fan him. Under your direction two other clients remove their PFDs and join the fan club. You direct a third client to keep pouring water over David, and a fourth client to massage David's arms and legs.

When the second raft arrives, you ask the second guide to call for help immediately on the radio waiting in the company van parked at the takeout. David's signs and symptoms show no improvement, and you know you'll be aggressively cooling him until medical assistance arrives. . . .

CHAPTER EIGHTEEN

Altitude Illnesses

You Should Be Able To:

1. Define and describe acclimatization.
2. Describe the signs and symptoms of acute mountain sickness, high-altitude pulmonary edema, and high-altitude cerebral edema.
3. Describe the treatment for acute mountain sickness, high-altitude pulmonary edema, and high-altitude cerebral edema.
4. Describe ways to prevent altitude illnesses.

It Could Happen to You

You're at the airport in Gunnison, Colorado. The plane brings clients from Florida, California, New York, and Texas—six excited teenagers. After a brief welcoming speech, you load the van and drive the group to the camp you work for near Crested Butte at an elevation of more than 9,000 feet above sea level.

On the morning of the second day of the week-long experience—after lessons in hiking and camping, outdoor cooking, and expedition behavior you lead your group out of camp toward the *really* high country. The hike is washed in sunshine, and the group stays together well—no stragglers. You set tents and establish a cooking area on bare ground near the edge of an alpine meadow at 11,200 feet.

On the third morning Ashley asks for a private conversation. She slept poorly. Her nausea erupted into vomit just before dawn. Her head "throbs."

"I can't go on," she says.

The medical problems collectively referred to as *altitude illnesses* are the result of *hypoxia*, insufficient oxygen in the blood for normal tissue function. Hypoxia can be a result of the decreased barometric pressure at higher altitudes. As you ascend, the barometric pressure decreases, the pressure of inhaled oxygen during each breath decreases, and the chance of altitude illness increases. Altitude problems affect either the brain or the lungs, or both.

Because there is a measurable increase in ventilation and decrease in aerobic exercise performance at elevations above 4,000 feet, "high altitude" can be said to start at that point. Complications seldom occur, however, below 8,000 feet. In defining terms, consider 8,000 to 12,000 feet as high altitude, 12,000 to 18,000 as very high altitude, and above 18,000 feet as extreme high altitude. At 18,000 feet barometric pressure is one half that of sea level, making the inspired pressure of oxygen also one half. You effectively get 50 percent less oxygen with each breath.

The human body adjusts to dramatic changes in barometric pressure, given enough time. Altitude illnesses which range from mildly disturbing to completely fatal—are determined, primarily, by four factors: (1) how high the person goes, (2) how fast the person attains a specific altitude, (3) the altitude at which the person slept (called "sleeping altitude"), and (4) predisposing factors, such as genetics and differences in individual physiology, that are not clearly understood. Children appear to have no more susceptibility to altitude illnesses than adults, but adults over the age of 50 appear to have less susceptibility than younger adults. Women and men appear to have the same susceptibility in some studies, but other studies indicate women are less prone to problems associated with the lungs.

Acclimatization

The rate of *acclimatization*—the process of physiologically adjusting to altitude change—differs with individual physiology and the specific altitude attained, but most people adjust enough to prevent illness if they spend 2 to 3 days in altitudes of 8,000 to 12,000 feet, and not gain more than 1,000 feet of sleeping altitude each successive day. For instance, if you slept at 14,000 feet last night, you can climb beyond 15,000 feet the next day, but you should drop back down to no more than 15,000 feet to sleep. It is not always practical, however, to gain no more than 1,000 feet per day, and gains of 2,000 feet per day of sleeping altitude are usually tolerated with a rest and acclimatization day after every 2,000 to 3,000 feet gained. Critical to acclimatization is adequate hydration and nutrition. Climbers in camp who mix rest with periods of light exercise seem to acclimatize faster than climbers who rest only.

Acclimatization is a complex physiological process involving, first and very importantly, an increased rate and depth of breathing. Heavy breathing increases the oxygen content of the blood, thus getting more oxygen to body tissues. It is OK to breathe fast as long as you are not overexerting yourself at the same time. This change starts immediately, with functional acclimatization for a specific altitude taking 6 to 8 days.

During acclimatization the heart rate speeds up and systemic blood pressure increases. After approximately 7 to 10 days at a specific altitude, heart rate and blood pressure decrease. Bone marrow is stimulated to produce more red blood cells, which increases the capacity of blood to carry more oxygen, and the final stage occurs on a cellular level, with capillary density increasing, muscle cells shrinking, and mitochondria—the intracellular "furnaces" where food is burned in the presence of oxygen to create energy—increasing. These cellular changes get more oxygen into action faster and more easily, but the changes take weeks. On the average, 80 percent of overall acclimatization is complete at 10 days, but 95 percent is not reached until 6 weeks.

Acclimatization is lost at approximately the same rate at which it is gained—significant loss at 2 weeks, most of it lost by 6 weeks—but the main point is this: If you take time, you will almost always adjust to higher altitudes.

Acute Mountain Sickness

The most common form of altitude illness—called *acute mountain sickness,* or AMS—is the first stage in illnesses associated with the brain. It produces symptoms that usually appear within 6 to 10 hours of arrival at altitude but that can appear in as little as 1 hour. The problems range in severity from mild to moderate in the grand scheme of debilitation.

Since the symptoms are nonspecific, it can be difficult to assess AMS with certainty. The signs and symptoms of AMS could also indicate dehydration, hypothermia, infection, carbon monoxide poisoning, a hangover, or a drug overdose, for instance. But if the patient recently arrived to an altitude of 8,000 feet or more, AMS should be your first guess. Headache is usually the first and by far the most common complaint. Other symptoms include *anorexia* (loss of appetite), nausea and vomiting, *insomnia* (inability to sleep), *lassitude* (weariness, exhaus-

tion), dizziness, and unusual fatigue. Patients are typically short of breath on exertion, a shortness of breath that quickly goes away with rest.

Peripheral edema, swelling of the face, hands, or feet, may or may not accompany AMS. A patient may show signs of peripheral edema without showing any significant other signs or symptoms of AMS.

> ## Signs and Symptoms of Acute Mountain Sickness
>
> 1. Headache.
> 2. Nausea, perhaps with vomiting.
> 3. Anorexia.
> 4. Insomnia.
> 5. Lassitude.
> 6. Unusual fatigue.
> 7. Dizziness or light-headedness.

High-Altitude Cerebral Edema

A worsening of AMS could indicate the patient is progressing toward high-altitude cerebral edema (HACE), a brain problem with a high potential for death. With HACE, fluid leaks out of capillaries in the brain and causes increased pressure inside the skull. The cause of death is the brain being squished by the increasing pressure. HACE develops progressively over hours to days. HACE and HAPE (high-altitude pulmonary edema) often occur in a patient at the same time.

HACE is announced by the onset of *ataxia* and/or an alteration in the level of consciousness. Ataxia may be dramatic with HACE patients. Even sitting up may be impossible. The dramatic altered mental status often includes disorientation, irritability, and combativeness, and you may see severe personality changes, such as hallucinations. Seizures are rare but not impossible. A severe headache may or may not be present in HACE. Lethargy (sluggishness, drowsiness), weakness, and vomiting are common.

> ## Signs and Symptoms of HACE
>
> 1. Ataxia.
> 2. Altered mental status.
> 3. Headache.
> 4. Lethargy.
> 5. Weakness.
> 6. Vomiting.

Ataxia

Ataxia is a loss of muscular control leading to difficulty in maintaining balance. A simple check for ataxia calls for the patient to stand straight up with his or her boots pressed together, with hands pressed into the sides of the thighs, and with eyes closed. If the patient wobbles or starts to fall and has to open his or her eyes to regain balance, the patient probably is ataxic. You may also ask the patient to walk a straight line touching the heal of the front boot to the toe of the back boot with each step. If the patient can't walk the line, the patient probably is ataxic. Because both acute mountain sickness and high-altitude cerebral edema are problems of the brain, and because balance is coordinated by the brain, *ataxia may be the single most useful sign indicating the patient is progressing from a mild or moderate altitude illness to a severe altitude illness.*

Treatment for AMS/HACE

Mild to moderate AMS does not cause physiological damage, but the signs and symptoms do indicate that the patient is not acclimatized to a specific altitude. Because you can't predict who will deteriorate from mild to severe altitude illness, you must stop the patient from ascending until the symptoms resolve. It is at this point the majority of life-threatening mistakes are made. The important rule is **don't ascend until the symptoms descend.** Fatal cases involving AMS or HACE almost universally occur among those who broke this rule.

Descent for AMS is mandatory if the symptoms do not resolve within 24 to 48 hours. Descent is critical if your assessment is HACE. A descent of as little as 1,500 feet can bring remarkably favorable results.

Ibuprofen may be given for headache. Light exercise increases respiratory drive and may relieve mild symptoms of AMS. Supplemental oxygen, however, is the treatment of choice for more serious problems. The combination of descent and oxygen should save just about everybody.

On the pharmacological scene, acetazolamide (Diamox) is a drug that speeds the resolution of mild to moderate AMS in about three out of four patients. The resolution may take up to 24 hours. The drug, available in the United States by prescription only, is also taken to prevent mild altitude illness, which it does effectively for most people. It increases the rate and depth of breathing, thus increasing arterial oxygen. Acetazolamide may also be used to aid sleeping, but drugs that suppress the respiratory drive should not be used. If you plan to carry and use acetazolamide, do so under the supervision of a medical adviser, and ask about side effects, such as numbness or tingling in the lips and fingertips. A patient treated successfully with drugs for mild to moderate AMS may continue to ascend, but the patient's symptoms should resolve before he or she continues the climb.

Patients treated with dexamethasone (Decadron) for more serious problems, tend to improve as fast as or faster than patients treated with acetazolamide. Dexamethasone, a steroid, may reduce pressure on the brain. The same recommendations for carrying and using a prescription drug apply here as they do for acetazolamide. The combination of acetazolamide and dexamethasone may be better yet but so far has not been fully tested; they are safe to use together but it is not certain that their synergy helps the patient improve faster.

Note: The use of medications, including supplemental oxygen, should never be used as a substitute for descent.

If descent is delayed, in addition to drug therapies, the patient may benefit greatly from being placed in a portable hyperbaric chamber, also known as a *Gamow Bag*. The device is an elongated bag made of sturdy nylon. The increased internal pressure, created by a simple foot pump, simulates a descent of several thousand feet, relative to the actual altitude. Although oxygen is not used to inflate the bag, the patient may benefit from supplemental oxygen while undergoing "descent" within the bag.

For any patient with ataxia and/or any other sign or symptom of HACE, immediate descent is mandatory. For any patient with a headache that grows worse despite rest and ibuprofen for pain, immediate descent is recommended.

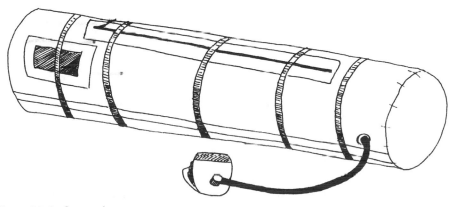

Figure 18-1: *Gamow bag.*

High-Altitude Pulmonary Edema

The most common severe form of altitude illness, the form most often causing death, is *high-altitude pulmonary edema* (HAPE), a problem of the lungs that typically shows up on the second night after reaching a specific altitude. (HAPE is uncommon beyond 4 days at a given altitude.) The pressure in the pulmonary (lung) arteries rises, and fluid, for reasons not totally understood, seeps out of the pulmonary capillaries and begins to fill the alveolar sacs. The patient begins to drown.

The earliest signs of HAPE are a decreased ability to exercise and a dry cough. A patient with HAPE develops increasing shortness of breath unrelieved by rest. In earlier stages of the illness, you may hear fluid in the lungs as "crackles" through a stethoscope. The sound appears most often in the right side of the chest beneath the armpit, but commonly on both sides, and without a stethoscope can sometimes be heard with an ear pressed against the chest wall of the patient. Late in the illness, breathing often is accompanied by gurgling sounds audible to the naked ear. As fluid continues to collect in the lungs, the patient develops a productive cough, eventually producing a frothy sputum that is pink with blood. The heart rate increases, and skin is typically cyanotic. Chest pain may be expected as a complaint. Fluid may build up to the point at which the patient suffocates to death. A low fever is common. The patient may have a decrease in level of consciousness. About half of all patients with HAPE also have signs and symptoms of AMS.

Treatment for HAPE

If your assessment is HAPE, immediate descent is critical. Descent should not overexert the patient because such exercise may increase pulmonary pressure and exacerbate the problem. In addition to descent, the patient will best be served by supplemental oxygen, which can reduce pulmonary arterial pressure 30 to 50 percent, enough to reverse the illness. It also immediately increases blood oxygen, protecting the brain.

When descent is delayed, a portable hyperbaric chamber (Gamow Bag) may also be of great benefit to a patient with HAPE. Supplemental oxygen may be used in concert with the Gamow Bag. Nifedipine (Procardia), a drug that reduces blood pressure in the pulmonary system—may be given according to prearranged instructions from a medical adviser, but should be considered only when descent is delayed.

Prevention of Altitude Illnesses

Most people adjust to altitude given enough time. Staged ascent is the key to acclimatization and, therefore, the key to preventing altitude illnesses. Above 10,000 feet, most people should gain no more than 1,000 to 1,500 feet of sleeping altitude per 24 hours. You can climb as high as you want during a day, but sleep as low as possible each night. As a group leader, pace your group to the speed of the slowest acclimatizers.

Adequate hydration is critical. Encourage all group members to drink enough water to keep urine output clear and copious.

A high-calorie diet is essential for the energy needed to ascend and acclimatize. A diet of 70 percent or greater carbohydrates, since they require less oxygen to metabolize, may aid in the prevention of altitude illness but probably will not serve much in terms of prevention under 16,000 to 17,000 feet.

Respiratory depressants, such as sleeping pills and alcohol, should be avoided, especially during the first 2 to 3 days at a specific altitude. Acetazolamide may be taken prophylactically, or dexamethasone for those intolerant of acetazolamide. Both are unarguably effective in preventing AMS, but *not* HACE or HAPE.

Training that is going to prepare you physiologically for altitude must be done at altitude, but physical fitness prior to ascent is a bonus in the game of safety and enjoyment. Fitness does not, however, protect against altitude illnesses. Fitter persons may actually be more susceptible, almost undoubtedly because they tend to go up too fast.

Evacuation Guidelines

Immediate descent should be initiated for any patient suffering altitude illness associated with ataxia, HAPE, and/or HACE. Further evacuation to definitive medical care depends on the level of the patient's recovery. Many patients who descend and recover from AMS and/or HAPE climb back up without complications—but it is not recommended.

Conclusion

Near the alpine meadow above Crested Butte, you ask Ashley to stand straight and tall, to press her palms into the sides of her thighs, to press her feet together, and to close her eyes. She easily accomplishes the task without the slightest loss of balance. Ashley has no difficulty breathing, and her breath sounds, to your naked ear, are normal.

Instead of gaining more altitude today, you decide to spend a second night at this camp, and use the day to teach skills scheduled for higher up. You encourage Ashley to hydrate, strongly suggesting she drink at least 4 liters of water during the day. You remind her how important it is for her to eat well, even though she may not feel hungry. You tell her she will feel better, and acclimatize faster, if she is active around camp today. You plan to monitor Ashley throughout the day, ready to descend if her condition deteriorates, but you are pretty confident this trip will continue.

Peter Hackett, MD, and Colin Grissom, MD, contributed their expertise to this chapter.

CHAPTER NINETEEN

Immersion and Submersion Incidents

You Should Be Able To:

1. Describe the safest approaches to an immersed or submerged person.
2. Describe the treatment for immersion hypothermia.
3. Demonstrate the treatment for a drowned person.
4. Define a near-drowned person and describe the treatment for a near-drowned person.

It Could Happen to You

Another oppressively hot, sultry, summer day, and you're enjoying the shade cast by an old oak along the shore of a quiet pond in southeastern Arkansas. Your cane pole extends out over the clear water, the bobber at the end of the line unmoving, the worm on the hook undoubtedly having gasped his last breath long ago.

Three boys, maybe high school age, maybe younger, have given up on the fish and now dive delightedly into the cool water. When you look after hearing a cry from across the water, you see two of the boys on shore, gesturing wildly.

About six body lengths from shore, turbulence in the still water catches your attention. A dash around the small pond, and you walk into the water, carrying a long stick you picked up from the edge of the forest. You'll reach with the stick, offering it to the boy if you find him capable of grasping it.

You find him, sure enough—just below the surface, his foot caught in a tangle of submerged roots. You detect no movement. Releasing the stick and diving in, you are able to free the boy and swim with him to shore, dragging him the last few yards of shallows. Face up on the shore, the boy remains unresponsive.

A patient whose nose remains above the surface of the water—a patient, in other words, who can breathe—is someone experiencing *immersion*. A patient whose nose has been below the surface of the water for a substantial amount of time—a patient who cannot breathe—is someone experiencing *submersion*. In recent years drownings, deaths following submersion incidents, have ended approximately 9,000 lives every 12 months in the United States, keeping it second or third as a cause of accidental death. Almost all the deaths had one thing in common: The victims never intended to be in the water. They planned to stay in the boat or on shore.

Several other factors related to deaths by drowning shed a bit of light on why some of the victims died.

1. Many of the dead were nonswimmers.
2. Most of the drowned victims were not wearing personal flotation devices (PFDs).
3. Some victims were whitewater paddlers not wearing helmets who hit their heads on the way down. Some of them *were* wearing helmets and hit their heads on the way down.
4. One study estimates that more than half of the dead had alcohol or some other mind-altering substance in their system.
5. *Immersion hypothermia,* the loss of core body temperature and/or the loss of coordination from being immersed in cold water, was a factor in a large number of drownings.
6. Males drown far more often than females, with males outnumbering females 12 to 1 in boating-related drownings.

Perhaps the most sobering estimate accompanying immersion and submersion incidents is this: Many of the dead or permanently debilitated—some experts guessing as many as six out of seven—could have been saved by immediate and proper actions by rescuers.

Few, if any, rescue scenes carry more risk for the rescuers than an immersion or submersion incident. The water threatening the life of your intended patient may suddenly threaten your life. The scene is not safe until all persons are safely out of the aquatic environment. An appropriate order of events determining how you attempt to rescue the patient from the water should be **reach, throw, row, tow, go.**

Reach first to a struggling immersed person from a secure position. If you can't reach him or her with your arm or leg, extend your reach with a stick, paddle, piece of clothing, or some other object. If your reach is not long enough, *throw* the person something that floats. He or she may be capable of swimming to shore with the aid of flotation. If throwing does not work, you may be able to *row* or paddle to the person in a stable water-craft, or you may be able to toss a line to the person and *tow* him or her to safety. To *go*, swimming to a drowning person, is to risk your life. Panic lends the struggling person great physical power and determination. Go only if you are well trained and capable.

Immersion

Sudden immersion in very cold water may precipitate an acute type of hypothermia, *immersion hypothermia*, that comes on rel-atively quickly, depending on the temperature of the water, the size of the patient, what the patient is wearing, and other factors. In general the patient should receive proper treatment for hypothermia (see Chapter 16: Cold-Induced Emergencies). In addi-tion to standard hypothermia treatment, the cold-water immersion patient should be handled gently, lifted from the water, and kept in a horizontal position. Even though he or she may be conscious and capable of walking, keep the patient flat to avoid stress on a pos-sibly cold heart. Dry the patient well, and insulate the patient from the environment before exposure to high external heat sources.

On rare occasions someone suddenly dies after being thrown into frigid water, a problem sometimes called *immersion syndrome*. The cause of death, not exactly understood, may be a reaction of the heart to rapid cooling, or the inhalation of icy water, or a combination.

Drowning

Those who die during a submersion incident typically go through a series of events that vary little from individual to individual. The person panics and struggles fiercely while holding his or her breath. The heart rate speeds up and the blood pressure rises. Involuntarily swallowing of water is common. Swallowed water may or may not cause vomiting. The drive to breathe becomes overpowering, and the person inhales water. Most people have an involuntary constriction of the muscles of the upper airway, a *laryngospasm*, which keeps the water out of the lungs. *Asphyxia*, an inadequate intake of oxygen, causes a loss of con-sciousness. Respiratory arrest and then cardiac arrest soon fol-low. At some point the laryngospasm relaxes, and water enters the lungs and begins crossing cell walls, entering the blood stream. Victims whose lungs are full of water are referred to as "wet" drownings, and they comprise the majority of the drowned. About 10 to 15 percent of drowning victims have a secondary laryngospasm, accounting for a "dry" drowning.

When the patient is pulled from the water, treatment must begin immediately. Artificial ventilation should start as soon as it is determined there is no breathing. Artificial breathing can even be done while the patient and rescuer are still in the water if the rescuer is stable, such as standing in shallows. Abdominal thrusts (the Heimlich maneuver) are not recommended as a means of clearing the airway prior to artificial breathing for drowned patients. Just start breathing for the patient (see Chapter 4: Airway and Breathing).

Once on a stable surface, such as land or a boat deck, determine if the patient's heart is beating. If it isn't, start chest compressions (see Chapter 5: Cardiopulmonary Resuscitation).

Cervical spine injury should be considered in some drowned patients, such as patients who have dived into shallow water, drowned while surfing, or drowned by flipping a kayak or canoe in whitewater. Use the jaw thrust to open the airway, and treat for spine damage if your CPR works (see Chapter 8: Spine Injuries).

Expect the patient to vomit. When vomit erupts, roll the patient immediately on his or her side, wipe the vomitus from the mouth, roll the patient back over, and continue CPR.

If you resuscitate the patient, treat for hypothermia. Hypothermia should be assumed in all immersion and submer-sion incidents, and the patient should be treated accordingly (see Chapter 16: Cold-Induced Emergencies).

The amount of time someone can remain under cold water and still recover is often surprising, especially if the patient is very young and the water is very cold. In reality these miracle saves are rare, but in July 1988, near Salt Lake City, Utah, a two-and-a-half-year-old girl was submerged for approximately 66 minutes in a frigid mountain creek. When she was found, CPR was started and continued en route to the hospital. With surgical intervention, she had a complete recovery. Treatment for nonbreathing patients should take precedence over treatment for hypothermia.

Survival Factors in Drownings

Length of submersion: The shorter, the better.
Temperature of the water: The colder, the better.
Contamination of the water: The less, the better.
Age: The younger, the better.
Struggle and trauma: The less, the better.

Near-Drowning

A *near-drowning* is a submersion incident in which the patient survives the underwater experience for at least 24 hours but does not necessarily go on to live a long, healthy life. Many near-drowning patients either swallowed and/or inhaled water. It takes very little water in the lungs to have profound effects on the patient. Vital body fluids are washed out, and unhealthy material in the water is absorbed into the body. Pneumonia, tears in weakened lung tissue, chemical imbalances in the blood, and other related problems may result in death days, weeks, or even months later. Some near-drowning patients suffer respiratory and/or cardiac arrest and are resuscitated. Near-drowners are probably hypothermic and should be treated accordingly. Anyone who has almost drowned should be hurried to a medical facility as soon as possible to be evaluated by a physician.

Prevention

Many drownings are preventable accidents. All you add to the water is a few drops of sound judgment. What you save the potential patient from, according to those who describe "near misses," is something mighty unpleasant.

1. Learn to swim.
2. Avoid swimming, and diving, in unsafe areas.
3. Never swim alone.
4. Wear a personal floatation device (PFD) while in a watercraft and, as a water-based trip leader, insist all party members also wear a PFD.
5. Do not ingest mind-altering substances prior to swimming or boating.
6. Cross wilderness rivers at safe fords with open run-outs, and (1) post someone downstream to aid anyone washed into the flow, (2) loosen pack straps and undo hip belts before crossing to make getting out of the pack easier for anyone who is washed into the flow, (3) wear boots or shoes for better balance, (4) use a long stick or pole as a "third leg" for balance, and (5) cross in linked groups.

Evacuation Guidelines

All patients should be evaluated by a physician following an accidental submersion incident when the patient was removed unconscious from the water, required resuscitation, exhibits difficulty breathing, complains of shortness of breath, has wet lung sounds, has a productive cough, complains of chest pain, and/or reveals a history of lung disease. A patient remaining unconscious after a submersion incident should be evacuated as soon as possible.

Conclusion

On the shore of the small pond in Arkansas, you look, listen, and feel for breathing in the young man. Finding none, you give 2 full breaths, mouth-to-mouth. A check for signs of circulation reveals evidence of a heartbeat. You return to artificial respirations, blowing in 1 breath after the patient's chest deflates following the preceding rescue breath. Within moments, it seems, the patient sucks in a gasping rush of air on his own and gags. You roll the young man onto his side just as vomit flies across the muddy shore.

Because he continues to breathe on his own, you maintain him in a stable side position. One of the other boys retrieves your day pack from near where your fishing pole still rests on the opposite bank. Your jacket, brought just in case, now covers the nearly drowned young man.

It's not far to your car, probably less than 200 yards. With the help of the other boys, you carry your patient to your car and drive him to the nearest hospital.

CHAPTER TWENTY

Lightning Injuries

You Should Be Able To:

1. Describe the mechanisms of lightning injuries, including direct strike, splash, contact, ground current, and blast effect.
2. Describe the most common lightning injuries.
3. Describe treatment for the most common lightning injuries.
4. Describe how to reduce the chance of lightning injury to a minimum.

It Could Happen to You

A July storm, rolling in an hour or so after midnight, brings heavy rain and booming claps of thunder to your campsite near Carter Notch in the White Mountains of New Hampshire. After a particularly close lightning strike, one that splits a tall tree in half, you make a quick tent-by-tent check on your young clients. As you move toward the last tent, a scream shifts you into high gear.

"Help! She's dead! She's dead!"

You crawl into the tent, to the side of a 13-year-old girl. The flash of the next strike paints an eerie mosaic of shadow and light on her immobile features. Her head and left shoulder still lie off her sleeping pad where they had been when the ground current swept under the tent.

Lightning occurs most often on hot days when warm, moist air rises rapidly to great heights, forming dark cumulonimbus clouds filled with static electricity. As a charge accumulates on the bottom of the cloud, an opposite charge develops on the top of the cloud and on the ground below the cloud. When the difference between charges reaches a potential greater than the ability of the air to insulate, lightning reaches out to equalize the difference.

Although it can run inside the cloud, cloud-to-cloud, and ground-to-cloud, the strike most likely to cause injury to humans runs cloud-to-ground. Most ground strikes occur immediately below a cloud. The strikes most often hit the nearest high point: sharp terrain features (such as mountaintops), tall trees, bushes in the desert, boats on water. Lightning also tends to strike long conductors such as metal fences, bridges, power lines, even wet ropes. On rare occasions, however, bolts of lightning have reached out as far as 10 miles ahead of a storm, and strikes a mile or so ahead of a storm are common. The bolt of electricity is a direct current that may reach 200 million volts and 300,000 amps, with a temperature of 14,432° F (8,000° C), and it may move through a channel only 8 centimeters wide.

Lightning flashes out approximately eight million times per day worldwide, or 100 times per second. Lightning kills more people in the United States every year than almost all other natural

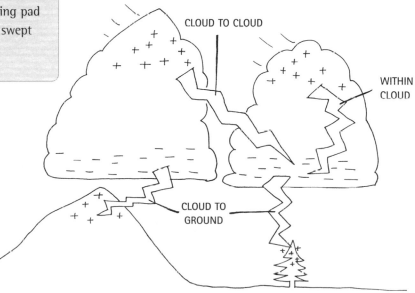

Figure 20-1: *Lightning strikes.*

disasters combined, placing second only to flash floods. Injuries usually occur from May to September, and those who die usually are working or playing outdoors.

Mechanisms of Injury

Injuries directly caused by a lightning strike may be classified in one of five mechanisms of injury:

Direct strike: As the name implies, the bolt of lightning hits someone directly. Most at risk is a person out in the open—hikers crossing an alpine meadow or anglers on the shore of an exposed lake. Unable to find shelter, the person may become the tallest "object" around. If a metal object—say, an ice ax—is carried higher than the head, the chance of attracting a direct strike is even greater—not because metal attracts lightning, but because of the increased height. Direct strikes carry a large potential for death.

Splash or side flash: The lightning strikes something more immediately appealing than the human, such as a tree or shelter, but "splashes" through the air to hit someone whose body offers less resistance to the electrical charge than the object receiving the direct hit. Side flashes may occur from person to person.

Contact: Lightning, either by direct strike or splash, hits something with which someone is in direct contact.

Ground current: The electrical charge radiates out from the strike point either along the surface of the ground, similar to the way waves spread in circles from a rock thrown in a lake, or along conducting routes. Conducting routes include long conductors, root systems, and drainages filled with running water. A person on the ground in line with the current may attract enough electricity to cause injury. Someone standing or walking with feet spread creates a *stride potential,* encouraging the charge to enter one leg and exit the other.

Blast effect: Although not an injury resulting directly from contact with electricity, the blast of the shock wave of air that explodes out from a lightning strike can throw someone with enough force to cause traumatic injury.

Types of Injuries

Lightning injuries are virtually unpredictable; the ultimate effect lightning has on a human body varies from minor to major, including death.

Cardiac arrest: The current of lightning upsets and sometimes stops the natural rhythm of the heart. If the heart is healthy, it often restarts on its own, but it may fail to start up again because it is too damaged, or it may restart and stop again if the heart suffers prolonged oxygen deprivation, that is, if the patient fails to start breathing again.

Respiratory arrest: The area of the brain that controls respiratory drive and the muscles used for breathing can be shut down by the charge of electricity. This arrest of breathing may be prolonged, leading to a second cardiac arrest when cardiac activity spontaneously returns after an initial cardiac arrest.

Other neurological injuries: The patient is most often knocked unconscious by the charge, and some patients will suffer temporary paralysis, especially in the lower extremities. Seizures and/or the inability to remember what happened may result. Post-lightning strike patients complain most often of ringing in the ears or loss of hearing that usually resolves in hours to days without permanent damage. Rupture of an eardrum, however, is common, and deafness is not impossible. Ears may bleed. Temporary loss of sight is not unusual, but permanent blindness is rare. Patients are often bothered by insignificant nausea and vomiting for a brief time.

Burns: It is rare to have serious skin and muscle burns after a lightning strike, but superficial burns—linear, feathery, and fern-like—are common. If the burn is deep and penetrating, the patient usually has much more to worry about than the burn. The treatment for lightning burns is the same as any burn treatment (see Chapter 15: Wilderness Wound Management).

Blast injuries: The impact of the blast effect can cause just about any trauma you can imagine: fractures, head injury, spinal injury, dislocations, or chest and abdominal injuries.

Management of the Patient

One factor in the management of patients struck by lightning is unique—the way *triage* is performed. Normally in the triage of multiple patients—the sorting of patients when a group has been injured—rescuers are taught to give priority to those still alive, and let the dead stay dead. It is the simple guideline of "the greatest good for the greatest number." After a lightning strike, however, the still, silent, dead patients are often recoverable and need to receive immediate attention. The moaning wounded can wait. If a patient does not suffer cardiopulmonary arrest, he or she is likely to survive. Cardiopulmonary resuscitation, done appropriately, may bring pulseless, breathless, post-lightning strike patients back to life (see Chapter 5: Cardiopulmonary Resuscitation). Even if the patient has a pulse, rescue breathing is often required, sometimes for a prolonged period, to save the patient (see Chapter 4: Airway and Breathing).

Aside from triage the basic management principles apply. When the scene is safe, perform an initial assessment followed by a focused assessment. Treat injuries as necessary.

Prevention of Lightning Injuries

Although wilderness lightning safety data remains relatively scarce, these suggestions provide what experts agree are the safest steps to take:

1. Know weather predictions and weather patterns for your intended region of wilderness travel. If storms are predicted and/or if storms typically strike, for instance, in mid-afternoon, avoid lightning-prone areas when storms are most likely to occur.

2. Time your activities. When you hear thunder, the storm lies 1 to 10 miles away, depending on the turbulence of the air. Calm air carries sound to you over the longest distance. Start looking for safe terrain as soon as you hear thunder.

3. Know where to find safe terrain. Peaks, ridges, and significantly higher ground are inherently more dangerous, but gently rolling hills are not attractive to lightning. A dry ravine or low spot in rolling hills provides excellent safety. Wide open spaces are dangerous because you might be the highest point. Standing in water is dangerous but naturally damp ground is not more dangerous than dry ground. Wet ground actually dissipates ground current faster than dry ground. Dry snow provides insulation, but wet snow is a conductor. Avoid cave entrances and shallow overhangs that may allow a charge to arc across the gap between the top and bottom. Deep in dry caves is safe. Avoid contact with long conductors. Trees attract lightning, especially lone trees. A forest of trees of relatively uniform size provides relative safety, but do not make contact with trees during the storm.

4. Paddlers, including sea kayakers, rafters, and other small craft operators should get off the water when they hear thunder. Then they should move well away from shorelines—at least 200 feet away—whenever possible. You may, however, find yourself in a narrow canyon without the possibility of moving well away from shore. If safe moorage is possible, it is still recommended to seek a shoreline because thunderstorms are typically accompanied by high and potentially dangerous winds. If safe moorage is not possible, the fact that the bottoms of canyons are seldom struck by lightning is in your favor.

5. Choose tent sites with the same care you give to choosing safe terrain during a storm. In a tent stay completely on your sleeping pad and out of contact with the sides of the tent during a storm.

6. Lightning, ground current especially, can injure entire "huddled" groups. Spread groups out, at least 50 feet between individuals when possible, to minimize the number who may be harmed by a strike and to maximize the number who may be able to respond to the injured.

7. During a storm, sit on your sleeping pad or some other non-conductive object and make yourself as small as possible. Huddled in a ball, keeping your feet close together, gives the least potential differences in the separate points of your body. No one has ever been seriously injured by lightning in this "lightning position."

Figure 20-2:
Lightning position.

Lightning Myths and Reality

1. Lightning *will* strike in the same place twice, and many more times than twice.
2. Lightning is *not* stored by a human body.
3. Lightning may strike *before, during,* and *after* a storm.
4. Lightning *can* cause serious internal injury while leaving no external signs of damage.
5. Lightning is *not* attracted to metal, but metal does conduct the charge extremely well.
6. Lightning *does* strike vehicles, and rubber tires on a car or truck do *nothing* to protect you. Because electricity stays on the outside of metal, taking the path of least resistance, being inside a car or truck with the windows rolled up *does* protect you.
7. Lightning *can* cause serious injury that doesn't show up until long after the strike.

Evacuation Guidelines

Evacuate survivors of lightning strikes even if no loss of consciousness occurred. Problems, especially problems involving the neurological system (mental and motor functions), can show up days later.

Conclusion

Near New Hampshire's Carter Notch, you kneel beside the 13-year-old girl, assessing for breathing, and finding none. You give her 2 full breaths. A check for signs of circulation reveals nothing. You initiate chest compressions.

You have no sense of time, but it seems after only minutes of ventilations and compressions, she regains a pulse but not spontaneous respirations. You continue rescue breathing for several minutes before she gasps out a ragged breath on her own. For the next hour, you are by her side, periodically giving artificial ventilations to supplement her weak breathing. Finally she starts to shift restlessly, blink her eyes, and moan.

The storm beats fiercely at the camp for an hour more. Once you assess the environment as safe, your associate group leader, who knows the area well, hikes out for help with two stronger members of the party. You wait in trepidation, glued to your patient's side. Storm clouds make a helicopter evacuation impossible, but a rescue team hikes in.

At approximately midday, the girl is packed on a litter and taken on her way to the nearest road. Although the strength of her breathing and heartbeat diminish during the morning and she lapses back into a coma, she remains alive, and fully recovers after a stay in the local hospital.

CHAPTER TWENTY-ONE

North American Bites and Stings

You Should Be Able To:

1. Describe the signs and symptoms of the most dangerous bites and stings common to North America.
2. Describe the treatment of the most dangerous bites and stings common to North America.
3. Describe the prevention of the most dangerous bites and stings common to North America.

It Could Happen to You

Cries from her tent bring you in a rush from where you're preparing dinner in the Sipsey Wilderness of northwestern Alabama. You find a member of your party, a young woman named Bethany, in extreme discomfort, curled into a fetal position, complaining of "the worst cramps y'all can imagine" in her abdomen and lower back. Sweat drenches her, and her skin appears flushed and feels warm. You suspect a fever. You open your mouth to ask what's wrong, but, before you can speak, she opens her mouth to splatter the tent wall with vomit. Her pulse races at 110 and seems weak to you.

Thinking appendicitis or a rare and fatal gastroenteritis, you quickly organize the group and abandon the campsite in favor of a mercifully short carry to the car and a rapid drive to the hospital.

The bites and stings humans receive from other animals vary greatly in severity depending on the species of animal, the reason for the attack, and the way the human acts during and after the confrontation. Despite the variations, similarities do exist in the damage done by bites and stings, and general principles have been established for the management of the wounds received from wild creatures.

In order of significance, treatment for bites and stings should include (1) making sure the scene is safe; (2) immediately stopping serious blood loss, when appropriate; (3) immediately cleaning the wound against the introduction of microorganisms; and (4) managing the patient and the wound to reduce the effects of envenomization, if the animal was poisonous.

Reptiles

Although venomous snakes may inflict a large number of bites on humans in the United States every year—perhaps more than 9,000 bites—deaths are unusual. In the period from 1983 to 1998, only 10 deaths by snakebite were reported to the Poison Control Centers of the United States. The reasons for the low mortality rate are more than likely to be the availability of effective antivenin and the relatively low toxicity of North American snakes. Those who die are usually very young or very old. Arizona, by the way, is the most likely place to die of snakebite, with Florida, Georgia, Texas, and Alabama filling out the top five.

Ninety-nine out of every 100 poisonous snakebites are received from *pit vipers:* rattlesnakes, copperheads, and water moccasins. The few remaining bites come from representatives of North America's only other venomous family, the *coral snakes,* or, occasionally, from exotic species kept as pets. Bites from lizards are unusual, and deaths are exceedingly rare.

Pit Vipers

They don't all have rattles, but all pit vipers do have distinctly triangular heads, catlike pupils, heat-sensitive pits between eyes and nostrils, and danger squirting from two very special teeth, hinged to swing downward at a 90-degree angle from the upper jaw. The jaw opens alarmingly wide, allowing the venom to be ejected down canals within the fangs and into the tissue of a prey or enemy. Venoms are not created equal. For instance, the poison of the Mojave rattlesnake is approximately forty-four times more potent than the Southern copperhead's.

Pit viper venom is generally yellowish, odorless, and slightly sweet to human taste buds. It evolved from the snake's saliva, which makes sense when you remember the purpose is to acquire and utilize food for the snake, not to deal death and destruction to large, inedible mammals who stumble over the low-lying reptiles. The venom attacks the nervous and circulatory systems of the snake's prey. In a mouse that can spell death in a few minutes.

How dangerous is this venom to a human? That depends on (1) the age, size, health, and emotional stability of the patient; (2) whether or not the patient is allergic to the venom; (3) where the patient was bitten (near vital organs being the most dangerous); (4) how deep the fangs go; (5) how upset the snake is; (6) the species and size of the snake; (7) the amount of venom injected; and (8) the first aid given to the patient.

The snake can control the amount of venom injected, and one pit viper bite in four carries no venom. The other three may vary from insignificant to mild to moderate to severe envenomation. Mild envenomations hurt, swell, and turn black and blue within 30 minutes. Moderate envenomations add swelling that moves *both* distal to the bite *and* up the arm or leg toward the heart. Numbness and swollen lymph nodes may follow. Substantial bleb (blister) formation is common, as are weakness, nausea, and perhaps vomiting. Sometimes patients report a "rubbery" taste in their mouths. A severe envenomation might add big jumps in pulse rates and breathing rates, profound swelling, blurred vision, headache, light-headedness, sweating, and chills. Death is possible, most often from respiratory or circulatory collapse. More importantly in most bitten humans, the venom starts to destroy tissue in the vicinity of the bite, with signs—the blebs—appearing in as little as 6 hours. Fingers or toes, hands or feet, even arms or legs could be lost to the "digestive" action of the venom. Antivenin and/or appropriate first-aid measures are very important.

■ TREATMENT FOR PIT VIPER BITES

1. Calm and reassure the patient. Agitation in the patient, both physical and emotional, can make an envenomation worse.
2. Keep the patient physically at rest, with the bitten extremity immobilized and kept at approximately the same level as the heart.
3. Remove rings, watches, or anything else that might reduce the circulation if swelling occurs.
4. Wash the wound.
5. Measure the circumference of the extremity at the site of the bite and at a couple of sites between the bite and the heart, and monitor swelling.

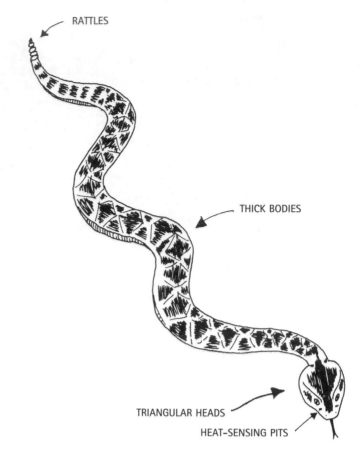

RATTLES

THICK BODIES

TRIANGULAR HEADS

HEAT-SENSING PITS

Figure 21-1: *Rattlesnake.*

6. Evacuate the patient by carrying, or going for help to carry, or, if the patient is stable, by slow walking with frequent rest breaks.
7. If the patient is kept still, keep her or him warm.
8. Keep the patient well hydrated with clear fluids unless he or she develops pronounced vomiting.
9. Do *not* cut and suck. Mechanical suction (not oral suction) may be valuable. The sooner mechanical suction starts, the better. Suction should be applied, for best results, for 30 minutes via a negative pressure venom extraction device (a Sawyer Extractor) and without cutting. To keep up the suction, you have to monitor the device and reapply it when the suction fails, which it will keep doing.
10. Do *not* give painkillers unless the patient is very stable, showing no signs of getting worse.
11. Do *not* apply ice or immerse the wound in cold water.
12. Do *not* apply a tourniquet.
13. Do *not* give the patient alcohol to drink.
14. Do *not* electrically shock the patient.
15. Attempt to identify the biting creature, but not if it puts you or anyone else at risk. If the species is known—and often even if it isn't—there is a high probability that antivenin is available if needed.

Coral Snakes

Brightly banded in red, yellow, and black, all coral snakes of the United States are described by a particular color sequence: "Red on black, venom lack; red on yellow, kill a fellow." In other words, red bands bordered by yellow bands equals dangerous. Short, fixed fangs in the front of a small mouth make it all but impossible for coral snakes to bite anything on humans other than a finger, a toe, or a fold in skin, and they often have to hang on and chew before they can inject venom to do damage. Immediate first aid, therefore, should include snatching off the chewing snake. Envenomation ranges from mild, with localized swelling, nausea, and vomiting, to severe, with dizziness, weakness, and respiratory difficulty. Since it takes up to 12 hours for signs and symptoms to reach the point where the patient wants help, early evacuation of known bitten persons is strongly advised. Walking out is OK. Although death by coral snake is rare, the venom is more potent than almost all pit viper venom. Antivenin should be started as soon as possible to be most effective. Little additional first aid works other than keeping the patient calm and cleaning the wound.

Pressure-Immobilization

Developed in Australia where numerous relatives of the coral snake live and bite, pressure-immobilization is a technique that involves securing elastic wraps (such as those used to secure sprained ankles) around the bite site, up the extremity, and back down the extremity, and followed with some type of splint. Properly applied—with about the same pressure as wrapping a sprained ankle—pressure-immobilization prevents the spread of some types of venom until a hospital can be found. It has proven very effective as a first aid for snakebites in Australia but remains untested on North American coral snake bites.

Lizards

Only two lizards in the world are considered venomous enough to end the life of a human. Unfortunately, meetings with both are possible in the southwestern United States and in Mexico. The Gila monster and the Mexican beaded lizard bite when they are picked up or stepped on. They have powerful jaws to compensate for primitive teeth and no means to inject the poison. They lock on while their venom drools into the wound. You may likely be required to heat the underside of their jaws with matches or a lighter to break the grip, or pry the mouth open with something rigid. There may be a great deal of local swelling and pain, but deaths in humans are rare. First aid is simply removing the attached lizard, cleaning the wound, and evacuating the patient for definitive assessment.

Prevention of Reptile Bites

1. Do not try to pick up or capture snakes or lizards.
2. Check places you intend to put your hands and feet before exposing your body part to a bite.
3. Gather firewood before dark, or do it carefully while using a flashlight.
4. In snake country, keep your tent zipped up.
5. Wear high, thick boots while traveling in snake country.
6. When passing a snake, stay out of striking range, which is about one half the snake's length.
7. If you hear the "buzz" of a rattler, freeze, find it with your eyes without moving your head, wait for it to relax the strike position, and back away slowly.

Spiders

Few people are gladdened to learn that almost all spiders, world-wide, carry venom that can be injected through nasty fangs. On the positive side, only a few dozen species on this planet have a bite harmful to humans because the spider injects too little venom and/or too impotent a venom or the spider's fangs cannot penetrate human skin.

Widows

One of the most venomous of spiders, the *widow* (*Latrodectus*), at least four species of which are common in the United States, bears the tag *cosmopolitan*, a spider found around the globe. Only the female, up to 1 inch in average length (2 to 2.5 centimeters) for the black widow, poses a threat to humans, and she packs more danger in every drop of venom than any other creature in North America. The shiny female black widow bears a typically red—but not always red—"hourglass" shape on her abdomen to help identify her. She has been found in every state but Alaska, secreting her tattered web under logs and large pieces of bark, in stone crevices, in trash heaps and outbuildings, or deep in clumps of heavy vegetation. Rarely aggressive, she may be touchy during springtime mating and egg-tending days.

Her drop of poison is tiny, a huge boon to bitten humans. Patients almost never feel the bite, although some have reported immediate sharp pain. There may be little or no redness and swelling at the site initially, but a small, red, slightly hard bump may form later. The bump may itch. Within 10 to 60 minutes symptoms usually begin to occur. Pain and anxiety become intense. Severe muscular cramping often centers in the abdomen and back. Burning or numbness characteristically disturbs the patient's feet. Headaches, nausea, vomiting, dizziness, and heavy sweating are all common reactions.

Even though patients claim it feels as though death is imminent, widows kill very few humans. In the 1960s, for instance, there were only four confirmed human deaths from black widow

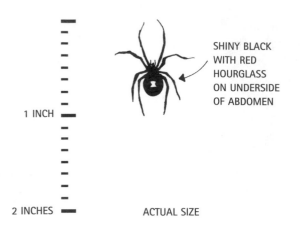

SHINY BLACK
WITH RED
HOURGLASS
ON UNDERSIDE
OF ABDOMEN

1 INCH

2 INCHES   ACTUAL SIZE

Figure 21-2: *Black widow.*

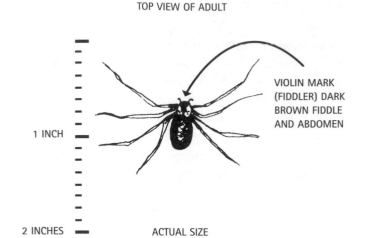

VIOLIN MARK
(FIDDLER) DARK
BROWN FIDDLE
AND ABDOMEN

1 INCH

2 INCHES   ACTUAL SIZE

Figure 21-3: *Brown recluse.*

bites in the United States. The dead are almost always the very young, the very old, or the very allergic.

You need to keep the patient as calm as possible, a course of action that "applies" perhaps your best first-aid treatment. If you can find the bite site—which may show up as a faint, red mark—wash it and apply an antiseptic such as povidone-iodine. Cooling the injury, with ice if possible or with water or wet compresses, will reduce the pain. Cold also reduces circulation, which slows down the spread of the venom. Medications for pain, if available, would be appropriate.

Evacuation to a medical facility is strongly advised, especially if you are unsure what is causing the symptoms, just in case complications arise. Most spider bite patients receive painkillers and 8 to 12 hours of observation at the hospital. Youngsters, older people, and the very sick may be admitted for longer. Antivenin is available if needed.

Recluses

The most common serious spider bite in the United States comes not from the venom of the black widow but from the solitude-seeking *recluse* (also known as the fiddleback or violin spider). Generally pale brown to reddish, with long, slender legs averaging about 1 inch (2 to 3 centimeters) in length, recluse spiders most often have the shape of a violin on the top front portion of their body. The head of this "fiddle" points toward the tail of the spider. Unlike the black widow, both sexes of recluses are dangerous.

The recluse (*Loxosceles*) prefers the dark and dry places of the South and southern Midwest but travels comfortably in the freight of trucks and trains and probably can be found in all 50 states. They don't shun the near company of humans, and set up housekeeping underneath furniture, within hanging curtains, and in the shadowed corners of closets. In the wild lands they hide in the daylight hours beneath rocks, dead logs, and pieces of bark in forests. They attack more readily in the warmer

months, usually at night, and only when disturbed. Curious children are their most frequent victims.

As with many spiders, the bite of the recluse is often painless. Having relatively dull fangs, the serious wounds they inflict are usually on tender areas of the human anatomy. Within 1 to 5 hours, a painful red blister appears where the fangs did their damage. Watch for the development of a bluish circle around the blister, and a red, irritated circle beyond that—the characteristic "bull's-eye" lesion of the recluse. The patient may suffer chills, fever, a generalized weakness, and a diffuse rash.

Sometimes the lesion resolves harmlessly over the next 1 to 2 weeks. Sometimes it spreads irregularly, as an enzyme in the spider's venom destroys the cells of the patient's skin and subcutaneous fat. This ulcerous tissue heals slowly and leaves a lasting scar. In a few children, death has occurred from severe complications in the circulatory system.

As with many spiders, without an eight-legged corpse as evidence, it is difficult to be sure what is causing the patient's problem. Initially, there is little to be done other than calming the patient, applying cold to the site of the bite for reduction of pain, and keeping the wound as clean as possible. Medications for pain, if available, would be appropriate. Any "volcanic" skin ulcers should be seen by a physician as soon as possible. Antibiotic therapy may be necessary to bring about healing of the wound.

Hobos

The *hobo spider* (*Tegenaria*) is an import from Europe that has spread at least across the Northwest. All hobo spiders are brown with gray markings and eight conspicuously hairy legs. The leg span reaches from 0.5 to 1.5 inches (1.25 to 3.75 centimeters). A herring-bone stripe pattern in brown, gray, and tan often appears on the abdomen. Hobo spiders have been mistaken for recluses, but hoboes lack the violin shape.

The bite of a hobo spider typically produces a blister that ulcerates and takes several months to heal. Approximately 50 percent of patients complain of headaches, muscle weakness, visual disturbances, and/or disorientation. Because the signs and symptoms are similar, hobo bites are often blamed on the recluse spider.

Bites from hobo spiders are rare. They tend to avoid large cities and congregate in small towns and rural communities. They like it under houses and deep in woodpiles and clumps of debris. Indoors they may lurk any place that is not regularly cleaned. You probably will never find one out in the far, untrammeled places. They don't bite unless trapped against the skin of an unsuspecting human with no way to escape.

First aid for hobo bites is the same as for recluse bites.

Tarantulas

Despite a fierce appearance, North American tarantulas are relatively harmless. Pain, seldom more than moderate, typically follows the bite. Later signs and symptoms are rare. Washing the bite site is encouraged. Cold and/or medications for pain are appropriate treatments.

Prevention of Spider Bites

1. Do not try to pick up or capture spiders.
2. Check places you intend to put your hands and feet before exposing your body part to a bite.
3. Gather firewood before dark, or do it carefully while using a flashlight.
4. In spider country, keep your tent zipped up.
5. If you must move around in the dark, wear boots or camp shoes and use a flashlight.
6. Take a look in your boots before you put them on.

Scorpions

As most spiders also do, scorpions hide by day and hunt at night. Scorpions all sting with the tip of their "tail"—the hindmost segments of their abdomen. From small species that reach maturity at 0.75 inch (about 2 centimeters) to humongous 9-inchers (23 centimeters), scorpions have crablike pincers used only to hold and tear apart their prey. Insects are their primary source of food.

Of interest to note, recent studies indicate many scorpions actually have two venoms. The first, a "pre-venom," is usually clear and injected when the scorpion is irritated or threatened. Not intended to kill prey, it causes pain and gives a warning: Stay away. The second, more milklike, is intended to kill, and carries more danger, of course, to the stung.

Most victims report no more pain than that inflicted by an irritated honeybee. An attack of the species *Centruroides* (Arizona bark scorpion) may be different. In North America only

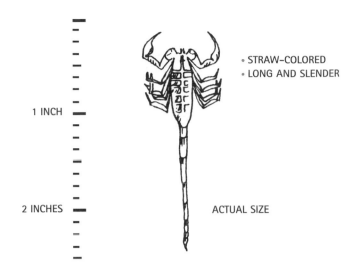

SCORPION-CENTRUROIDES

1 INCH

2 INCHES

- STRAW-COLORED
- LONG AND SLENDER

ACTUAL SIZE

Figure 21-4: *Scorpion.*

the *Centruroides* is a known killer of humans. They are usually old-straw-yellow or yellow with dark longitudinal stripes, and reach from about 1 to 3 inches (2 to 7.5 centimeters) in length. Their pincers are long and slender as opposed to bulky and lobsterlike. The sting, immediately and exquisitely painful, is increased by a light tap on the site. Deaths have almost exclusively been in small children, the elderly, and the severely allergic. This scorpion is found only in Mexico and the extreme southwestern United States.

First aid for any scorpion sting should involve cooling the wound, which allows the body to more easily break down the molecular structure of the venom. Cooling also reduces pain. Use ice or cool running water if available. On a warm night, a wet compress will help. Keep the patient calm and still. Panic and activity speed up the venom's spread. If the scorpion was *Centruroides*, post-sting manifestations may include heavy sweating, difficulty swallowing, blurred vision, incontinence (loss of bowel control), jerky muscular reflexes, and respiratory distress. These serious signs are cause for quick evacuation to a medical facility. Antivenins are available in many areas where dangerous scorpions live.

In the Southwest, most stings occur May through August. The victims are usually putting on clothing, walking barefoot in the dark, or picking up objects (like firewood) off the ground. The simple rules of spider avoidance should be followed to avoid scorpions.

Zoonoses

Zoonoses are animal-borne diseases that are communicable to humans under natural conditions. They include diseases borne by ticks and mosquitoes as well as those borne by mammals.

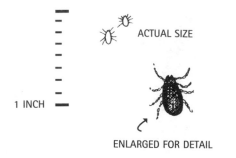

Figure 21-5: *Tick.*

Ticks

A relative of the spider, the tick crawls around on its unsuspecting host on eight tiny legs, looking for the right spot to settle down for a few days. It may search for hours. With specialized pincer-like organs, it digs a small wound in his host. Into the wound goes a feeding apparatus called a *hypostome;* its relatively powerful sucking mechanism allows the tick to feed on the blood of the host. Anchored firmly in the wound, it feeds for an average of 2 to 5 days, sometimes longer, depending on the species, and drops off weighing hundreds of times more than when it first arrived. In the host it often leaves a reminder of its visit—disease-causing microorganisms. Worldwide, only the mosquito spreads more illness to humans than the tick.

■ LYME DISEASE

First diagnosed in Lyme, Connecticut, the corkscrew-shaped bacteria *Borrelia burgdorferi,* a spirochete carried by some ticks, has spread to almost all states, still in heaviest concentrations in the Northeast and upper Midwest and along the coast of northern California. The tick must be attached for approximately 48 hours to spread Lyme disease.

The warning signs of Lyme disease include the following:

1. Two days to 5 weeks after the bite, a well-defined rash appears—a ragged bull's-eye, red surrounded by lighter shades—in 60 to 80 percent of patients. The rash appears anywhere it pleases, unrelated to the bite site, in several places at once, disappearing to reappear in different spots.
2. Flu-like symptoms, with a fever, headache, fatigue, and a stiff neck, occur before or after the rash. Rash and illness disappear in a few days to weeks.
3. Extreme, chronic fatigue with irregular heartbeat and partial numbness or paralysis can occur weeks to months later in a few sufferers. This stage can be very serious.
4. Swelling and pain in joints, especially the knees, may start up to 2 years later. Any joint can be affected, and the arthritis

may move from joint to joint with periods of remission. A blood test will show if you have the disease, but only in later stages.

Lyme disease is not fatal, although it is possible to die from cardiac complications associated with the disease. Without antibiotic treatment, however, it can lead to lifelong arthritic problems.

■ ROCKY MOUNTAIN SPOTTED FEVER

Montana's Bitterroot Mountains were the first site of the identification of Rocky Mountain spotted fever, caused by the parasite *Rickettsia rickettsii,* but it spread from coast to coast. After feeding for approximately 3 hours, infected ticks may pass the disease. Three to 12 days later, the patient develops a spotty rash, usually beginning on the hands, feet, wrists, and ankles. The spots migrate over the arms, legs, face, and abdomen. Severe headaches are common, with stiff neck and back and general muscle aches. The characteristic fever rises during the first days, and remains high. If untreated, approximately 20 percent of the victims die. Almost everyone recovers with antibiotic treatment, however.

■ COLORADO TICK FEVER

All the Rocky Mountain states have recorded patients with Colorado tick fever, as well as western Canada and the Dakotas. A virus produces the sudden fever with muscle aches and headache that develops 3 to 6 days after the bite of the tick. Diarrhea, vomiting, and stomachaches are common signs and symptoms. The patient often recovers and relapses several times in the course of the illness. It is very rarely serious, although sufferers report feeling so bad they wish they could die. There is no specific treatment, but patients should receive supportive care.

■ TULAREMIA

Another bacteria, often borne by ticks of the South and Southwest, causes the high fever and flu-like symptoms of tularemia. A decaying wound at the site of the bite is common. Antibiotics defeat the bacteria.

■ TICK PARALYSIS

At least forty-three species of ticks worldwide have been known to cause tick paralysis, with more cases showing up in North America than anywhere else. A venom in tick saliva causes the problem, which appears to be a block to nerve messages; children are affected more often than adults. The patient may first be restless and irritable, with complaints of numbness or tingling in hands and feet. Ascending paralysis develops over the next 24 to 48 hours. Once the tick is removed, the patient recovers, almost always without complications. Any patient with an ascending paralysis and without a known cause should be inspected carefully for an embedded tick.

OTHER TICK-BORNE DISEASES

Several other tick-borne diseases have been reported in the United States. Any patient from whom an embedded tick has been removed who develops an otherwise unexplained illness should be treated by a physician as soon as possible.

TICK REMOVAL

Quick tick removal is necessary to reduce the chance of disease transmission. Forget those ineffective and possibly dangerous ways of smearing them with petroleum jelly, nail polish, or gasoline, or burning them out with a match. Using bare fingers to pull them out works, but it's not the best way—you can crush the tick, propelling the tick's juices into the patient. A pair of tweezers should be in your first-aid kit. If the tweezers are fine-pointed, all the better. Grasp the tick near the skin, parallel to the longitudinal axis of its body, and pull out gently. No twisting, yanking, or squeezing. After the tick is out, scrub the area gently with alcohol, an antibiotic ointment, or soap and water. What if your removal technique leaves tick mouthparts in the skin? If your pull is straight and slow, the mouthparts very rarely detach from the tick. Even if they do, the chance of getting a disease from just the mouthparts may be nonexistent. Save the tick, preferably without touching it with your hand, if you would like to have it checked for disease. Make note of the time and your location to keep with the removed tick.

PREVENTION OF TICK BITES

1. Tick checks, several times a day if you're active outdoors in tick season, are extremely important. Obviously, considering a tick's inclination to tuck itself into hard-to-see areas, a partner will add much to an adequate check. Free-roaming ticks on your skin can be easily and safely lifted off. Embedded ticks should be immediately removed.

2. If you wear light-colored clothing and tuck your pants into your socks, you stand a great chance of recognizing and removing the dark-colored tick before it reaches your skin.

3. Repellents have proven effective in keeping ticks off. Tested and proven is DEET, the active ingredient in most repellents. Also proven is permethrin, a spray repellent and insecticide that is applied to clothing. A third repellent shown to be effective is lemon eucalyptus oil, available commercially in products, although it has to be applied more often that DEET or permethrin.

Mosquitoes

Experts estimate that as many as one out of every 17 humans currently alive will die from the bite of a mosquito. The reason: Mosquitoes aggressively pass devastating diseases including malaria, dengue fever, yellow fever, and West Nile virus. As of this date, only West Nile virus poses a threat in the United States,

and that threat so far is minimal. In addition, mosquitoes have done as much or more to ruin a wilderness experience than anything else known to man or woman.

You can apply an ice pack to minimize the swelling and itching in the first few minutes after a bite, or use a topical itch-reducing product. Use of an antihistamine, such as diphenhydramine at 25 to 50 milligrams every 6 hours, is recommended for more intense reactions.

Prevention of mosquito bites (and the bites of other small insects) is most easily handled with a repellent. The most effective repellent is DEET, and a concentration of 30 percent is all that is needed. Higher concentrations last longer but do not increase repellency. DEET should be washed off as soon as possible after exposure to insects has passed. DEET may be applied to clothing, such as collars, cuffs, and hats, to repel insects. Permethrin, originally extracted from chrysanthemum flowers, is a potent insect neurotoxin currently synthesized for human use as an insect repellent that is applied to clothing. It's not really a repellent—it's an insect killer. Within minutes after contact with permethrin-treated clothing, the insect dies. Permethrin bonds strongly to the fibers of clothing and, depending on the concentration and application process, can withstand numerous washings, remaining active, in some cases, for years. Permethrin is colorless and odorless and does no harm to vinyl, plastic, or other fabrics. It can be applied to mosquito netting on tents, to sleeping bags, even to window screens at home. It should not be applied to human skin. After many tests, experts agree it apparently does no harm to humans, but it quickly loses its efficacy when applied to skin. A third repellent, lemon eucalyptus oil, keeps insects off humans as effectively as low concentrations of DEET, according to studies.

Loose clothing helps prevent bites from bugs that can reach their biting apparatus through tight clothing. Smoke keeps some insects away. In desperation, mud smeared on exposed skin prevents insects from biting.

Some insects, especially mosquitoes, feed primarily at dawn and dusk and sometimes throughout the night. Use a tent with adequate insect netting to sleep safe from bites. Set camp in a high and dry place where winds may keep some insects away. Avoid low and wet areas where mosquitoes breed.

WEST NILE VIRUS

The first case of West Nile virus was identified in Uganda in 1937, on the banks of the West Nile River. But no known case appeared in the United States until 1999. Since that first victim in the New York City area, the disease has been reported in almost 40 states.

Mosquitoes seem to get the virus from infected birds and maintain it in their salivary glands—as with other mosquito-borne diseases—then spread it to humans when the insects bite and feed. West Nile virus has also been found in horses, cats, bats, chipmunks, squirrels, skunks, and domestic rabbits, but there is no evidence that humans get the virus from those animals without a mosquito serving as the go-between. Humans cannot pass the disease to other humans.

Cases in humans are definitely on the rise. The signs and symptoms are almost always mild and flu-like, and may include fever, headache, muscle aches, and, occasionally, a rash on the trunk of the body and swollen lymph glands. These signs and symptoms eventually go away harmlessly, usually within a few days.

In rare cases, however, the virus enters the brain, with potentially deadly results. Serious signs and symptoms may include headache, high fever, neck stiffness, stupor, disorientation, coma, tremors, convulsions, muscle weakness, and paralysis. In severe cases problems may persist for weeks, and neurological effects may be permanent. The chance of death is greatest for patients over 50 years of age and/or the immunocompromised.

If you have been in an area where West Nile virus could be carried by mosquitoes, and you have been bitten by mosquitoes, and you think you could have the virus, you should see a physician as soon as possible. A blood test can confirm the presence of the disease. Unfortunately there exists no specific treatment for West Nile virus, but supportive care leads to a complete recovery in most severe cases. And once you get West Nile virus, you probably can never get it again.

Current West Nile Virus Facts

1. Experts estimate that less than 1 percent of mosquitoes in virus-prevalent areas carry West Nile virus.
2. Less than one in five people who contract the disease develop any indication of illness.
3. In approximately one patient in 150 the virus crosses the blood-brain barrier and causes a serious inflammation of the brain (known as West Nile encephalitis), a serious inflammation of the membranes surrounding the brain and the spinal cord (known as West Nile meningitis), or a serious inflammation of the brain and its surrounding membranes (known as West Nile miningoencephalitis).
4. Less than 1 percent of the people in the United States who have been proven to have the disease have died from it.

Hantavirus

Strong evidence points to the deer mouse as the primary reservoir for hantavirus. It's been found in pinyon mice, brush mice, western chipmunks, and other rodents, primarily in the southwestern United States. Rodents don't get sick, but they carry the germs in their saliva, urine, and feces for weeks. Inhaling aerosolized microscopic particles of dried rodent saliva, urine, or feces gets the virus into humans, where it causes an extremely dangerous respiratory syndrome. So far, approximately 50 percent of diagnosed cases have died. Rodent bites may possibly transfer the disease, but human-to-human transference does not occur. There has been no known insect-to-human transference. Most inhalations occur in dwellings where rodent droppings have collected.

Infected humans appear to have the flu—fever, ache-ridden muscles, headache, cough—then, suddenly, their lungs fill with fluid and they suffer respiratory failure, usually within 2 to 6 days.

The following steps should be taken to avoid hantavirus:

1. Avoid contact with all rodents and their burrows.
2. Do not use enclosed shelters unless they have been cleaned and disinfected, and wear a protective mask if you're doing the cleaning.
3. Do not pitch tents or place sleeping bags near rodent burrows.
4. Use tents with floors or sleep on ground tarps that extend 2 to 3 feet beyond sleeping bags.
5. Store food away from rodent contact.
6. Promptly and appropriately dispose of all trash and garbage.

Rabies

Only mammals get rabies, and it acts in this way: The virus is in the saliva of infected animals. An infected animal bites a noninfected animal. The virus travels at a constant speed from the bite site to the spinal cord, and up the cord to the medulla, where it replicates. (Once it replicates in the brain of a human, death is assured.) It then travels back out along the nervous system of the newly infected. The second animal becomes infectious after the rabies virus collects in its saliva glands.

Infected animals can spread the disease without biting. If their saliva contacts a mucous membrane (the inner surface of lips or the eye) or an already open wound, the disease may result. There are several well-documented cases of transmission by corneal transplant from an infected human.

The incubation period in humans—the time from bite to brain—varies with individuals and with the bite site. Ten days is probably the minimum, but some patients have presented years after the inoculation. An average individual would incubate the virus for about 3 weeks after a bite on the face and about 7 weeks after a bite on the foot.

Annually in the United States about one million people are treated for animal bites, including those bitten by other humans. Somewhere between 20,000 and 25,000 of those people will be treated for rabies. Each year somewhere between zero and three cases of human rabies are actually diagnosed, and the diagnosis is made postmortem (after death).

Statistics from the United States, Canada, and Mexico give evidence of the primary reservoirs of rabies. The major hosts are currently skunks, raccoons, bats, cattle, cats, and dogs. The rest of the carriers were a few wolves, bobcats, coyotes, groundhogs, muskrats, weasels, woodchucks, foxes, horses, and a rare human. The significant statistic for outdoor enthusiasts to note is that an estimated 96 percent of the rabies virus in the United States is carried by wild animals.

Rabies produces general early signs and symptoms: headache and fever, cough and sore throat, loss of appetite, fatigue, abdominal pain, nausea, vomiting, and diarrhea. Patients also report a tingling sensation at the bite site. Once in the central nervous system, the virus makes the patient anxious, irritable, depressed, disoriented, unable to sleep, and prone to hallucinations. The patient may complain of stiff neck, double vision, and visual sensitivity to light. You may notice muscle twitching and, later, seizures. The final stage includes bizarre behavior like aimless unreasonable activity, biting at those who approach, and drooling. Painful muscle spasms in the throat result when the patient tries to eat or drink. The pain leads to avoidance of swallowing, and thus the drooling. Some patients have throat spasms at the sight of water, which is why the name *hydrophobia* (fear of water) became attached to rabies. As the nervous system deteriorates, patients become paralyzed, slip into a coma, fail to breathe adequately, and die.

Immediate care for the bite of any animal, including rabid animals, needs to concentrate on cleaning the wound. Wash it vigorously with soap and water. Aggressive washing can deactivate the rabies virus and does more than any other first-aid treatment. Rinse the wound thoroughly. The sooner cleaning takes place, the better.

As soon as possible, have the wound checked by a physician. Tetanus shots and/or antibiotics may be recommended to prevent bacterial infections. The doctor may suggest the rabies vaccine. Fortunately, the current human diploid cell vaccine against rabies works every time in minor exposures. Human antirabies immunoglobulin is also required in more significant exposures. A series of five relatively painless shots has replaced the old painful abdominal injections. To be most effective, the shots should be started as soon as possible—within 24 hours would be best.

How do you decide if the biting animal had rabies? There is only one sure way: have the brain of the animal tested for presence of the virus. This is often impossible. Without the head, five other questions should be asked:

1. Was the animal one of the highly suspect species?
2. Was the attack provoked? Rabid animals tend to attack without provocation. Trying to pick up or feed a wild animal and having it take a nip out of your finger is a very natural and unsuspect action. Having it leap from the shadows at your throat is an unprovoked and suspect attack.
3. If the animal was a pet, what is its vaccination status?
4. What is the geographical incidence of rabies in the area of the attack?
5. How did the animal behave in general? Loss of natural timidity and weird behavior are important nonclinical signs of rabies infection in animals.

When in doubt, get the expensive shots. If you make the wrong choice, you have accepted the death sentence. Once the clinical symptoms of rabies begin, you will almost assuredly die.

Plague

From 1347 to 1350, the Black Death, caused by the bacteria *Yersinia pestis*, began somewhere in Asia and eventually killed about 25 million Europeans (roughly one-third the population), including 90 percent of the population of England. Before those devastating years, even in ancient times, reports of the ravages of plague were known and feared.

In recent years plague has been on the rise in the western United States. Carried by rodents and passed primarily by the bite of rodent fleas, both rodent and flea are killed by the bacteria, an unusual aspect of this disease. Black rats are especially susceptible, and *Rattus rattus* is blamed for the Black Death of fourteenth-century Europe. In the United States, deer mice and various voles maintain the bacteria. It is amplified in prairie dogs and ground squirrels. Other possible carriers include chipmunks, marmots, wood rats, rabbits, and hares.

Hikers and campers are at mild risk if they hang around rodent-infested areas. Coyotes and bobcats are known to have transmitted plague to humans after the animals were dead and the humans were skinning them. Skunks, raccoons, and badgers are suspect. **Sick people transmit plague readily to other people.** Meat-eating pets that eat infected rodents (or get bitten by infected fleas) can acquire plague. Dogs don't get very sick, but cats do. There is only one known case of plague being passed to a human by a dog, but several cats have passed the disease to humans by biting them, coughing on them, or carrying their fleas to them.

Several forms of plague exist, but the three most common are *bubonic, septicemic,* and *pneumonic*. Buboes are inflamed, enlarged lymph nodes, and they give bubonic plague its name. After an incubation period of 2 to 6 days, patients usually suffer fever, chills, malaise, muscle aches, and headaches. Blackened, bleeding skin sores gave the name to the Black Death. The septicemic form plague may appear similar but does not give rise to buboes. Gastrointestinal pain with nausea, vomiting and diarrhea is common. The pneumonic form results most often from inhaling droplets that contain the bacteria, but it can develop from bacteria that get into the bloodstream. Coughing often produces blood in the sputum.

If plague is suspected, the patient should be isolated. Use body substance isolation. Do not inhale air the patient has exhaled. Transport the patient immediately. Fatalities are common. Antibiotics are required, and the use of them almost always prevents death. Prevention includes avoidance of rodents, avoiding touching sick or dead animals (if you must touch them, wear rubber gloves), and restraining dogs and cats while traveling in infected areas.

Hymenoptera

Bees, wasps (paper wasps, yellowjackets, and hornets, for instance), and fire ants (all *Hymenoptera*) are related largely due to their habit of injecting venom when they sting. Most humans find the pain extremely annoying, and that's the end of the story. But every year, for an estimated 50 to 100 people in the United States, the sting causes death, usually in less than an hour. Some experts guess the fatality rate runs even higher. Death almost always results from *anaphylaxis*, a severe allergic reaction (see Chapter 28: Allergic Reactions and Anaphylaxis).

Stings cause immediate pain, followed by redness and swelling, followed by itching. With any of these insects, ice packs generally ease the pain and swelling. Mild to moderate allergic reactions can be treated with an oral antihistamine. If severe breathing difficulty results, only an injectable drug, epinephrine—available by prescription in preloaded syringes—reverses the reaction.

Honeybees, nonaggressive by nature, lead the swarm as a source of fatalities in humans. As many as 15 percent of all humans may have some sensitivity, often mild, to bee venom, and the bee, unlike other *Hymenopterans*, has a barbed stinger that rips out of the insect and stays in human skin, continuing to pump venom for up to 20 minutes. The famed "killer" bees, near relatives of honeybees, are edging into the extreme southern U. S. from Latin America, where they've attacked several hundred people with fatal results. Killer-bee venom is no more potent than honeybee venom, but killers are noted for mass attacks by hundreds of individual bees.

Evidence suggests there is no significant difference in the amount of envenomization whether the stinger is scraped out or pinched and pulled out. The amount of envenomization does increase the longer the stinger is left in; therefore, bee stingers should be removed as rapidly as possible. Bee venom is for defense, and not for killing prey, so it is not particularly dangerous to the nonallergic in small doses. Simultaneous stings, however, numbering 40 to 50 could cause a systemic reaction: vomiting, diarrhea, headache, fever, muscle spasms, breathing difficulty, and maybe even convulsions. Depending on the individual who got stung, it takes somewhere between about 500 and 1,400 simultaneous stings to cause death by toxicity. Interesting to note, neither honeybees nor killer bees are native to North America, having been imported from Europe and Africa, respectively.

Wasp, unlike bees, are predators and scavengers who are attracted to meat and decaying matter. Their dirty stingers have a higher rate of infection, and the site of their stings should be thoroughly cleaned. Their venom, intended to kill prey, is more potent that bee venom, and 10 or more stings can cause vomiting, diarrhea, headache, fever, muscle spasms, breathing difficulty, or convulsions; 100 or more could possibly cause cardiac arrest. One of these insects is capable of multiple stings, but deaths in humans rarely occur.

Step on a fire-ant bed, which typically contains up to 25,000 ants, and 30 seconds later you're covered in hundreds. As other ants do, fire ants bite, but that's only the first part of the attack. Holding on with mandibles, fire ants arch their backs, jab in their stingers located in the end of their abdomens, release their venom, pull out their stinger, rotate, and jab in their stinger again. Unless you intervene, each ant will create a ring of burning stings. The venom causes local tissue destruction that produces, within 24 hours, a fluid filled bump that itches horribly for a week or more. Oral antihistamines may provide some relief. Fire ants, another imported insect, are well established in at least 11 southern states, and fire-ant victims seeking medical attention have soared in number to as high as 85,000 in one year. Deaths, due to anaphylaxis, may run as high as 30 per year.

If confronted by a bee or two, or a wasp, stay calm and back away slowly. They don't like rapid movements, especially swatting movements. If attacked by a swarm, run for dense cover, lie face down, and cover your head with your hands. If attacked by a hoard of fire ants, running and swatting are both approved methods of fighting back. Brightly colored summer clothing seems to attract winged insects. Tan, light brown, white, and light green appear to have no special appeal. Food left uncovered around camp is a no-no. Insect repellents will not work.

Centipedes

Centipedes have one pair of legs per body segment while millipedes have more than one pair per body segment. Some centipedes, but no millipedes, have venom glands and fangs that can penetrate human skin. Burning pain and swelling may result. Redness and discomfort may persist for a couple of weeks, perhaps more. Although most centipede bites heal without specific treatment, oral antihistamines and/or topical hydrocortisone may relieve the symptoms. Centipede bites are not considered life threatening, but infection rates in poorly cleaned wounds are high.

Bears

There are three recognized species of bears in the United States and Canada. In the far north lives the great white polar bear (*Ursus maritimus*). The most widespread is the black bear (*Ursus americana*). Black bears may be jet black to creamy yellow, average 300 pounds but may reach 600 pounds, and have no hump and a straight profile. The third is the brown bear (*Ursus arctus*), currently divided into two subspecies: the Kodiak brown (*Ursus arctus middendorffi*) and the grizzly (*Ursus arctus horribilis*). Brown bears may be dark brown to blond, average 500 to 900 pounds but may reach 1,400 pounds, and have a prominent shoulder hump and a dish-face profile.

Statistically, your chances of being killed by a bear are slim, but, to reduce the chances to an absolute minimum, here are three basic rules concerning bears:

1. *Hike and camp in a manner designed to avoid bears.* In known bear country, avoid areas that are used often by bears—trails with bear tracks and bear droppings, trails along salmon streams and through berry patches, and trails through dense brush and thick forest, especially at night. Avoid areas that smell of decaying meat: Bears like to cover their uneaten food lightly and camp nearby to finish it off later. If possible, camp in the open. Cook food at least 100 yards from sleeping sites. If camping near a river, sleep upriver from the cooking site. Night breezes tend to blow downriver, pushing the food smells away from the sleeping bags. Camp cleanly. Avoid wiping food-stained hands and utensils on clothing. Avoid spilling food on the ground. Avoid fish and greasy food. Pack food and other odorous products, such as toothpaste and soap, in a separate bag so the smells don't get into pack and clothing. Hang the food at night, if possible, out of bear reach from the ground and away from the trunk of the tree. All food residue should be packed with the food, not burned. Dishes should be cleaned even further away from the tents than the cooking site and kept packed with the food. Although no evidence indicates bears are attracted particularly to menstruating women, menstrual wastes should still be stored away from camp, possibly bagged securely in plastic and hung with other odorous material.

2. *Travel in a group large enough to ensure a measure of safety.* Bears like a measure of safety, too, and have not attacked a group of four or more in recent history.

3. *Strive to never surprise a bear.* Bears, particularly grizzlies, do not accommodate people up close and personal. Mother bears need even more room. Push their comfort zone, and they tend either to attack or run away. Once smelling or hearing humans, almost all bears will run away. Traveling with the wind and making noise can help make bears aware of your proximity.

A surprised bear that has seen the surpriser cannot be predicted to act in a certain way. If you're lucky, it will turn and run. If the bear doesn't run, speak in a calm, quiet voice. Back away slowly, but do not run. Running encourages the bear to play chase, a game bears win. If the group of people is four or more,

it usually works best to maximize the threat to the bear. The people should stand close together, raise their arms, speak in a reasonably loud and assured voice, identifying themselves as humans, the age-old nemesis to the bear. Statistics say the bear will retreat.

Bears who feel threatened turn to the side, displaying their size. They will often *woof* aggressively. They may charge toward the threat, and suddenly stop. These are invitations for the human to retreat. It is advisable to do so, but, remember, no running. It is best to back off slowly and keep speaking in a calm voice (which may be difficult by now). Humans without backpacks may benefit from turning to the side while backing off, an act that makes you look smaller and less intimidating. Climbing a tree is seldom worth the effort. Black bears climb like squirrels, and grizzlies can climb into the lower branches at a very fast rate. Avoid eye contact with the bear, an act of aggression.

If the bear actually attacks, different tactics are called for depending on the species of bear. Black bears seldom attack seriously unless they are hungry. They are not used to having food fight back, and it is statistically best to counterattack, doing all possible to convince the black bear to dine elsewhere. If the bear is a grizzly, assuming a least-threatening posture—playing dead—is the best tactic. Curl up to guard vital parts, clasping hands protectively behind neck. Fainting is not a bad idea. The brown bear may take a few bites, but then leave you alone. Those playing dead should remain so until the bear is well away. There seems to be no reasonable response to a polar bear's attack.

Patients surviving bear attacks should have serious blood loss stopped. The wound should be adequately cleaned and bandaged and evaluated by a physician (see Chapter 15: Wilderness Wound Management).

Dangerous Marine Life

Below the dancing surface of the ocean, which covers about 71 percent of the earth, live 80 percent of the world's living organisms. Some of these creatures are potentially dangerous to humans. Generally, risky aquatic life can be divided into three categories: (1) those with big teeth, (2) those that envenomate, and (3) those that sting with nematocysts.

Those with Big Teeth

Of all marine life worldwide, sharks elicit the greatest fear. Their triangular teeth are continually migrating forward to fall out and be replaced by new ones growing in the back of their crescent-shaped mouths. The result is an orifice always full of razor-keen weapons capable of removing large amounts of tissue from a prey's body. Sharks account for the most human deaths annually from aggressive marine life. Even so, the number of shark attacks each year averages only from 50 to 100 worldwide. It seems safe to go in the water considering the ratio of sharks to people who swim off shore.

If you work or play in the oceans of the United States, your chance of being shark food is perhaps one in five million. There are steps you can take to reduce the risk of attack even more:

1. Swim in groups.
2. Do not swim at dusk or during the night in infested waters. Those are the times sharks feed.
3. Wear bright or gaudy-colored swimsuits or wet suits to reduce the chance of being mistaken for something sharks like to eat.
4. Avoid turbid water, where shark eyesight will be impaired.
5. Avoid thrashing and splashing and other erratic movement and noise that attract attention. Sharks hunt with vibration receptors as well as their eyes.
6. Be watchful if you are a successful spear fisher. Sharks also use their sense of smell and are attracted to blood and other body fluids in the water.

If a shark seems interested in you, face it and swim away slowly. Avoid the crawl stroke, the butterfly, and other quick swimming styles. Stick to a breast stroke or easy side stroke. If the shark charges, curl into a ball. If it seems intent on taking a bite out of you, kick and punch at the eyes, nose, and gills.

In warmer waters, swimmers sometimes run the risk of toothy contact with barracudas or moray eels. Barracudas seldom attack humans, but when they do the attack is swift and fierce and usually in murky water where they have become confused. Their big, V-shaped mouths, with long pointed teeth, can do much damage. The glitter of shiny paraphernalia and jerky swimming movements may attract the barracuda. Moray eels do not bite unless threatened. Their sharp teeth and powerful jaws tend to hold a victim so strongly that the eel may have to be killed to be removed. Divers are at risk if they stick their hands into coral holes and rock crevices where eels like to lurk in wait for tastier morsels than humans.

Patients surviving an attack by sea life with big teeth should have blood loss stopped and wounds thoroughly cleaned. Humans tend to deal poorly with marine bacteria, and wounds from marine bites carry a high risk of infection. Any wound that significantly breaks the skin should be seen by a physician as soon as possible.

Those that Envenomate

Venomous sea creatures are almost always nonaggressive organisms that bite after being stepped on or handled unknowingly. The most likely of these animals to become the victims of human carelessness are stingrays and sea urchins. Contact usually produces local burning pain, followed by redness, swelling, and aching (and bleeding in the case of the stingray that has large barbs on the end of its tail).

The wound should be cleaned immediately with sea water. Explore the clean wound and remove any loose piece of the creature that may be stuck in the patient, a typical problem with stingrays and sea urchins. Cleaning also removes some of the venom, and, if the water is cold, relieves some of the initial pain. Scrubbing the wound with a povidone-iodine solution can help reduce the effects of the venom. As soon as possible, soak the wound in hot water. This may further remove venom and pain. Once the early pain has eased, the application of heat should not increase pain. Check again for debris in the wound. These injuries carry a high risk of infection. When the wound has dried, cover it with a sterile dressing and bandage it loosely. With all marine-related cuts, a loose bandage is best, allowing the wound to drain. Monitor the patient for signs of infection (see Chapter 15: Wilderness Wound Management). Prophylactic antibiotics are typically recommended due to the high incidence of infection.

Hawaii is the only state with venomous sea snakes. Reptiles in every sense of the word, they breathe air but can remain submerged for hours. They commonly range from 3 to 4 feet in length, but individuals have been measured to 9 feet. They flatten out toward the rear and propel themselves forward and backward with undulating motions, like delicate ribbons waving in sea currents. Attacks are rare, unless they are provoked. Their short fangs deliver a relatively mild bite, but the venom is very toxic. Developing in minutes to hours, the patient complains of nausea and nonspecific feelings of illness. Anxiety increases. Pain grows, with partial paralysis possible. Breathing difficulty may follow. Heavily envenomed patients develop a growing intensity of signs and symptoms and may lapse into a coma. One in four victims dies without antivenin.

Treatment in the field consists of removing the patient from the water and calming him or her down. Hysteria and movement increase the activity of the poison. Place a light constricting band, not a tourniquet, close to the bite and between the wound and the patient's heart. Cutting and sucking is dangerous and nonproductive. Evacuate the patient as soon as possible to a medical facility.

Those with Nematocysts

Coelenterates, a group that includes all the jellyfish, form an enormous phylum of sea creatures, with around 9,000 named species. Some of them are harmful to humans, with common names that sometimes reflect the risk they pose: fire coral, stinging medusa, Portuguese man-of-war, sea wasp, sea nettle, hairy stinger, and stinging anemone. They all share a stinging organelle, called a *nematocyst*, that lives encapsulated in a cnidoblast on the tentacles of the coelenterate. The cnidoblast has a trap door (the operculum) with a trigger (the cnidocil) that opens the door. Filled with venom, the nematocyst has a sharp, coiled, thread-like appendage that springs out when the trigger is touched, lodging in the patient.

The reaction in humans ranges from mild to severe. Mild reactions are immediate, and include stinging, burning, itching, and sometimes local numbness to touch. The stung area may acquire a bruised appearance that lingers for days. A moderate reaction progresses to headache, nausea, and vomiting. Severely reactive patients have breathing difficulty that may lead to loss of consciousness, convulsions, and the possibility of death.

Because the leading cause of death is drowning in the ensuing panic, removal of the patient from the aquatic environment is of immediate importance. Once safely out of the water, the irritated area of the patient's skin should be rinsed with copious amounts of water or vinegar. Physically lift off any tentacles that still cling to the patient, but do so without touching the tentacles with a naked hand.

The final step in treatment is to remove the embedded nematocysts. They can be rubbed off with sand and sea water, shaved off with a knife, or scraped off gently with anything that has an appropriate edge. Persistent itching can be relieved with a topical corticosteroid, and oral antihistamines have proven to give some relief. The patient should be watched over the next 24 hours for signs of a severe allergic reaction.

Evacuation Guidelines

A patient treated for a bite or sting from a known venomous animal should be evacuated to definitive care. Any patient developing an illness subsequent to a bite should be evacuated. A patient treated for anaphylaxis, even though he or she seems to have recovered, should be evacuated. Any suspicion of rabies calls for an evacuation of the patient. Bite wounds from large animals call for an evacuation of the patient.

Conclusion

When the bite occurred to the young woman in the Sipsey Wilderness was never known. It caused no immediate pain. What bit her was never established beyond the shadow of a medical doubt. The physician in charge discovered a red, hardened bump on her right lower leg and recorded the high probability of a black widow envenomation. The patient's stay in the hospital included medication for the pain, a watchful eye for more severe reactions such as breathing difficulty, and nothing more as far as treatment went. After a few days, she was released with complete recovery.

CHAPTER TWENTY-TWO

Diving Emergencies

You Should Be Able To:

1. Describe the signs and symptoms of the most common scuba diving emergencies including barotitis, barosinusitis, pulmonary overinflation syndrome, arterial gas embolism, and decompression sickness.
2. Describe the appropriate care for the most common scuba diving emergencies.
3. Describe prevention of the most common scuba diving emergencies.

It Could Happen to You

You're learning to scuba dive in the warm waters off Kona, Hawaii. With successful completion of the basic course behind you, you're about 40 feet down, about a quarter mile from land, a student in an Open Water Diver course. Near you a second student floats suspended beneath the waves. The instructor is on the boat somewhere above, waiting at a pre-arranged pickup spot.

You take a compass bearing from the device on your right wrist and kick off in a southwesterly direction. The second student heads in a different direction. He will follow a prearranged underwater course, as you are doing, to arrive at the same pickup spot.

A moray eel pokes its head from a crevice in a rock wall, and jerks back in as you swim past. Colorful fish of a dozen species scurry away at your approach.

Nearing the anchor chain of the instructor's boat, the point where you'll slowly ascend to the surface, you see the second student swimming toward the same spot when he suddenly stops, stares at you a moment, takes a quick look at the air pressure gauge on his left wrist, and begins to swim rapidly upward. The first words you hear when your head breaks into sunshine come from the second student and sound distressed. He complains of chest pain and breathing difficulty. The instructor leans over the side, telling you, in an urgent whisper, to hurry on board.

Scuba—the word is so well known many people have forgotten it stands for "self-contained underwater breathing apparatus." Scuba diving has swelled in popularity; there are well over three million active recreational divers in the United States, and more than 300,000 new divers are trained each year.

For healthy, well-trained, and well-equipped individuals, scuba diving is a very safe activity. It is, however, an activity with some inherent risks. The most common risks are associated with the often rapid changes in air pressure divers must deal with, sometimes descending, sometimes ascending. Improperly treated, some of the medical emergencies associated with scuba diving can cause permanent disability or death.

Physical Principles of Diving

When a scuba diver descends beneath the surface, the ambient pressure increases because of the weight of the water. Water, much more dense than air, creates pressure changes that are substantial even for small changes in depth. At a depth of only 33 feet, the pressure is 2 atmospheres, or twice the pressure at sea level.

Body tissues consist primarily of water, which is not compressible, so body tissues are not significantly affected by the changes in pressure that occur at depths down to 100 feet, the maximum depth at which most scuba diving is done. Gases, however, are compressible, so the gas-filled spaces of the body are directly affected by changes in pressure.

Gas Laws

Knowledge of three gas laws is fundamental to understanding why pressure-related diving medical emergencies occur. Knowledge of these laws is *not* a critical part of knowing how to handle the emergencies, but the laws are presented here for reference.

1. *Boyle's law* states that the volume and pressure of a gas are inversely related to its pressure at a constant temperature. In other words, when the pressure on a gas increases, the volume decreases, and the converse is true. This law explains the basic mechanism for all types of *barotrauma*.
2. *Dalton's law* states that the pressure exerted by each of the individual gases in a mixture of gases is the same as it would exert if it occupied the same volume alone. Alternatively, this

law states that the total pressure of a mixture of gases is equal to the sum of the partial pressures of the component gases. Since the biological effects of a gas depend on partial pressure, Dalton's law is fundamental to an appreciation of *decompression sickness.*

3. *Henry's law* states that the amount of gas dissolved in a fluid is proportional to the partial pressure of that gas in contact with the fluid. This law explains why a more inert gas, such as nitrogen, dissolves in a diver's body during descent and, conversely, is released from tissue with ascent.

Types of Barotrauma

Air pressure within the middle ear, sinuses, lungs, and other gas-filled spaces of the body is normally in equilibrium with the environment. If something obstructs the free passage of air into and out of these spaces, then a loss of equilibrium will develop. Tissue injury may occur. Such injuries are collectively referred to as *barotrauma.*

Overall, barotrauma is the most common problem for scuba divers. Barotrauma can be categorized according to whether it occurs during descent or ascent.

Barotrauma of Descent

Barotrauma of descent, or *squeeze,* as it is known to most divers, results from the compression of gas in enclosed spaces as ambient pressure increases with descent underwater. The ears and nasal sinuses are most often affected. More than one type of barotrauma may be present at the same time.

■ BAROTITIS

Ear squeeze, or *barotitis,* affects essentially all divers at one time or another and is the most frequent type of barotrauma. Although the external ear can be affected, this is unusual unless the patient was wearing a hood or earplugs, making the ear canal a closed space. Much more common is *middle ear squeeze,* or *barotitis media,* which results when the diver fails to equalize pressure in the middle ear because of closure or dysfunction of the eustachian tube (the tube from the middle ear to the throat that allows for compensation in pressure changes). Divers usually notice a sense of fullness in the ear, followed by increasingly severe pain. *Vertigo* (dizziness caused by a disturbance in equilibrium) is a common complaint. Blood may leak from the ear, and the *tympanic membrane* (eardrum) may rupture, resulting in *tinnitus* (ringing in the ear) or hearing loss, depending on the severity of the injury. Nausea may be a complaint.

The field treatment for middle ear squeeze is abstinence from diving or other pressure exposures until the condition has resolved. The use of oral decongestants, such as pseudoephedrine, is recommended to shrink swollen mucus mem-branes and help open up the eustachian tube. A combination of an oral decongestant and a long-lasting nasal spray, at least for the first 2 to 3 days, is usually most effective. Analgesics (painkillers) may be given as needed.

Although far less common, *inner ear barotrauma* is much more serious than middle ear barotrauma because of a possible permanently disabling injury. Inner ear barotrauma typically results from the sudden or rapid development of markedly different pressures between the middle and inner ear. These different pressures may result from an overly forceful Valsalva maneuver, a common and typically safe procedure to equalize pressure in the middle ear by pinching the nose shut, holding the mouth shut, and forcefully expelling air, or from an exceptionally rapid descent, during which the middle ear pressure is not equalized.

The classic triad of symptoms indicating inner ear barotrauma is roaring tinnitus, vertigo, and deafness. A feeling of fullness or "blockage" of the affected ear, nausea, vomiting, pallor, sweating, disorientation, and/or loss of coordination may be present. The onset of these symptoms may occur soon after the injury or may be delayed many hours, depending on the specific type of inner ear injury and the diver's activities during and after the dive. The essential point, however, is that any scuba diver with a hearing loss should be considered to have inner ear barotrauma until shown otherwise and should be seen by a physician as soon as possible.

■ BAROSINUSITIS

Just as with the ears, the nasal sinuses may also fail to equalize pressure during descent, thus resulting in *barosinusitis,* or *sinus squeeze.* The frontal and maxillary sinuses are most often affected. Sinus squeeze usually causes pain in the affected sinus, and the diver may notice bleeding from the nose or the mouth. Your examination may be unremarkable or may elicit tenderness when you lightly tap over the affected sinus.

Treatment of sinus squeeze is similar to that for middle ear squeeze and consists of the use of decongestants, analgesics, and abstinence from diving until the condition is resolved. Antibiotics are usually indicated in cases of frontal sinus squeeze because of concerns about complications resulting from frontal sinusitis. Consult a physician.

Signs and Symptoms of Ear/Sinus Barotrauma

1. Pain in the affected area.
2. Bloody or unusual fluid leaking from the ear or nose or mouth.
3. Vertigo.
4. Tinnitus and/or hearing loss.
5. Nausea and/or vomiting.

Barotrauma of Ascent

Although less common than squeezes, scuba divers may also suffer barotrauma of ascent, which is the reverse process of what happens in the squeeze syndromes.

■ PULMONARY OVERINFLATION SYNDROME

The only type of barotrauma of ascent that occurs with any real frequency is *pulmonary overinflation syndrome*. A dramatic demonstration of Boyle's law, pulmonary overinflation syndrome results from the expansion of entrapped air in the lungs. Air entrapment usually occurs because of breath-holding during ascent, as when a diver runs out of air and/or the diver panics, forgetting to breathe in both cases. The net effect leads to alveolar rupture, with air escaping to fill spaces outside the alveoli. Pulmonary overinflation syndrome usually is indicated by gradually increasing substernal chest pain and difficulty in breathing and swallowing. *Subcutaneous emphysema*—bubbles of air trapped beneath the skin—may be present. You may be able to hear decreased breath sounds.

The treatment for pulmonary overinflation syndrome usually consists only of observation and abstinence from further diving for 4 to 6 weeks after the condition has resolved. In severe cases hospitalization may be necessary. Administration of supplemental oxygen may hasten resolution of the condition. An important point, except in exceedingly rare situations, is that recompression—hyperbaric oxygen treatment—is not recommended because of the fear of causing further pulmonary barotrauma.

■ ARTERIAL GAS EMBOLISM

Air bubbles entering the bloodstream, forming an *arterial gas embolism* (AGE), are the most serious complication of pulmonary barotrauma of ascent and, indeed, the most dramatic and serious medical emergency associated with scuba diving. Next to drowning, a gas embolism entering the brain is the leading cause of death in scuba divers. An arterial gas embolism typically manifests itself just before or within 10 minutes of surfacing from a scuba dive.

Manifestations of AGE are many, but they tend to be dramatic, the brain being affected most often. Loss of consciousness, seizures, blindness or other visual disturbances, inability to speak, confusion, vertigo, headache, weakness (including *hemiplegia*, full or partial paralysis of one side), and various sensory disturbances are the most common signs and symptoms. The patient may appear to be having a stroke (see Chapter 25: Neurological Emergencies). Sudden loss of consciousness in any scuba diver within 10 minutes of surfacing should always be considered to be attributable to AGE until proven otherwise. If gas bubbles affect the heart, the patient may appear to be having a heart attack (see Chapter 23: Cardiac Emergencies).

All patients suspected of suffering a gas embolism must be referred for recompression—hyperbaric oxygen treatment—as rapidly as possible. This is the primary and essential treatment. As soon as possible, all patients should be given high-flow/high-concentration supplemental oxygen.

The majority of AGE patients have definite signs and symptoms of neurological injury when first evaluated, but a significant number manifest spontaneous recovery, at least initially. Nonetheless, they must still be transported for recompression treatment because it is impossible to fully exclude damage, and many of these patients deteriorate over the next few hours, often to a worse condition than at first.

During evacuation, patients should be maintained in a supine position. Placement in the left-side/head-down position is not recommended because of the uncertain benefit of such a maneuver, concerns about causing or aggravating cerebral edema (especially if left in the head-down position for longer than 30 to 60 minutes), and increased respiratory difficulty associated with being in such a position.

Other Diving Disorders
Decompression Sickness

Decompression sickness (DCS), or more often called *the bends,* is a multisystem disorder that results when the nitrogen from breathing compressed air that accumulated on descent—explained by Henry's law—collects in the blood in a greater concentration than can be returned to a gas and breathed out on rapid ascent. Gas bubbles then form in the blood and other tissues. Bubbles cause mechanical effects, such as blockages in blood vessels. The overall effect is to decrease tissue perfusion.

The manifestations of DCS are many, with the musculoskeletal system and central nervous system being most often affected. Joint pain is the single most common symptom of DCS and occurs in about 75 percent of patients. The joints most often affected are the shoulders and elbows, although any joint may be involved. The pain is usually described as dull and is usually located deep in the joint. Movement of the joint worsens the pain. The patient keeps her or his joints bent—thus the origin of "the bends."

Neurological manifestations of DCS are less common. Because of the random manner in which bubbles may affect the central nervous system, essentially any symptom is compatible with neurological DCS, but the lower thoracic, lumbar, and sacral portions of the spinal cord are most often affected. Consequently *paraplegia* (paralysis of the lower extremities) or *paraparesis* (partial paralysis of the lower extremities), lower extremity *paresthesia* (numbness or tingling), and bladder or bowel dysfunction are the most common symptoms of neurological DCS.

Anyone who manifests symptoms after a scuba dive that

cannot be adequately explained by other conditions should be presumed to have DCS until proven otherwise. Such a patient should be transported for recompression treatment without delay. All patients suspected of having DCS should be started on high-flow/high-concentration supplemental oxygen as soon as possible, and other life-support measures should be given according to the patient's specific condition.

Nitrogen Narcosis: Rapture of the Deep

Several diving-related problems may develop as a result of breathing gases at a higher than normal atmospheric pressure. Among these is *nitrogen narcosis,* a result of the anesthetic effect of nitrogen at elevated partial pressures. There is considerable variability in susceptibility to nitrogen narcosis, with symptoms usually becoming evident at approximately 3 atmospheres, or 100 feet down. At depths of 200 feet, divers suffering nitrogen narcosis are usually so impaired that they can do little or no useful work. Loss of consciousness begins to occur at depths greater than 300 feet.

Nitrogen narcosis is a reversible condition that has manifestations similar to alcohol intoxication, including impaired judgment, giddiness, poor concentration, loss of coordination, and slowed motor response. Nitrogen narcosis completely resolves with ascent to shallower depths, and divers often do not notice or recall the adverse effects.

The real importance of nitrogen narcosis is not in its occurrence per se, but in its ability to impair a diver's judgment and memory, thereby possibly precipitating an accident. This possible confounding factor always must be considered when taking a diving accident history, especially when there is a history of diving deeper than 100 feet.

Prevention of Diving Emergencies

1. Use Valsalva maneuvers to equalize pressure in ears and sinuses.
2. Refrain at all times from breath-holding.
3. Monitor carefully the volume of air in your tanks.
4. Monitor your depth carefully, and do not exceed prescribed times at specific depths.
5. Avoid rapid ascents. Base ascent rates on depth of dive and time at depth.

Evacuation Guidelines

Any patient assessed with arterial gas embolism or decompression sickness should be evacuated as soon as possible. Any patient suffering increasing difficulty breathing and/or sudden unconsciousness should be evacuated as soon as possible.

Conclusion

By the time you've removed your mask, fins, and tank, and stored them on board the boat lying off Kona, the instructor has the second student in a position of comfort, breathing 100 percent oxygen from the emergency cylinder he keeps below deck. The patient's condition appears stable, but he tells you he'd rather not take any chances. Heading back toward shore means drawing closer to a hospital. Better safe than sorry.

Mark Crawford, EMT-P, United States Air Force Pararescue, Retired, contributed his expertise to this chapter.

Section Five

Medical Emergencies

CHAPTER TWENTY-THREE

Cardiac Emergencies

You Should Be Able To:

1. Describe cardiovascular disease and the risk factors related to cardiovascular disease.
2. Define and describe the treatment for angina, myocardial infarction, and congestive heart failure.

It Could Happen to You

The snowshoe trips you lead in the snowy Cascades east of Seattle leave after breakfast and return to the lodge in time for dinner. All a client has to do is show up. No prerequisites. No medical forms to fill out. The equipment is supplied by the lodge that pays your salary.

Robert, "call me Bob," seems like a nice fellow and appears, generally, to be in good health. But by mid-morning, Bob has fallen behind the group. You drop back to check on him.

It's cold, but not cold enough to explain the paleness of his face, and not cold enough to justify the pain in his eyes.

"Feels like someone parked their snowmobile on my chest," he pants out in a ragged whisper. "Must be that rich breakfast they fed us at the lodge."

Heart attacks lead to more deaths than any other cause in the United States, and wilderness enthusiasm does not provide immunity. While the incidence of cardiac disease increases dramatically after age 50, younger people are having heart attacks, and vigorous elders are venturing far into wild lands.

Coronary arteries, the vessels that leave the aorta and supply blood to the heart muscle itself, are very susceptible to injury to their lining. The general term is *arteriosclerosis,* several conditions in which the arteries thicken, harden, and otherwise lose their elasticity. *Atherosclerosis,* probably the most common form of coronary artery disease, is a gradual clogging of the arteries from fatty deposits and other debris. As the arteries narrow, less blood passes through to reach the heart. Decreased blood flow may lead to several cardiac emergencies, including *myocardial infarction* (heart attack).

Encouragement for clogging comes from dietary cholesterol, cigarette smoking, lack of exercise, obesity, and *hypertension* (high blood pressure). Heredity and aging are factors. Diabetes predisposes a person to coronary artery disease. For women over age 35 who smoke and have high blood pressure, oral contraceptives seem to encourage heart attacks even more.

You'll notice some of these factors are uncontrollable: heredity, age, and gender. On the positive side of things, however, many of the factors are controllable. With awareness and dedication to prevention, any individual can reduce his or her chance of coronary artery disease.

Types of Cardiac Emergencies

Angina Pectoris

Pain caused by interruptions in adequate blood flow to the *myocardium* (heart muscle), a result of narrowed arteries, is called *angina pectoris.* It usually comes on with physical or emotional stress or with heat or cold stress, events that increase the workload on the heart. Angina pectoris, however, can occur without unusual stress.

Angina pectoris means "pain in the chest," and is more often referred to simply as *angina.* But sudden chest pain, ranging from a mild ache to crushing pressure (often described as "squeezing," "tightness," "constricting," or sometimes "burning") is only one of several typical signs and symptoms. The patient may report the pain radiates to the left side of his or her body: shoulder, arm, neck, or jaw. Shortness of breath, nausea, and vomiting are not uncommon. Patients typically deny anything serious is happening even though they often appear anxious. Indigestion is routinely suspected by the patient. Heavy sweating, pale cool skin, and complaints of dizziness or light-headedness may be a part of the assessment. You may find a rapid, slow, weak and/or irregular heart rate.

Signs and Symptoms of Angina

1. Chest pain.
2. Shortness of breath.
3. Denial and anxiety.
4. Nausea and vomiting.
5. Light-headedness or dizziness.
6. Rapid, slow, weak and/or irregular heart rate.
7. Pale, cool, sweaty skin.
8. Pain relieved by rest and/or medication.

Since the heart is starved for oxygen, administering high-flow/high-concentration supplemental oxygen would be of great benefit. Do what you can to calm and reassure the patient, keeping him or her comfortable and physically at rest. A comfortable position is almost always sitting or reclining, rarely lying down. Comfort includes maintenance of normal body core temperature.

Ask about medications. Angina sufferers may be carrying nitroglycerin in tablets or spray. Nitroglycerin works by reducing the demands on the heart and dilating the coronary arteries, allowing more blood to pass through. If it has been prescribed to your patient, and if the expiration date has not been reached, help the patient use the drug. Most patients carry tablets, and most prescriptions call for one tablet taken under the tongue. Classic symptoms indicating the nitro is working include a burning under the tongue and a headache. In 5 minutes, if the chest pain remains, a second tablet may be given and, if that doesn't work, a third, as long as the patient's blood pressure remains high enough. Nitro should not be given if the patient's systolic blood pressure has dropped below approximately 90 to 100. The medication might drop the blood pressure even more and precipitate shock. Because of the possibility of a drop in blood pressure, always give nitro with the patient sitting down in a comfortable, stable position. As long as the patient's radial pulse remains strong, the blood pressure may be considered high enough. In addition to nitro, the patient will benefit from taking aspirin, even if he or she is currently taking aspirin prophylactically (see "Aspirin," this chapter).

With rest and/or nitroglycerin, the pain of angina usually goes away, leaving no permanent damage from the episode. In almost every case, however, you'll want to evacuate the patient, even when the patient appears to have fully recovered, so that he or she can be evaluated by a physician. This is especially true if the angina is new, if it takes longer to resolve than usual, or if the pattern of the pain (the onset, duration, quality) is different from the norm for the patient. If you work as a trip leader, it will prove beneficial if (1) you've screened the person prior to the trip, (2) you've discussed the outcome should the person suffer an attack of angina on the trip, and (3) you've ensured the patient is carrying nitroglycerin. If the patient continues to show signs and symptoms after 15 to 20 minutes, it is recommended to assume a myocardial infarction and initiate an evacuation.

Myocardial Infarction

Angina, a sign of narrowed arteries and underlying heart disease, may slowly progress to a *myocardial infarction* (MI), or an MI may happen suddenly and with little warning. A myocardial infarction results when a clot blocks a narrowed coronary artery or reduces the blood flow through that artery enough to cause the permanent death of the heart muscle cells fed by that artery. Although heart attacks can be painless and referred to as "silent MIs," probably more than 80 percent of MI patients complain of center-chest discomfort: crushing, squeezing pain, or heavy pressure. Pain may radiate to the shoulder, down the arm, and/or into the jaw, predominantly on the left side. Pain may wax and wane, but, in contrast to angina, it does not go away. Rapid, shallow respirations are common, as are complaints of shortness of breath. Pulse may be weak, rapid or slow, regular or irregular. Nausea, vomiting, weakness, and light-headedness are common. Skin usually turns pale and cool with heavy sweating, and cyanosis is possible. Patients often deny the possibility that this could be "the big one"—a heart attack—but expect high levels of anxiety.

Signs and Symptoms of Myocardial Infarction

1. Sudden, crushing chest pain.
2. Shortness of breath.
3. Denial and anxiety.
4. Nausea and vomiting.
5. Light-headedness or dizziness.
6. Rapid, slow, weak, and/or irregular heart rate.
7. Pale, cool, sweaty skin.
8. Pain *not* relieved by rest and/or medication.

Myocardial infarctions create several possibilities for the patient:

1. MIs can cause sudden death or death within a couple of hours of the onset of signs and symptoms, depending on the extent of damage to the heart. Cardiopulmonary resuscitation should be initiated in hopes of salvaging the patient who is breathless and pulseless (see Chapter 5: Cardiopulmonary Resuscitation).
2. If the left ventricle sustains extensive damage, the heart will fail to adequately pump blood to the body, perfusion drops, and cardiogenic shock follows. As many as eight out of 10 patients who develop shock from a heart attack die, usually within 24 hours. Treatment for shock, in this case, should be initiated with one exception to the rules: **Do not elevate the patient's legs** (see Chapter 7: Shock).
3. MIs may cause congestive heart failure.
4. The patient can live a long, healthy life if he or she takes steps to prevent further MIs.

In the wilderness there is little else to be done for the patient suspected of having an MI other than keeping the patient physically and emotionally calm, in a position of comfort, and insulated from a cold environment. Activity could make the patient much worse. The patient would benefit, often very much, from high-flow/high-concentration supplemental oxygen, which is rarely available in the wild outdoors. If the patient has been prescribed nitroglycerin, follow the directions for providing aid. If aspirin is available, the patient should take approximately 160 milligrams immediately. The patient requires immediate transport—no walking—to a medical facility.

Aspirin

Aspirin is an extremely effective antiplatelet agent, a means to prevent further blood clot formation. For that reason patients suffering angina or heart attack may greatly reduce the damage done to their hearts by taking aspirin as soon as possible. Aspirin, in other words, could save a life. Although the dosage remains somewhat controversial, research indicates the dose should be approximately 160 milligrams orally to start—about one half of an adult aspirin or two "baby" aspirins—and approximately 80 to 160 milligrams each subsequent day until you turn the patient over to definitive care. The aspirin acts faster if it is chewed before swallowing.

Congestive Heart Failure

Heart attack, coronary artery disease, and other forms of damage to the heart may result in a myocardium no longer able to meet the demands for blood that a body makes. A weak myocardium

may result in *congestive heart failure* (CHF), a somewhat complex process in which blood and tissue fluids congest (collect excessively) due to the heart's inadequacy. Failure of the right side of the heart, the side that takes blood from the body and pumps it to the lungs, creates congestion in the periphery of the body, usually most noted in swollen ankles. More seriously, failure of the left side of the heart, the side that takes blood from the lungs and pumps it to the body, creates congestion in the lungs—pulmonary edema—which can rapidly turn into a threat to life.

It is extremely rare to encounter a CHF patient in the wilderness, with the most likely chance occurring, possibly, in a heart attack patient who cannot be evacuated quickly. The most important indications are breathing difficulty and the terror such difficulty produces. A patient typically refuses to lie flat, preferring to sit upright, a position that makes it easier to breathe. Pulse is elevated. Skin may be cyanotic. The patient may cough productively, revealing pink sputum. Death may follow.

Signs and Symptoms of Congestive Heart Failure

1. Anxiety.
2. Rapid pulse.
3. Difficulty breathing and/or rapid breathing.
4. Cyanosis.
5. Productive cough.
6. Swollen legs and ankles.

Assist the patient in sitting upright, a position that can prevent some blood from congesting in the lungs. Give high-flow/high-concentration supplemental oxygen when available. Other than that, your emergency care is limited to the techniques used for a patient experiencing a myocardial infarction.

General Treatment for Cardiac Emergencies

1. Maintain an open airway (see Chapter 4: Airway and Breathing).
2. Keep the patient inactive, physically and emotionally calm, and in whatever position is most comfortable for the patient, almost always sitting or semi-reclining.
3. Deliver high-flow/high-concentration supplemental oxygen, if available.
4. Assist in the maintenance of body core temperature.
5. Assist the patient in following the instructions for prescription medications, if applicable.
6. Give aspirin.
7. Initiate a rapid evacuation that involves no physical effort by the patient.

Evacuation Guidelines

A patient suffering from an attack of angina should be, in almost every case, evacuated. Patients who recover from an angina attack may walk. A patient assessed with myocardial infarction and/or congestive heart failure should be evacuated as soon as possible by a means of transport that requires no exercise by the patient. An evacuation accomplished within 4 to 6 hours gives the patient the best chance of improvement via drug intervention.

Conclusion

Although your assessment reveals Bob has a history of angina and has been prescribed nitroglycerin tablets, he forgot his medication at the lodge. You reassure your patient. You dig out your emergency foamlite pad and extra clothing from your pack, stamp out a quick seat in a nearby snow bank, and make Bob as comfortable as possible. As the rest of the group gathers, you announce today's trip will be cut short. Within 10 minutes, Bob has greatly improved. With you carrying your patient's light day pack, the group tramps slowly back to the lodge, where you intend to suggest a medical form be filled out by all clients prior to future trips.

CHAPTER TWENTY-FOUR

Respiratory Emergencies

You Should Be Able To:

1. Describe the signs and symptoms of the most common respiratory emergencies, including hyperventilation syndrome, pulmonary embolism, upper respiratory infection, pneumonia, chronic obstructive pulmonary disease, and asthma.
2. Describe the treatment for the most common respiratory emergencies.

It Could Happen to You

The trail into the Great Smoky Mountains rises toward a cloudless sky, and the summer heat mingles with humidity high enough to drown a small child. While you lead the group of backpackers up the steep path, your co-instructor brings up the rear. Somewhere in the middle of the group, unseen by you or your co-instructor, a voice lifts in a cry for help.

Hurrying back down, you find the other instructor leaning over the collapsed form of one of the young women on the trip. Although it's only the second day of the hike, you've already decided the woman on the ground, Jen, is strong-willed, with an unusual drive to prove herself. You are worried.

"What happened?" you ask the cluster of clients.

"She just fell over," says one.

"She hit her head," says another.

"I think she's dying," suggests a third.

Jen lies on her side, knees drawn up, curved almost into the shape of a crescent moon. Her upper arms are held tight to her body. Her hands, twisted into odd spasms near her chin, are curled back toward her forearms. Her skin looks pale. Sweat moistens her face. She breathes very rapidly and deeply, apparently desperate for air.

Comprised of five large and spongy lobes, the lungs contain millions of air sacs—150 million to 300 million of them—called *alveoli*. Alveoli are, essentially, microscopic spaces at the end of small airways called *bronchioles*. Lungs also have an extraordinarily dense arrangement of tiny blood vessels called *pulmonary capillaries*. The elastic walls of the alveoli are only one cell thick; oxygen passes through these thin walls to enter the blood and bind onto oxygen-depleted red blood cells, while carbon dioxide leaves the blood to be breathed out on the next exhale. Without a steady diffusion of these gases, oxygen in and carbon dioxide out, life fails to go on. There is simply no more important action for a Wilderness First Responder than ensuring the patient keeps breathing. And few, if any, medical emergencies cause the patient, and rescuers, more distress than *dyspnea*—air hunger resulting in breathing difficulty.

Types of Respiratory Emergencies

Hyperventilation

Not a disease itself, *hyperventilation*—breathing unusually fast and/or deeply—is a human response to some type of stress to the body, triggered by emotion, exercise, pain, a sudden dip in icy water, or, sometimes, disease. Hyperventilation can also be a voluntary act. To treat hyperventilation, therefore, you have to assess and treat the *cause* of hyperventilation.

First and foremost in the assessment of a hyperventilating patient is determining if there are any immediate threats to life, which you do during your initial survey. Finding none, you should move to the focused assessment, which may shed light on why the patient is breathing fast. A cause discovered and treated may reduce hyperventilation.

Hyperventilation syndrome, defined as breathing fast without another cause, occurs relatively often in otherwise healthy patients as a result of psychological stress. The patient, like most hyperventilating patients, has a high level of anxiety or panic. Rapid breathing causes more carbon dioxide than normal to be exhaled, altering the pH of blood by raising it, a condition called *respiratory alkalosis*. Alkalosis may cause numbness and tingling in the mouth, fingers, and toes. Alkalosis may also cause *carpopedal spasms*, muscular spasms in the hands and feet. Rapid breathing may cause the slow onset of chest pain from overtaxed chest muscles. Light-headedness and dizziness are not uncommon. Skin color may vary from pale to flushed. These physiological responses to hyperventilation do nothing to calm the psychologically stressed patient. Prolonged hyperventilation causes constriction of the blood vessels in the brain, which may lead to fainting. During the early part of the fainting spell, the patient may not breathe because carbon dioxide levels are low, and breathing is based on the carbon dioxide level in the blood rather than the oxygen level—and carbon dioxide is reduced by the act of hyperventilating. Consciousness will return.

As soon as you know the patient's life is not threatened, you should begin to help the patient regain control of breathing. Stay calm. Clear the immediate area of other people. State clearly to the patient that he or she feels terrible *because of* fast breathing. Ask the patient to breathe more slowly. Ask the patient to hold his or her breath for 3 seconds, longer if possible, and you do the counting. Ask the patient to breathe through the nose. Ask the patient to breathe along with you in a slow, calm manner. Be supportive in your attitude and words. Do not ask the patient to breathe into a bag. It does not work faster than calming the patient down, and it may cause more anxiety for the patient. When the patient returns to normal breathing, and the signs and symptoms go away—almost always within 20 to 30 minutes—recovery is complete.

Pulmonary Embolism

An *embolism* is an obstruction that lodges in an artery, preventing the normal flow of blood. The embolism can be a "foreign" substance or a blood clot, and it is most often a blood clot. When the obstruction occurs in an artery of the lungs, the result is a *pulmonary embolism,* an emergency that ends the lives of an estimated 50,000 or more in the United States each year.

In approximately three out of four patients, the embolus formation takes place in a deep vein in the legs before it breaks loose and flows into the lungs. High-altitude climbers are susceptible if they are dehydrated and tent-bound for an extended period of time, such as during a storm. A patient with a history of a recent illness or surgery that kept him or her bedridden may be susceptible. Smokers are more at risk than nonsmokers. Oral contraceptives, especially taken by women over 40, predispose women to pulmonary embolisms. Trauma to the legs is also a risk factor.

The pulmonary embolism patient complains of the sudden onset of breathing difficulty. The patient is typically apprehensive and complains of chest pain that may be described as "sharp." With a stethoscope you hear normal breath sounds. Many patients develop a cough, and about one patient in four develops *hemoptysis* (the coughing up of blood). The signs and symptoms are typical of shock. A high flow of supplemental oxygen and rapid transport to a medical facility are the only treatments available for use by the Wilderness First Responder.

Upper Respiratory Infection

An upper respiratory infection (URI) is a viral or bacterial infection of the *mucosa* of the respiratory tract: the nasopharynx, pharynx, larynx, sinuses, and/or bronchi. Patients often complain they have a "cold," but this one is worse than usual. Signs and symptoms may include increased mucous production (with the accompanying "runny nose"), a cough that may be productive, sore throat, muscle aches, a "run-down" feeling, and a fever. Patients need rest and hydration. Decongestants may be given for congestion, and acetaminophen or ibuprofen relieve the aches and fever. Patients who improve with treatment can stay in the wilderness, as long as they want to, but patients who continue to deteriorate may have a bacterial infection requiring antibiotic therapy—and these patients should be evacuated. It is not unusual for a serious URI to become a pneumonia.

Pneumonia

Pneumonia is an infection in the lungs caused by a wide variety of agents, including bacteria, viruses, protozoa, and fungi. The disease process can cause swelling and fluid collection in the lungs, leading to breathing difficulty. Unlike chronic obstructive pulmonary disease, pneumonia is often curable, especially if the patient is young and healthy before the illness. Unfortunately and despite antibiotic therapy, pneumonia still is the fifth- or sixth-leading cause of death in the United States each year.

A patient most in need of emergency care has an infection, most likely bacterial, that produces any or all of six classic signs and symptoms: shortness of breath (the most important), chest pain, chills, fever, increased sputum production, and a productive cough that may produce sputum colored from yellowish to greenish, even brownish. With a stethoscope, you may hear wet breath sounds. The patient is typically exhausted.

The pneumonia patient requires specific antibiotic treatment and should be speedily transported to definitive medical care, preferably while breathing a high flow of supplemental oxygen. In the wilderness the patient may be treated with antibiotics, such as erythromycin, during evacuation. As a group leader you would do well to ask a physician for recommendations and a prescription. The patient may be given *antipyretic* (anti-fever) drugs and should be kept well hydrated. Encourage the patient to breathe deeply from time to time and to cough up the sputum collecting in his or her upper airway.

Chronic Obstructive Pulmonary Disease

The term *chronic obstructive pulmonary disease*, or COPD, covers a collection of diseases sharing the common symptoms of airway obstruction in the small to medium airways, excessive secretions, and/or constriction of the bronchial tubes. Cigarette smoking is the cause of most COPD, with air pollution, genetic factors, and some occupational exposures (such as to coal dust) sometimes playing a role. Because it takes so many years of inhaling "bad stuff" to produce the signs and symptoms of COPD, patients are seldom less than 50 years old. The most widespread of chronic obstructive pulmonary diseases are, by far, *chronic bronchitis* and *emphysema*—and most COPD patients have components of both diseases.

Chronic bronchitis is assessed after a patient has had at least 2 successive years of a productive cough lasting at least 3 months per year. The cough develops because, to be brief, the patient has inhaled so much smoke he or she has damaged the ciliated cells of the bronchial tubes. The *cilia*—hairlike processes—normally sweep excess mucus produced by the bronchial tubes up and out. Once cilia are damaged, excess mucus collects more and more in the tubes, forcing patients to cough continually to clear the airway. Ventilatory effort becomes less efficient, and more and more carbon dioxide is retained. In later stages patients with chronic bronchitis may assume a mild but permanent cyanotic complexion.

Although virtually unheard of in a wilderness context, you may one day find yourself facing a patient with chronic bronchitis who needs assistance due to unusual respiratory distress. Almost undoubtedly the patient can relate the nature of the problem, having been diagnosed years before. The patient requires definitive medical care. Your transport should be rapid. Allow the patient to assume a position that makes it most comfortable for her or him to breathe. If possible, start a flow of supplemental oxygen high enough to meet the patient's needs.

Emphysema results from destruction of the alveoli, destruction that is irreversible. In a normal lung, the appearance of alveoli might be compared to a microscopic bunch of grapes. After years of inhaling a harmful substance, such as cigarette smoke, the alveoli of the emphysematous patient have eroded into rounded spheres, sort of like one larger grape instead of a bunch of smaller ones. Deterioration of the alveoli means a loss of surface area for gas exchange, which means less and less efficient breathing. The alveoli also start to collapse. To keep the alveoli as open as possible, the patient tends to sit up straight and breathe with chest muscles which eventually deforms the chest into a more circular shape instead of the normal oval shape. The chest shape of advanced emphysema is known as "barrel chest." In attempts to breathe more easily, emphysema patients often purse their lips to exhale against resistance, a technique that creates back pressure in the lungs, which helps keep the alveoli from collapsing. Emphysema leaves the patient physically wasted and in a constant state of fatigue.

The patient may require emergency care when, for some reason, the ability to breathe suddenly decreases. With a stethoscope you hear diminished or absent breath sounds on the affected side of the chest. Field treatment is the same as for chronic bronchitis.

Asthma

Asthma is often considered an obstructive pulmonary disease because it does involve airway swelling, increased mucus (very sticky mucus) production, and spasms of the lower airway, all of which lead to breathing difficulty. Asthma, however, is a *reactive airway disease:* Instead of having a disease that deteriorates the ability to breathe, a patient with asthma has an airway that reacts to something by narrowing. Millions suffer from asthma, and an estimated 5,000 die from the disease each year in the United States.

Asthma can be divided into two types. *External asthma* (extrinsic, allergic asthma) is a reaction to substances such as dust, pollen, spores, mold, or animal dander. *Internal asthma* (intrinsic, nonallergic asthma) is a reaction to internal stress caused by such factors as infection, cold air, emotion, or exercise. Asthma is not contagious and not curable. No one knows why some people have asthma, but heredity and environmental factors seem to play a part. Some people have it all their lives, some grow out of it, and some develop it as adults.

Sufferers with a relatively mild form of asthma usually care for themselves, often with an over-the-counter inhaler. Such inhalers release epinephrine in a mist, dilating the airway. Moderate asthma calls for stronger inhaled drugs available by prescription only. Some of these inhalers are for prevention and some are for an acute attack, and it is important to know which is which. A few severe asthmatics may be taking an oral drug to help control their disease. If an asthmatic is taking steroids (prednisone) for management of the disease, he or she is probably more at risk for a severe reaction.

Asthma attacks narrow a section of the airway, often leaving the patient able to inhale with relative ease or actively suck air in past the closure, but not exhale with ease or passively blow air out past the closure. This is sometimes called an *air trapping syndrome*. Breathing during an attack produces wheezes, sometimes audible to the naked ear, almost always audible through a stethoscope, as the patient moves air past the partial closure of the airway. The patient may be using accessory muscles to aid breathing, expanding the chest and lifting the shoulders.

It is of great importance to identify and treat an asthma attack early in hopes of keeping it on the mild to moderate side. The patient experiencing a moderate attack may need your help to shorten the attack. Treatment includes calm encouragement of the patient to relax and breathe with control. Change the environment—move away from known environmental factors, such as pollen. Also, breathing against the back pressure of pursed lips may help open the airway. The patient may require assistance finding and using her or his inhaler. The inhaler should be held with the mouthpiece on the bottom, shaken, and placed between the lips. The patient should exhale before the bottle is depressed into the mouthpiece, releasing the mist. As the mist is released, it needs to be inhaled. The mist must be sucked deep into the airway to work. After inhaling, the patient should hold her or his breath for 5 to 10 seconds before exhaling. The usual dose is 2 puffs, the second one administered after the first one has been inhaled and held.

Severe asthma attacks are uncommon and are very unlikely in a patient who does not have a history of severe asthma. The patient experiencing a severe attack may need your help to survive. Severe attacks usually prevent the patient from speaking except in one- to two-word clusters: "I [gasp] can't [gasp] breathe [gasp]!" The lungs that were full of diffuse wheezes may have now gone silent. If the patient seems extremely tired or tells you he or she is "tired," or "too exhausted to breathe," respiratory collapse may be imminent. When the patient cannot inhale and/or the inhaler is not working, an injection of epinephrine usually is the only life-saving treatment available in the wilderness. The dose is the same as for anaphylaxis, and beesting kits may be used (see Chapter 28: Allergic Reactions and Anaphylaxis). The use of epinephrine for severe asthma falls outside the scope of practice for Wilderness First Responders—unless the WFR is acting under specific protocols established by and managed by a medical adviser. Trip leaders are well advised to ask an asthmatic at the start of a trip if he or she has ever had a severe attack, and if so, how it developed and how it was treated. Going prepared can save a life. When the patient of a severe attack can breathe again, high-flow/high-concentration supplemental oxygen would be of great benefit.

Status asthmaticus is a prolonged asthma attack unrelieved by conventional treatment. Even an injection of epinephrine may not help. Only rapid transport to definitive medical care offers the patient a chance at life.

Care of asthma patients after an attack includes giving copious amounts of drinking water to keep the patient well hydrated. Adequate hydration helps the patient move the sticky mucus out of the airway. If the attack has an external cause, do all you can to prevent future contact with the offending agent.

An asthmatic should carry medications at all times, especially on wilderness trips. Twice as much as necessary should be carried into the wilderness, one set of medications carried by the patient, and one set carried by someone else. On water-based trips, the sets of medications should be carried in separate boats. If one set is lost or ruined, the second set will remain.

Figure 24-1: *Patient inhaling medication.*

Auto-Inhalers

Some asthmatic patients carry auto-inhalers, devices that deliver the medication when the patient inhales. Although useful in ordinary circumstances, an auto-inhaler will not work in a critical situation when the patient cannot inhale.

Evacuation Guidelines

Any patient treated for a medical emergency involving breathing difficulty—pulmonary embolism, pneumonia, or unresolved asthma—should be evacuated as soon as possible, with the exception of a patient successfully treated for mild to moderate asthma or hyperventilation syndrome. During evacuation allow the patient to assume a position that gives the most comfort.

Conclusion

At Jen's side in the Great Smoky Mountains, you do an initial assessment and discover no threats to your patient's life. Your focused assessment uncovers no apparent injuries. The carpopedal spasms indicate the possibility of hyperventilation syndrome.

Asking your co-instructor to move the group back and give you "lots of breathing room," you speak quietly and confidently to Jen, asking her to hold her breath for a count of three. She ignores you. You speak more firmly, telling her she feels sick because she's breathing so fast. She ignores you. Once again, you repeat your instructions. She stops breathing—and faints.

With Jen already on her side, you ensure that her airway is clear and open. It seems like forever, but she starts to breathe again. Her respiratory rate immediately starts to increase. You speak to her, going over your instructions one more time. Now she hears you and tries to control her breathing. It takes about 30 minutes before Jen seems to be completely in control again, not an unreasonable amount of time to recover from hyperventilation syndrome. Several hours pass before she feels like her old self again.

Jen, you learn, was pushing herself very hard to keep up, breathing hard and fast, suffering from the heat and humidity. She began to feel dizzy. Numbness crept into her lips and tongue. Fearing some kind of serious ailment, she began to breathe even faster. She collapsed to the ground, unable to continue.

You decide to slow the pace of the group, stopping more often for rest breaks. Jen, you are confident, will be able to finish the trip.

CHAPTER TWENTY-FIVE

Neurological Emergencies

You Should Be Able To:

1. Describe the assessment and emergency care of an unconscious patient.
2. Describe the signs and symptoms of the most common neurological emergencies, including cerebrovascular accident and seizure.
3. Describe the treatment of the most common neurological emergencies, including cerebrovascular accident and seizure.

It Could Happen to You

In Oregon one of Smith Rock's almost 1,000 named routes is perfect for introducing young students to rock climbing. From a secure anchor at the top, you belay the men and women up the face. From below, your co-instructor coaches them over the trickier moves. It is a warm, bright fall day, full of sunshine, devoid of clouds.

With all the students happily at the top, you tie a static line to a second bombproof anchor and begin to teach rappelling. The general excitement among the students is apparently not shared by Ben. Some years ago Ben suffered a severe traumatic head injury that required surgical intervention. However, his doctor reported that he is fully recovered. Today Ben ascended the rock slowly, requiring extra encouragement to finally make it.

When Ben's turn comes to descend, you check his harness and knots, and he steps back to the edge of the cliff. His nervousness alarms you, and you suggest he consider walking down the easy path to the bottom. Ben declines. He wants to overcome his fear.

As Ben leans out over the rock face, his body stiffens. You calmly remind him to relax and trust the rope. You put tension on the belay rope to remind him you're there providing security. Instead of relaxing, he grows stiffer, his eyes roll up in his head showing nothing to you but white. Ben's feet slip off the edge, and his body begins to jerk in violent convulsions.

The brain, the spinal cord, and the peripheral nerves, known collectively as the *neurological system*—or *nervous system*—may be further divided into two distinct physiological components: (1) the *central nervous system* (CNS), or the brain and spinal cord; and (2) the *peripheral nervous system* (PNS), all the other nerves. As far as performance goes, the brain—in charge of thought, vital body functions, and emotion—uses the rest of the nervous system to send its messages to all parts of the body.

Although a critical aspect of life, the PNS receives little emergency care. You should be able to recognize early signs of damage to the PNS, such as loss of circulation, sensation, and motion, but other than reducing a dislocation or an angulated fracture, Wilderness First Responders primarily document PNS changes and report them to physicians. On the other rescuing hand, you should gear your medical efforts toward preserving and protecting the CNS.

Basic Anatomy of the Brain

The *cerebrum*, the largest part of the brain, is divided into right and left *hemispheres*, the right side controlling the right side of the head and the left side of the body, and the left side of the brain controlling the left side of the head and the right side of the body. The cerebrum houses the speech center near the middle of the brain, the vision center in the back of the brain, and the hearing centers on both sides. In the front of the brain, creativity, abstract thought, and personality traits are housed. Voluntary movement and skilled movement are controlled by the areas of the brain near the top.

A firm, spongy bundle of nerves weighing an average of only 3 pounds, the brain receives approximately 20 percent of the body's supply of blood and oxygen. Despite being made of so many nerve cells, the brain itself, interestingly, cannot feel pain, heat, or even touch, but deprive it of blood and oxygen for more than a few moments and brain cells begin to die. Unlike other cells of the body, nerve cells cannot be repaired.

Unconscious States

Unconsciousness ranks among the more ambiguous terms. The patient could be simply daydreaming, unaware of immediate happenings, on one end of unconsciousness, or completely unresponsive—U on the AVPU scale (see Chapter 3: Patient Assessment)—on the other end. Determining the level of consciousness (LOC) is an early part of all patient assessments. For patients with an altered LOC, you should do your best to determine the reason for the alterations, especially when the patient lies in a wilderness environment, far from definitive care.

Common causes of unconscious states or, to say it another way, things that commonly cause the brain **To STOP** are:

To: Toxins (such as alcohol or carbon monoxide poisoning).

S: Sugar (not enough) and Seizures.

T: Temperature (too high or too low).

O: Oxygen deprivation.

P: Pressure (too much, such as high-altitude cerebral edema and increasing intracranial pressure).

AEIOU TIPS is another helpful mnemonic for determining, with more specificity, why a patient is unconscious (or has changes in mental status):

Allergies: Did an allergic reaction cause unconsciousness (see Chapter 28: Allergic Reactions and Anaphylaxis)?

Epilepsy: Did a seizure cause unconsciousness?

Insulin: Did a diabetic reaction cause unconsciousness (See Chapter 26: Diabetic Emergencies)?

Overdose: Did an overdose of alcohol or legal or illegal drugs cause unconsciousness?

Underdose: Did an underdose of medication cause unconsciousness?

Trauma: Did a traumatic injury, such as a blow to the head, cause a pressure increase leading to unconsciousness (See Chapter 9: Head Injuries)?

Infection: Did a body-wide infection cause unconsciousness?

Psychological/Poison: Did a psychologically traumatic event cause unconsciousness? Is the patient faking unconsciousness? Or is the patient unconscious due to poisoning (See Chapter 27: Poisoning Emergencies)?

Stroke: Did a stroke cause unconsciousness?

The immediate emergency care for all unconscious patients, despite the cause, is generally the same:

1. Protect the airway. Consider rolling the patient into the recovery position to help maintain the airway.
2. Protect the spine unless you do not need to take spinal precautions (see Chapter 8: Spine Injuries).
3. Perform a thorough patient assessment. You will have to be more clever with patients who are unconscious for unknown reasons. Check carefully for medical alert tags. Check the pockets of the clothing for clues. Check the backpack. Ask bystanders who may know the patient. Look for clues near the patient, such as empty bottles or partially eaten substances.

4. Protect the patient from the environment.
5. Monitor the patient for changes, especially changes in LOC.
6. Plan and carry out an evacuation.

Specific treatment depends, of course, on the cause of unconsciousness, but consider giving sugar to patients unconscious for unknown reasons (see Chapter 26: Diabetic Emergencies).

Types of Neurological Emergencies

Cerebrovascular Accident *Stroke*

A *cerebrovascular accident* (CVA)—also called a *stroke*—is an interruption of normal blood flow to a part of the brain. The interruption could be caused by a blood clot forming in a cerebral artery, an embolus circulating into the brain from another part of the body to form a clot, or a hemorrhage from a cerebral artery. Similar to a heart attack, a stroke may be called a "brain attack." As a rescuer, you won't be able to tell the type of stroke involved, but you don't need to: The signs and symptoms are all similar, as is the treatment.

Although not impossible in younger patients, strokes are far more likely in older patients with a history of atherosclerotic disease (see Chapter 23: Cardiac Emergencies).

Signs and symptoms vary, depending on the part of the brain affected by the interruption in blood flow. Alterations in mental status are common—often described by ambiguous terms such as "confused," "stuporous," "semiconscious," or "unconscious"—and may leave the patient unable to adequately manage his or her own airway. An open airway, as always, demands priority attention. Don't be surprised by a conscious patient who has lost the ability to speak, or who has slurred speech, or who complains of memory loss. Breathing irregularities are not uncommon. Changes in heart rate and blood pressure give no accurate assessment clues for a CVA, but if you find a slowing heart rate accompanied by a rising blood pressure, suspect increasing intracranial pressure. Indications of increasing intracranial pressure (ICP) are ominous signs found late in the patient's deterioration (see Chapter 9: Head Injuries). Facial paralysis or facial drooping may be apparent, usually on one side of the face. *Hemiparesis* or *hemiplegia* (weakness or paralysis affecting one side of the body) is likely. Loss of reactivity in one pupil is not uncommon. Blurred or decreased vision in one eye may be a complaint. Incontinence may occur.

The signs and symptoms may last for less than 24 hours, often less than 10 minutes, in which case the patient suffered a *transient ischemic attack* (TIA), also known as a "temporary stroke." A TIA sounds a loud alarm. Patients who have had a TIA are at a relatively high risk for having as stroke relatively soon, often within 90 days.

Signs and Symptoms of Stroke

1. Altered mental status.
2. Inability to speak or slurred speech.
3. Memory loss.
4. Hemiparesis or hemiplegia.
5. Facial droop/paralysis.
6. Incontinence.
7. Vision changes.

Evaluating the Stroke Patient

When evaluating a patient for possible stroke, the Cincinnati Prehospital Stroke scale can be used to identify a high percentage of acute stroke patients with only three physical findings:

1. *Facial droop.* Have the patient smile or show his or her teeth. Look to see if one side of the face does not move as well as the other.
2. *Motor arm drift.* Have the patient hold both arms out straight for 10 seconds with eyes closed. Look to see if one arm does not move or drifts downward compared with the other.
3. *Speech difficulties.* Have the patient repeat back a phrase such as, "You can't teach an old dog new tricks." Look to see if the patient slurs, uses the wrong words, or is unable to speak.

Interpretation: If any one of these three signs is abnormal, there is a 72 percent chance that the patient has had a stroke, according to the Cincinnati Prehospital Stroke scale.

■ TREATMENT OF THE STROKE PATIENT

As far as treatment goes, the Wilderness First Responder is somewhat limited. Emotional reassurance is of great importance, even if you think patients can't hear or understand you. The fact is they often can. Allow the patient to assume a position of comfort unless the patient is unable to do so. Patients unable to assume a position of comfort should be placed on the affected side, the weak or paralyzed side, to protect the airway. High-flow/high-concentration supplemental oxygen should be started, if available. Evacuation should be immediate.

If you suspect a TIA, evacuation also should be immediate. The use of aspirin after a stroke remains controversial, but patients who have had a TIA benefit from aspirin. The National Stroke Association recommends aspirin at a dose of 325 milligrams per day.

■ COMMUNICATING WITH THE STROKE PATIENT

You may be able to create a communication system with a stroke patient who cannot speak by asking him or her questions with "yes" and "no" for answers. Encourage the patient to communicate by either nodding or shaking the head or by blinking the eyes (one blink for "yes," two blinks for "no"). Most patients will feel great relief if you are able to establish some form of communication. Stroke is a devastating event that many patients consider worse than death.

Seizures

Suddenly, there is a great discharge of uncontrolled electrical activity in the *cerebral cortex,* the outer layer of the cerebrum—the "peach pit" part of the brain. There is an episode of involuntary behavior, which may or may not be associated with an altered mental state. Your patient has had a *seizure.*

Although more than 20 different types of seizures have been identified, they can be classified as being either partial or generalized. For simplicity, *partial seizures* may be described as seizures affecting a localized part of the brain, typically causing no loss of consciousness. Violent or jerking motions, if present, will be limited, most often, to a single extremity. The patient may stare blindly, wander aimlessly, or repeat a simple motor movement such as lip smacking. No emergency care is required other than careful monitoring of the patient to make sure the seizure does not progress and/or the patient does not endanger himself or herself.

Generalized seizures involve widespread firing of cerebral neurons. Some patients describe an aura prior to their seizure—a "funny" feeling, a sick feeling, an odd taste or smell. Patients experiencing auras know they're going to have a seizure. Not all generalized seizures cause *tonic-clonic activity,* but, when they do, they present the most dramatic form of seizure. The patient assumes a *tonic* (rigid) posture. The patient's head may turn to one side or be forced backward. Within less than a minute, typically, the patient enters the *clonic* (jerking) phase, rapidly thrashing around, striking the head and extremities on whatever happens to be in the way. Collapse and unconsciousness may occur before, during, or after the tonic-clonic phase. The patient may bite his or her tongue, drool, or lose bowel and/or bladder control. Breathing may sound like snoring or may stop. Skin turns pale or cyanotic. The seizure seldom lasts more than 2 minutes, but it may last longer. Following the seizure is a *postictal* period, a period of confusion, disorientation, drowsiness, and fatigue.

Although known seizure sufferers are commonly referred to as epileptics, it is more correct to say the patient suffers a seizure disorder. The problem, however, may be caused acutely by head injury, heat stroke, high-altitude cerebral edema, diabetes, fever, brain tumor, infection, eclampsia, alcohol or drug withdrawal, and assorted other overstimulations of the brain. Diagnosed patients often take drugs to suppress seizures, such as Dilantin or Tegretol.

■ TREATMENT OF THE SEIZURE PATIENT

A seizure must run its course once it has begun. You can't stop it, but you can protect the patient during the episode. Do not restrain the patient. Move objects away that might cause damage if hit. Try to place a cushion under the patient's head using a shirt or parka. The patient cannot swallow his or her tongue, so **do not try to put anything into the mouth.** Many seizure sufferers are harmed by misdirected aid.

If consciousness does not return immediately after the event, roll the patient gently into a stable side position to maintain an adequate airway. High-flow/high-concentration supplemental oxygen, if available, should be started. Take some notes on what happened: what time the seizure occurred, how long it lasted, focal areas if you noticed any, and triggering events if known. Stay calm and comfort the patient once a normal level of consciousness returns. Protect the patient's dignity. Allow time and privacy for changing of clothes, if required. Most often the patient does not remember the event. Don't let the patient drink or move around at first. Check for injuries that could have occurred.

The greatest danger to a seizure patient is *status epilepticus*, a persistent seizure or series of seizures with no time for adequate breathing, a true emergency. If seizures go on for too long, permanent damage or death may result. The patient needs medical intervention immediately, but, unfortunately, there is little or nothing to be done in the wilderness for this person other than rapid evacuation.

A person diagnosed with a condition that causes seizures does not need to be immediately evacuated if a seizure occurs, as long as he or she is dealing with the situation, and the intended wilderness pursuits do not jeopardize his or her life. Patients with first-time seizures or who have seizures for unknown reasons should be evacuated for medical attention, but there is seldom reason for a rapid evacuation. Once a seizure has occurred, thoughtful consideration should be given to continuation of the wilderness trip, and seizures for some known reasons, such as heatstroke or high-altitude cerebral edema, do merit a rapid evacuation.

A question often arises: Should known seizure sufferers risk participating in an extended wilderness venture? Many outdoor programs, on the advice of their physicians, recommend that someone with a seizure disorder be seizure-free for at least 1 year prior to a trip. Seizure-suppressing drugs, prescribed to the patient, should be carried in twice the amount thought needed. A full set of drugs should be carried by someone other than the patient, in case of loss or ruin of one set.

Evacuation Guidelines

In almost all cases, a patient with a significantly altered mental status should be evacuated, especially if the cause cannot be ascertained and/or adequately treated. A patient assessed with a stroke or TIA should be evacuated as soon as possible. A patient suffering a seizure should be evacuated right away if the seizure is a first-time occurrence, the seizure occurred for unknown reasons, or if multiple seizures occur in a short period of time.

Conclusion

At Smith Rock in Oregon, you are empowered by a rush of adrenaline, and you pull Ben to the top of the cliff via the belay rope. Dragging him quickly away from the edge, you pad beneath his head with a coil of the rope. In less than 2 minutes, his seizure subsides.

Ben seems confused and drowsy. You roll him into a stable side position and proceed with a patient assessment that reveals no injuries. Drool drains from Ben's mouth, and he has wet his pants. You ask the members of the group that remain nearby to descend and return with Ben's day pack.

By the time the other students return, Ben has regained a relatively normal level of consciousness. He changes into clean shorts. Keeping a watchful eye on Ben, you lead the group back to the foot of the cliff and prepare for the ride home.

Diabetic Emergencies

Diabetes was known by early Greek physicians as a disease producing copious, sweet-tasting urine. It is not recorded who did the taste test, but the name *diabetes mellitus* (Greek for "a passing through of sweetness or honey") stuck. Diabetes is often thought of as a disturbance in sugar balance in the blood. It's actually much more than that. Defects in the action of the hormone *insulin* affect fat and protein as well as sugar metabolism. In the short term diabetes can manifest itself as too much or too little sugar in the blood. Long-term complications of diabetes include high blood pressure; heart, blood vessel, and kidney disease; blindness; and poor wound healing.

There are two main types of diabetes. In *insulin-dependent diabetes,* or *Type 1 diabetes,* (5 to 10 percent of cases), the body's immune system destroys the pancreatic cells that make insulin. People with this form of diabetes need insulin injections to survive. *Noninsulin-dependent diabetes*, or *Type 2 diabetes*, (90 to 95 percent of cases), is caused when cells become resistant to insulin or the pancreas produces too little insulin. Noninsulin-dependent diabetics may be able to manage their blood sugar level with diet, exercise, and oral noninsulin medications.

According to current estimates, diabetes effects 6 percent of the U.S. population, or 17 million people, and about half of those are undiagnosed. The past decade has seen a dramatic rise in the incidence of diabetes mellitus. If you're leading or participating in a wilderness trip or camp or perhaps responding as a member of a rescue group, you're likely to work with a person with diabetes.

Normal Physiology

Glucose, a form of sugar, is the fundamental fuel for metabolism. It's burned to produce energy. After eating, food is digested; a lot of it is processed into glucose, and the blood glucose level rises. In nondiabetic people the pancreas secretes insulin into the blood, carefully matching the amount of insulin to the blood glucose level. The insulin helps the glucose diffuse across cell membranes; without insulin the glucose stays in the blood (except in the brain, where cells take in glucose without insulin). The cells then burn the glucose to drive life-sustaining chemical reactions. As the blood glucose level decreases, the amount of insulin produced by the pancreas decreases. The human body stores the glucose it doesn't use as glycogen in muscles and the liver.

Diabetic Physiology

In people with diabetes the blood sugar rises after eating, but the pancreas is unable to secrete sufficient insulin, or the insulin it does secrete is inefficient. Without insulin too much glucose remains in the blood and outside the cells. The cells are starved for glucose. The kidneys attempt to excrete the excess sugar (thus the sweet-tasting urine). Without insulin patients urinate more, and as a result the patient becomes dehydrated and develops electrolyte imbalances.

Hyperglycemia

Diabetics who have defective or insufficient insulin can develop *hyperglycemia* (high blood sugar). Additionally, when diabetics become ill they are in danger of developing hyperglycemia as the body responds to the infection by increasing blood sugar levels. On wilderness trips there is also the possibility of losing the insulin. Without insulin, as mentioned earlier, the kidneys work to excrete the abundance of sugar in the blood, along with water and electrolytes, and the result is dehydration, electrolyte problems, and altered metabolism.

Hyperglycemia develops slowly. The first symptoms are increased hunger, increased thirst, and an increased volume of urine output, and there may be nausea and vomiting. The patient's breath may have a fruity odor from the metabolism of fats as an energy source, the result of the body's demand for nutrition. There may be abdominal cramps and signs of dehydration that include flushed, dry skin.

The brain, which needs glucose almost as much as it needs oxygen, absorbs glucose freely from the blood. Because glucose is able to cross the brain-blood barrier without insulin, the onset of neurological changes is delayed, often taking days to appear. Loss of consciousness, therefore, is a late and very serious sign—although you can expect to see changes in mental function, such as attitude alterations (alterations that often make the patient appear "drunk") by the time the breath smells fruity.

Signs and Symptoms of Hyperglycemia

1. Increased hunger, increased thirst, increased urine output, fatigue.
2. Restless, "drunken" level of consciousness
3. Weak, rapid heart rate.
4. Increased respiration rate.
5. Skin is warm, pink, and dry.
6. Breath odor is sweet.

Treatment for Hyperglycemia

The hyperglycemic patient has a complex physical disturbance, is dehydrated with an electrolyte imbalance, and needs the care of a physician—unless you can access the patient's insulin and she or he is capable of administering the correct injection. Treatment is supportive: airway maintenance, vital signs monitoring, and treatment for shock. *Do not give insulin if the patient cannot self-administer.* Without a thorough knowledge of how much, what kind, and when to inject insulin, the patient can be seriously harmed, even killed. If the patient is alert, give oral fluids to combat dehydration.

Hypoglycemia

Hypoglycemia (low blood sugar) can result from the treatment of diabetes, not the diabetes itself. Without a pancreas releasing insulin automatically to match the blood sugar level, insulin-dependent diabetics must monitor their blood sugar (with a glucometer) and match injections of synthetic insulin to their food and activity level. Ideally, their blood sugar, in the diabetic under control of his or her disease, stays within normal limits. Hypoglycemia can occur if the diabetic skips a meal but takes the usual insulin dose, takes more than the normal insulin dose, exercises strenuously and fails to eat, or vomits a meal after taking insulin. If a diabetic takes too much insulin and not enough food, the blood sugar level will be greatly depleted, insufficient to maintain normal brain function.

In contrast to hyperglycemia, the onset of signs and symptoms for hypoglycemia is usually rapid and reflects the sensitivity of the brain to low blood sugar. Altered mental status—confusion, slurred speech, dizziness, nervousness, weakness, and irritability—are typical early signs and symptoms. Other vital signs changes will include a rapid heart rate, normal or shallow respiratory rate, and pale, cool and clammy skin. As the blood sugar drops, disorientation and unconsciousness can result. *Diabetic hypoglycemia is a true threat to life and requires immediate treatment.*

Signs and Symptoms of Hypoglycemia

1. Level of consciousness may range from disoriented to irritable to combative to unconscious; seizures are possible.
2. Heart rate is rapid.
3. Respiratory rate is normal or shallow.
4. Skin is pale, cool, and clammy.
5. Blood pressure may be unchanged or slightly elevated.
6. Breath odor does not change.

Nondiabetic Hypoglycemia

Hypoglycemia may also be a diagnosis made after a series of laboratory tests that reveal a condition in which blood glucose is persistently lower than normal. This is not diabetes, and the patient should have been given precise dietary and exercise instructions concerning the management of his or her glucose levels. The body is amazingly adept at maintaining a normal level of glucose, even after extended periods of exercise, and nondiabetic hypoglycemia is a rare condition.

Treatment for Hypoglycemia

Brain cells can suffer permanent damage quickly from low blood sugar levels. Treat hypoglycemia by administering sugar. If the patient is conscious, a sugary drink or candy bar (simple sugars) will help increase blood sugar level. If the patient is unconscious, establish an airway by placing the patient in a stable side position, and then carefully rub sugar, or a sugary substance, into the patient's gums. Sugar is absorbed through the oral mucosa. Improvement may not occur quickly, and you could find yourself rubbing in sugar for quite awhile. But don't give up.

The diabetic may have an injectable medication, glucagon, which mobilizes sugar reserves stored in the muscles and liver as glycogen. If the patient is an expedition companion, ideally you are well informed about his or her diabetes and know when and how to use the glucagon.

If you cannot return the patient to a normal level of consciousness, a rapid evacuation is necessary.

Hypoglycemia or Hyperglycemia?

How can you tell the difference between hypoglycemia and hyperglycemia?

Hypoglycemia usually has a rapid onset. The patient is pale, cool, and clammy and has obvious disturbances in behavior and/or level of consciousness. Hyperglycemia has a gradual onset. Often, the patient acts "drunken" and rarely may be found in an unexplained coma, with flushed, dry skin. The fruity breath odor may be present. The patient in hypoglycemia will respond to sugar, and the hyperglycemic will not—but the extra sugar will cause no harm.

If you get there in time, ask two questions of the diabetic patient: "Have you eaten today?" and "Have you taken your insulin today?" If the patient has taken his or her insulin but has not eaten, you should suspect hypoglycemia. If the patient has eaten but has not taken insulin, hyperglycemia should be suspected. If you remain unsure, give sugar.

Diabetics in the Wilderness

In the wild outdoors the physical and emotional stress, the new physical and social environment, the heavy packs, the altitude, the sun, and the battle against dehydration may be new challenges for the diabetic, but they are ones that can be managed through care and education. People with diabetes routinely participate in extended wilderness expeditions.

Discuss the illness beforehand with the diabetic person undertaking the expedition. Most diabetics are very knowledgeable about their reactions and intuitively know if they are getting into trouble. Make sure you both understand the disease, the timing and side effects of any medications, the appropriate emergency treatment, and any other of the person's health needs. It is important to inform the rest of the group—especially the person's tent mates—about the condition and how to deal with it in an emergency. It is important for diabetics to eat at regular intervals. Carry twice the amount of insulin and syringes that you and the diabetic think he or she will need, and carry them in separate packs or drybags, in case one set is lost. Carry extra batteries for the glucometer.

Prevention of Diabetic Emergencies

If you are a leader on a wilderness trip, and the group has one or more members with diabetes, there are precautions you can take to reduce the chance of a diabetic emergency to a minimum:

1. Many outdoor programs, under the advice of their physicians, ask anyone with diabetes to be free of "crashes" (the need for emergency care) for at least 1 year, an indication that he or she understands the illness and how to cope with it.
2. Ask anyone with diabetes to practice on overnight and weekend trips before undertaking longer wilderness adventures. The diabetic must be able to make adjustments due to the changes she or he will experience in normal routine, changes such as exercise level, variations in diet, cold stress, and possible illnesses, such as diarrhea or vomiting.
3. Encourage anyone with diabetes to stay in shape, a high level of physical fitness being an extremely important factor in health maintenance.
4. Refresh your memory concerning diabetic emergencies and know how to give injections.
5. Remind anyone with diabetes to stay well hydrated.
6. Carry a readily available carbohydrate source, such as glucose tablets or gel, and make sure everybody traveling with you knows where the glucose is and how to best help get it into the patient if the need arises.
7. Ask anyone with diabetes to carry twice as much insulin, glucagon, and syringes as he or she thinks will be needed. Divide the supplies so that they are not all in one pack or canoe bag, just in case a set of supplies is lost.

8. Remember that insulin must be carried near the body in extremes of cold to prevent freezing.

9. Remember to protect insulin from overheating in hot climates.

10. Learn how to use the glucometer. Suggest that anyone with diabetes check his or her blood glucose often the first few days of the trip. If readings are high, suggest reducing the amount of carbohydrates being taken in before increasing the insulin dose.

11. Suggest that anyone with diabetes limit the use of concentrated sources of simple glucoses. For snacks, suggest nutrient-dense foods such as nuts and dried fruits.

12. Remind anyone with diabetes to not skip meals.

13. Work with anyone with diabetes to plan each day the night before, including wake-up time, activities, and meals. Sleeping late can be adjusted for by merely pushing all meals back. Adjust for late dinners by switching the bedtime snack with the regular dinner time or by reducing the morning insulin dose and taking an evening injection of regular insulin just prior to the evening meal.

Evacuation Guidelines

Evacuation of a hypoglycemic patient depends on whether or not the patient returns to a normal condition and whether or not the patient wishes to remain in the field. Hypoglycemic patients that maintain an altered level of consciousness require a rapid evacuation. If treatment of the hyperglycemic patient is not working, the patient requires evacuation to a medical facility.

Conclusion

Aware of your client's diabetic condition, you recognize the onset of hypoglycemia. Grabbing a full water bottle, you throw in a handful of presweetened Kool-Aid and add a couple of teaspoons of sugar. The man eagerly accepts the bottle and shakily drinks until the container is empty.

As your patient begins to calm down and return to normal, you serve dinner—the fine trout you caught today. The evening discussion reveals not only that your client has been exercising much more strenuously but also that he skipped lunch in his eagerness to catch fish. He vows to be more careful tomorrow.

CHAPTER TWENTY-SEVEN

Poisoning Emergencies

You Should Be Able To:

1. Describe the ways in which poison can get into the human body.
2. Describe the treatment for the most common wilderness poisoning emergencies.

It Could Happen to You

Unseasonable snow keeps you and your partner, Daryl, huddled in the tent where you've been for more than 24 hours, waiting for a break in the weather to continue your ascent of Mount Rainier. Hunger and thirst lead to firing up the stove under the vestibule, but a harsh wind convinces you to move it inside where water now simmers, melting the snow you periodically dump into the pot.

At first you leave the tent's door about one third open, but rushes of icy air lead you to close the opening. Eventually you're zipped up inside the tent without adequate ventilation. You develop a mild headache, pain that resolves when you step outside for a few minutes to urinate.

Back in the tent, and several cups of tea and a freeze-dried dinner later, your mild headache returns and reaches toward throbbing proportions, and your stomach has you thinking your culinary efforts might resurface on the tent floor. Daryl, complaining of head pain earlier, now seems irritable and increasingly confused. He periodically gasps for air, grumbling that he can't seem to get a full breath.

Any substance you ingest, inhale, absorb through your skin, or get injected into your body that causes a malfunction in normal biological processes is called a *poison*. Most deaths from poisoning occur in homes to small children who eat something "bad." In the wilderness fatal poisonings are rare. When they do happen, it is usually the result of cooking in an inadequately ventilated tent or snow cave that lets carbon monoxide build up or of ingesting a poisonous plant, usually a deadly fungus. Absorbed poisonings seldom occur outside of industrial or farm settings where strong chemicals are used. Injected poisons in the wilderness are almost always from a bite or sting (see Chapter 21: North American Bites and Stings). In cases of serious poisonings proper intervention by the Wilderness First Responder may save a life.

Ingested Poisons

Mushrooms

Almost all deaths from mushroom ingestion have been in adults who mistakenly ingested poisonous fungi for dinner or for a hallucinogenic high.

The mushroom most likely to kill? The *Amanita* species (death cap, death angel, destroying angel) is responsible for 90 to 95 percent of all human deaths by mushrooms. Amanitas contain cyclopeptide amatoxins and can produce fatal liver and kidney failure in 2 to 3 days. Typically growing under deciduous trees in the United States, *Amanitas* show a yellowish to white cap 1.5 to 6.5 inches (4 to 16 centimeters) in diameter and a thick stalk 2 to 7 inches (5 to 18 centimeters) in length, with a large bulb at the base. The gills under the cap are usually easily visible and white to green in color.

Onset of gastrointestinal distress—severe nausea, vomiting, abdominal cramps, and diarrhea—usually occurs within 6 to 12 hours after ingestion of *Amanita*, as well as all other potentially lethal mushrooms. As a general rule remember this: If symptoms develop within approximately 2 hours of ingestion, it is unlikely that the mushroom is one of the potentially lethal varieties. In other words, if stomach discomfort soon follows mushroom munching, the chance of serious mushroom poisoning is extremely slim.

Note: By the time signs and symptoms show up in serious mushroom poisonings, it's too late to do anything except hurry to a hospital where supportive care might save the life of the patient. That means if you think someone has eaten a bad mushroom, *start treatment quickly*. If you're in doubt, *start treatment quickly*. You don't want to wait for signs and symptoms. Each moment that passes lets more and more poison be absorbed into your patient's system (see "Treatment for Ingested Poisons," this chapter).

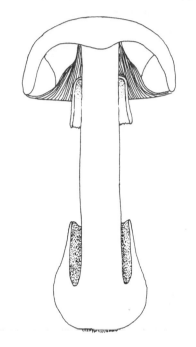

Figure 27-1: *Cross section of death cap.*

Jimsonweed

Hallucinations and deaths in the wilderness have been attributed to a pot of tea—a tea brewed voluntarily by the patient from jimsonweed, *Datura stramonium*, a plant that grows all over North America, sometimes known in differing locales as stinkweed, thorn apple, Indian apple, angel's trumpet, sacred datura, and belladonna. The genus *Datura* is responsible for a large percentage of plant poisonings that show up each year in hospitals of the United States.

Of the twenty-five species of *Datura* growing worldwide, most are shrubs that grow close to the ground. A few are treelike and reach a height exceeding 35 feet (11 meters). But common to all are trumpet-shaped flowers in shades of red, pink, yellow, or, by far most familiar, white. Native American cultures used *Datura* to treat venomous snakebites, insect bites and spider bites, asthma, sore throat, nasal congestion, and bruises. But most often it has been, and still is, used to produce a state of euphoria. It is intentional ingestions by seekers of a hallucination that now account for most poisonings. The leaves, roots, stems, seeds, and fruit of

jimsonweed contain a combination of alkaloids including atropine and scopolamine. These alkaloids are *anticholinergics* that affect the nervous system, and in proper doses they are very useful. Scopolamine is the active ingredient in transdermal patches, the little sticky things that go behind your ear to alleviate motion sickness. In large doses the chemicals produce delirium, fast heart rates, dilated pupils, and fever. In larger doses anticholinergics may produce coma, seizure, respiratory failure, and death.

Water Hemlock

Water hemlock is considered the most violently poisonous plant in the northern temperate regions of Earth. Growing predominantly along waterways, hemlocks are tall perennial herbs, members of the carrot family. The entire plant is poisonous, from the flat-topped clusters of small white flowers, down past the narrow, toothed, pointed leaves to the roots that exude a gummy yellow juice when cut. One mouthful of root will kill a large adult. *Cicutoxin*, the poison in hemlock, attacks the central nervous system, causing tremors, spasms, convulsions, paralysis, and, often, death. On the way to death, extreme stomach pain, diarrhea and vomiting are common.

Solanums

An interesting group of plants, containing around 1,500 species, is the genus *Solanum*. This genus includes the deadly nightshade (*S. nigrum*), climbing nightshade (*S. dulcumara*), Jerusalem cherry (*S. pseudocapsicum*), the potato (*S. tuberosum*), and the eggplant (*S. melongena*). Each of these has caused poisoning in humans. The poison is a group of alkaloids lumped together under the name *solanine*. It occurs in different parts of different Solanums. The vines, for instance, of the potato and eggplant contain large amounts of solanine, but the ripened tubers and fruits are perfectly wonderful to eat. Green potatoes and immature eggplants contain substantial amounts of solanine and should, of course, be avoided. Symptoms of solanine poisoning include nausea, vomiting, fever, weakness, and, sometimes, paralysis.

Seeds

Other sources of plant poisonings not suspected by those people who regularly eat them include the seeds of the apple, apricot, plum, peach, and cherry. These seeds contain *amygdalin*, a cyanogenic glucoside. When broken down in the digestive tract, the seeds release cyanide. Although it is extremely rare for someone to eat enough of the seeds to cause harm, fatal ingestions have been well documented.

Medications

Not all ingested poisons in the wilderness are plant poisons. Students, especially young students, on outdoor programs may intentionally or unintentionally ingest harmful levels of medications from their own or the group first-aid kit. Acetaminophen (Tylenol), for example, can be toxic to the liver when taken in large doses. Ibuprofen (Advil) can be destructive of the kidneys when taken in large doses over a long period of time.

Treatment for Ingested Poisons

There is no such thing as a Wilderness Poison Control Center, but cell phones may allow you to call the American Association of Poison Control Centers at (800) 222-1222. The people there can provide immediate and exact information to guide you.

In their absence, you are left with some management principles. The most important is that *limiting the absorption of the poison from the gastrointestinal tract is the prime goal of field management.* There are three practical ways to do this in the wilderness:

1. Dilute the poison with as much water as the patient will drink.
2. Induce vomiting.
3. Bind the toxin with activated charcoal.

If vomiting can be induced early, within 1 hour of ingestion, it may be very beneficial, especially when a mushroom poisoning is suspected. Stimulation of the gag reflex typically works as well, or better, than anything else to induce vomiting. Lean the patient forward, ask him or her to stick the tongue out as far as possible, gently reach into the mouth with a finger, and tickle the back of the throat. You might want to stand back or off to the side.

Do *not* induce vomiting if:
1. The patient is losing consciousness.
2. The patient has a seizure disorder or heart problems.
3. The patient has swallowed corrosive acids or bases, both of which can increase damage as they come up.
4. The patient has swallowed petroleum products that can cause serious pneumonia if even a small amount is breathed into the lungs.

In the case of ingestion of corrosive chemicals or petroleum products, get the patient to drink a liter of water. Diluting the poison reduces its effects. If someone takes an accidental swallow of white gas, the petroleum product you brought for the stove, do not worry. White gas ingestion can almost always be managed with dilution and without any harm to the patient.

Activated charcoal is postcombustion carbon residue treated to increase absorbency. With most poisons, even if your care will be short term, binding the toxins with charcoal may provide an even better treatment than inducing vomiting. Why? By the time you realize you have a poisoned patient, much of the toxin has already passed out of the stomach, and there are *no* contraindications for the use of activated charcoal. Charcoal may also be administered postvomiting. The usual dose is 50 to 100 grams for adults and 20 to 50 grams for children. Although it is odorless and tasteless, swallowing the slurry of fine black powder may prove a chore. It can be added to flavored drinks (such as fruit drinks), but it should *not* be mixed with milk or milk products.

Activated charcoal can be found in almost any pharmacy and can be bought without a prescription. Brands such as Sorbitol enhance the speed with which contents pass through the bowels, which reduces the time the poison has to be absorbed. Activated charcoal, however, is an impractical inclusion in most wilderness medical kits due to its size and weight and the fact that it's seldom needed. You might find it at expedition base camps, on raft trips, or in established backcountry camps.

If the patient goes unconscious, evacuation to a medical facility is probably what is going to save a life. Keep the patient on her or his side during the evacuation to maintain the airway.

Prevention of Poisonings by Ingestion

Poisonous plants grow abundantly in some wilderness areas, where they occur naturally or have escaped captivity and gone wild. The potential for harm is created when the cardinal rule of wild edibles is forgotten: **Absolutely positive identification of wild edibles must precede consumption.**

Inhaled Poisons

Carbon monoxide (CO) poisonings account for approximately 50 percent of the deaths by poison in the United States every year, making CO the leading cause of fatal poisonings. This invisible, odorless, tasteless, nonirritating gas creates one of the few serious poison threats on wilderness ventures. CO is the result of incomplete combustion of any carbon-based fuel—gasoline, kerosene, natural gas, charcoal, and wood. Outdoor stoves, for example, burn inefficiently in an enclosed space with inadequate oxygen. Higher altitudes increase the chance of poor combustion of the fuel being burned, and CO poisoning may be mistaken for altitude illness (see Chapter 18: Altitude Illnesses).

Once inhaled, CO enters the blood of the patient, where it is approximately 200 to 250 times more bondable than oxygen to the hemoglobin of red blood cells. Hemoglobin normally carries oxygen to the cells of the body. With CO attached to it, hemoglobin can't carry as much oxygen and can't release what is attached as efficiently. The brain and heart, the organs most in need of a constant flow of oxygen, begin to deteriorate. Tissue death can occur rapidly and lead to the death of the patient.

As the amount of attached carbon monoxide increases in the body to as little as 10 percent of the maximum potential, the patient develops a terrible headache, nausea, vomiting, and a loss of manual dexterity. At 30 percent of maximum potential, the level of consciousness descends into irritability, impaired judgment, and confusion. It is increasingly difficult for the patient to get a full breath, and he or she grows drowsy. At 40 to 60 percent of maximum potential, the patient lapses into a coma. Levels above 60 percent are usually fatal. The cherry-red skin often associated with the terminal stage of CO poisoning is, in truth, very rarely seen. Death typically results from heart failure, not the pleasant drift into slumber depicted in some movies.

In the field treatment of CO poisoning is simple: Move the patient to fresh air. If the patient has been exposed to low concentrations of CO (or high concentrations for a short time), he or she will probably recover completely in a few hours. The half-life of CO attached to hemoglobin is about 5.5 hours. If the concentrations have been high, the patient will not get better and may die even removed from the source of the gas. A high concentration of supplemental oxygen is the most important treatment for severe CO poisoning. Rapid evacuation to a high-pressure (hyperbaric) chamber may prove beneficial. Although controversial, use of a Gamow Bag will not harm the patient and may prove of some benefit (see Chapter 18: Altitude Illnesses). Patients unconscious from CO poisoning will need to have their airway maintained during the evacuation.

Treatment for Inhaled Poisons

For any inhaled poison, including the intentional inhalation of poisons (such as glue) for a hallucinogenic high, the following steps should be taken:

1. Remove the patient from exposure to the poison.
2. Administer high-flow/high-concentration supplemental oxygen, if available.
3. Evacuate the patient with serious CO poisoning to a hyperbaric chamber.

Prevention of Poisonings by Inhalation

Do not burn any carbon-based fuel inside a tent or snow cave, even if the shelter is well ventilated. Operating a stove under the open vestibule of a tent, however, is usually safe for short periods of time.

Absorbed Poisons

Rare in the wilderness, dry poisons should be brushed off, and wet poisons should be washed out and off.

Injected Poisons

In urban settings injections are a common source of poisonings due to the use of "recreational" drugs. In the wilderness injected poisons are most often the result of the bite or sting of a venomous creature (see Chapter 21: North American Bites and Stings).

General Treatment Guidelines for Poisonings

With any suspicion of a poisoned patient, ask early about nausea and vomiting, abdominal cramps, diarrhea, loss of visual acuity, muscle cramps, or anything else unusual. Especially indicative of serious poisonings are changes in the level of consciousness of the patient and changes in the respiratory drive of the patient. All vital signs should be monitored with alertness to changes that indicate shock.

The historical evidence you gather may be extremely helpful in assessing and dealing with the problem:

1. What was ingested, inhaled, absorbed, or injected?
2. How much?
3. When?
4. Who (age, sex, body size of the patient)?
5. With ingested poisons, when did the patient last eat? What else is in the stomach?
6. If more than one person suffers, what possible poison do they have in common?
7. Was contact with the poison accidental or intentional?

Evacuation Guidelines

Any patient suspected of the ingestion, inhalation, absorption, or injection of a potentially lethal dose of a poison should be evacuated as soon as possible. Rapid evacuation is critical for patients with changes in level of consciousness and changes in respiratory drive.

Conclusion

Suddenly the light of realization shines on your tent on Mount Rainier: carbon monoxide poisoning. You turn off the stove and open wide both tent doors. A blast of cold air sends a shiver down your back. After approximately 20 minutes, you notice the ill effects are beginning to wear off.

Daryl, whose exposure to the gas was constant, is no worse—but no better, either. To attempt a descent in this storm, which has strengthened in intensity, is ridiculous. There's nothing to do other than keep the tent ventilated and hope for the best.

Daryl drifts in and out of slumber during the evening, falling fast asleep some time well after dark. His breathing, you are overjoyed to note, becomes more and more regular and easy. Early the next morning he awakes, not exactly feeling great, but definitely well along the road to full recovery.

CHAPTER TWENTY-EIGHT

Allergic Reactions and Anaphylaxis

You Should Be Able To:

1. Describe an allergic reaction.
2. Describe the treatment for an allergic reaction.
3. Describe an anaphylactic reaction.
4. Describe the treatment of anaphylaxis, including the importance and methods of using epinephrine.

It Could Happen to You

It's a fine spring day, plants blooming in a riot of color, and your group of budding naturalists is busy identifying flowers near Crested Butte, Colorado. The alpine meadow, bordered by a deep creek, is a picture of serenity until a scream pierces the air. You see one of your students running across the meadow engulfed in a swarm of bees that angrily spit from a disturbed nest. The student runs to the creek and immerses herself to escape. By the time you reach the water most of the bees have given up. You dodge the last few buzzers to reach her side. She sobs hysterically from pain and fright. You coax her onto shore, noticing at least a couple of dozen red, raised welts from the stings.

The next few minutes pass so quickly you can hardly believe what is happening. Before your eyes her face begins to redden and swell. Suddenly she complains not of pain but of increasing difficulty catching her breath. Her breathing grows ragged and gasping. She can only choke out one- and two-word sentences. Less than 5 minutes after you help her from the creek, she collapses on the ground, her skin flushed, her swollen tongue protruding slightly from her mouth.

An *allergic reaction* results from an acquired hypersensitivity to a substance that causes no reaction in the great majority of humans. The allergy-causing substance is called an *allergen*. Someone with an allergy has, essentially, an immune system disorder that develops *after* that individual has been exposed to the allergen. You have to contact the allergen at least once to develop the hypersensitivity. Virtually anything that can be swallowed, rubbed on the skin, injected through the skin, or inhaled can result in an allergic reaction.

Types of allergic reactions range from relatively mild and delayed to immediate and severe. Reactions can be unpleasant, frightening, and in the most severe case of immediate reaction, potentially fatal.

Allergens that commonly cause allergic reactions include foods, environmental agents (such as pollen), animal dander, plant oils, medications taken either orally or by injection, and animal bites and stings, especially insect stings.

Some Specific Allergens

Foods: Shellfish and other seafood, nuts, berries, preservatives and food additives (such as MSG, or monosodium glutamate).

Environmental agents: Pine pollen, cottonwood pollen, ragweed.

Medications: Penicillin and other antibiotics, aspirin and other anti-inflammatory drugs, anesthetics.

Animals: Hymenoptera (bees, hornets, wasps, yellow jackets, fire ants), jellyfish, spiders, caterpillars.

Allergic Reactions

Allergic reactions are caused by an excessive release of histamines and other substances from the body's immune system in response to the presence of allergens. A relatively small release of histamine causes mild to moderate signs and symptoms. A massive histamine release causes, among other reactions, increased dilation of capillaries, which could lead to shock in a severe reaction, and constriction of the smooth muscles of the bronchial tubes, which could lead to asphyxiation in a severe reaction. These reactions can occur rapidly (within seconds) following exposure, or they may take several hours to develop.

On the mild to moderate end of things, allergic reactions are exceedingly common. An immediate local reaction—a red, swollen area (sometimes a *large* red, swollen area)—encountered in the wilderness, for instance, is generally from stings or puncture wounds that introduce an allergy-causing substance under the skin. Bees are probably thought of most often in this type of reaction. Cold packs or cold compresses may alleviate some of the local pain. The immediate use of a nonprescription antihistamine, such as diphenhydramine (Benadryl), may be helpful in minimizing the allergic reaction.

Even more common are the patients who suffer seasonal "hay fevers." It begins as an itch in the nose—gradually or abruptly—just after the onset of the pollen season. It spreads, like ants fanning out from a disturbed anthill on multitudes of tiny feet, to the eyes, the roof of the mouth, and the back of the throat. Eyes begin to water, and the patient begins to sneeze. Hacking cough, headache, rash, loss of appetite, lack of sleep, depression, and irritability are all possible symptoms. There might be a feeling of tightness in the chest and a whistling sound—a wheeze—when the patient takes a breath.

In spring allergies are often caused by pollen from trees, including oak, elm, maple, alder, birch, and cottonwood. In summer allergies are likely to be triggered by grass pollens or weed pollens. Fall allergies may result from weeds and, occasionally, from airborne fungus spores. Fungi sometimes cause allergic reactions in winter. All of these allergies are often lumped under the general category of "hay fever."

About 95 percent of all hay fever medications contain the same ingredients, and they can be divided into two categories: *antihistamines* and *decongestants*. It doesn't matter what you take, as long as it works for you.

Antihistamines work by blocking the body's normal response to histamines. That usually makes the symptoms ease off. All over-the-counter antihistamines also tend to cause drowsiness. You can put the drowsiness to good use if you need a mild sleeping aid, but it can be dangerous if you're hiking a rough trail in steep terrain and debilitating if you're trying to cast a fly precisely. Now available over-the-counter, clemastine (sold as Tavist-D) does not cause drowsiness in many people who take it.

Decongestants work by decreasing blood flow to nasal tissue, shrinking mucous membranes and opening up the airway. Sometimes a decongestant puts the user's nerves on edge, so the medication should be tried during the day before taking it at bedtime. Anyone with high blood pressure, heart disease, thyroid disease, or diabetes should use decongestants with caution.

Anyone who has a tendency to react to allergens with wheezing and chest tightness should carry an over-the-counter pocket inhaler. The inhaler contains a drug, usually epinephrine, that dilates constricted bronchial tubes. The mist is squirted into the mouth while the user is inhaling, and the airway opens up almost immediately.

Another way to fight allergies is to visit a physician who specializes in the problem. First the doctor figures out what an individual is specifically allergic to. There could be a vaccination program ("allergy shots") to cure the particular sensitivity. Allergy shots do not pump medications into the body. They are, instead, small doses of what an individual is allergic to, given over a long period of time, until that person develops a natural immunity to the allergen.

Immediate generalized reactions may come from an allergen. The result may be increased allergic symptoms (nasal and chest congestion, itchy eyes)—plus *bronchospasm* (spasms in the muscles of the bronchi), *urticaria* (hives), and *angioedema* (swelling of the mucous membranes of the lips, mouth, or other parts of the respiratory system). These general reactions usually do not result in true anaphylaxis, but they can be quite severe and cause breathing difficulty that is potentially dangerous. Late generalized allergic reactions can occur starting 6 hours after exposure, but serious reactions causing breathing difficulty have not been known to develop after 1 hour.

Anaphylaxis

Anaphylaxis is a true, life-threatening emergency. Although substantial data remains elusive, it is thought that *anaphylaxis*, the most severe of allergic reactions, does not occur often on wilderness trips. When it does occur, it often begins like a general reaction but rapidly results in respiratory and/or circulatory collapse. The breathing difficulties (from airway constriction) and *anaphylactic shock* (from rapidly dilating blood vessels) that result in a true anaphylactic reaction require rapid field treatment. The onset of the signs and symptoms of anaphylaxis typically occur within minutes of a bite or sting and within 30 to 60 minutes of ingestion of an allergen.

The signs and symptoms of anaphylaxis may involve any combination of reactions in the following body systems:

Integumentary system (skin): Flushing, itching, burning, and swelling, especially of the face; cyanosis of the lips; swelling of the tongue; hives that itch severely.

Circulatory system: Signs and symptoms of shock.

Respiratory system: Painful tightness in the chest; upper airway obstruction; breathing difficulty; coughing; wheezing.

Neurological system: Restlessness; lightheadedness; convulsions; confusion; loss of consciousness.

Gastrointestinal system: Nausea; vomiting; abdominal cramps; diarrhea.

Treatment for Anaphylaxis

If you suspect a person susceptible to anaphylaxis has been exposed to a known allergen, give an oral antihistamine as soon as possible, a treatment that might forestall the anaphylactic reaction. The antihistamine may be chewed and swallowed for faster action, but remind the patient that it will probably taste bad and numb the tongue.

Because deaths due to anaphylaxis occur primarily from loss of airway and general circulatory collapse (shock) and because most fatalities occur within minutes to 1 hour of onset of symptoms, the ability to respond immediately typically means the difference between life and death. The ability to reverse fatal anaphylaxis requires the administration of epinephrine.

There are two ways in which the Wilderness First Responder can assist with the administration of epinephrine:

1. Because many of the severe reactions involve bronchospasm and pulmonary congestion of a typical asthmatic nature, the use of inhaled epinephrine (such as Primatene) to relax the airway can be life-saving. This is a nonprescription medication. Unfortunately, inhaled epinephrine won't work if the patient can't inhale, and it does not help significantly in the case of cardiovascular collapse in severe anaphylaxis.

2. The most specific and valuable treatment is the use of injectable epinephrine that works to alleviate both bronchospasm and anaphylactic shock. This prescription product is available in kit form. The WFR should be aware of how to help the patient administer this medication if it is available and consider acquiring a prescription and written protocols for its use from a medical adviser prior to a wilderness journey.

In the past two kit systems have been available. One is EpiPen, with a single 0.3-milliliter bolus of 0.3 milligram of epinephrine per injection (EpiPen Jr, for young children, has 0.15-milligram of epinephrine). The other is the Ana-Kit or Ana-Guard (with dual 0.3-milliliter injections of 0.3 milligram of epinephrine per kit). Many patients with a history of insect sting allergy have one of these kits prescribed for their use by their physician.

Dangers to the Person Injecting Epinephrine

With either EpiPen or Ana-Kit/Ana-Guard, you must take great care to prevent injecting yourself in a finger or hand. The vasoconstricting effect of epinephrine is strong enough to cause loss of small, low circulation areas, especially a fingertip.

■ USE OF EPIPEN

EpiPen is an auto-injection system. Using EpiPen involves three simple steps:

1. Pull off the safety cap.
2. Place the black tip on the outer thigh, preferably against the skin, but it can be used through thin clothing.
3. Push the unit against the thigh until it clicks, and hold it in place for several seconds. It is important to hold the unit in place until all the epinephrine has been injected. You can count to ten to be safe.

■ USE OF ANA-KIT/ANA-GUARD

Ana-Kit and Ana-Guard are manual injection systems utilizing the same type of syringe. Ana-Kit contains additional items, including oral antihistamine tablets. Ana-Guard includes only the syringe. Using the syringe involves the following steps:

1. Remove the red rubber needle cover.
2. Hold the syringe upright and push the plunger until it stops to expel air and excess epinephrine.
3. Rotate the plunger one-quarter turn until the crossbar aligns with the slot in the barrel of the syringe.
4. Insert the needle straight into the muscle of the upper thigh or upper arm.
5. Push the plunger steadily, but not quickly, until it stops.
6. Remove the needle from the muscle and protect the needle in case the second injection is needed. The syringe barrel has graduations so that smaller doses may be measured for young children. By rotating the plunger once again, the second 0.3-milliliter injection may be administered.

■ USE OF AMPULES AND VIALS

Currently Ana-Kits are not available, though many are still in use. Outdoor programs, such as the National Outdoor Leadership School, have turned to carrying epinephrine in single-dose ampules and multidose vials—along with syringes. Ampules and vials require the rescuer to draw into the syringe the 0.3-milliliter dose of epinephrine prior to the injection. In all cases a prescription and a written protocol from a medical adviser are the first steps in the use of epinephrine.

When to Inject Epinephrine

Waiting for respiratory distress to treat anaphylaxis may be waiting too long. Watch for the following symptoms:

1. Large areas of swelling, typically involving face, lips, hands, and feet.
2. Swallowing difficulty.
3. Signs and symptoms of shock.
4. Respiratory distress—the patient is unable to speak in more than one- or two-word clusters.

Treatment After Epinephrine

Sometimes one injection is not enough, and rebound or recurrent reactions can occur up to 24 hours after the original incident. A second injection should be given in approximately 5 minutes if the condition of the patient does not improve. Some physicians recommend carrying up to 3 total doses, and some recommend more. It is recommended that the WFR should have at least 3 injections of epinephrine at his or her disposal. There is, however, no endpoint for the administration of epinephrine. If you have it, and the patient needs it, you should give it.

The patient will frequently have blanching of the skin around the injection site. This is a normal response to the local vasoconstriction caused by this medication.

The response to the administration of epinephrine in a normal, healthy person is the development of *tachycardia* (increased heart rate). The anaphylactic patient already has tachycardia, and the use of epinephrine may lower this rapid heart rate as it reverses the vasodilatation and bronchospasm of anaphylaxis.

Epinephrine is rapidly inactivated in the body and becomes subtherapeutic (too inactivated to be effective) within 20 minutes. Repeat dosage, therefore, may be required.

Oxygen is the second most useful drug in treating anaphylaxis because *hypoxia* (oxygen deficiency) can cause rapid cardiovascular collapse during this crisis. If supplemental oxygen is available, immediately start a low flow via a face mask or nasal cannula for both psychological and physiological support. In cases of severe respiratory distress, increase the flow to high (see Appendix B: Oxygen and Mechanical Aids to Breathing).

Ana-Kit includes four tablets of an oral antihistamine that should be given as soon as the patient can accept the tablets and swallow. Many over-the-counter antihistamines, such as diphenhydramine, may be in a wilderness medical kit and should be given. The recommended dose of diphenhydramine is 50 to 100 milligrams to start, and 25 to 50 milligrams every 4 to 6 hours until the patient is turned over to definitive medical care. As soon as the patient can swallow, adequate hydration must also be maintained.

If you don't have epinephrine, there are other prescription medications the patient, or someone in the party, may be carrying that can be useful in treating anaphylaxis. Albuterol, sold under the names Ventolin and Proventil, is prescribed to treat asthma and may be given in 2 puffs inhaled every 3 hours to help with the respiratory component of anaphylaxis. Some asthmatic patients have access to an ipratropium (Atrovent) inhaler, which can be life-saving, especially in recurring episodes of breathing difficulty. The treatment dosage may have to be fairly high, 15 to 30 puffs per 4-hour period for maximum results. This is a larger dose than an asthmatic person has generally been instructed to use by his or her physician. Under normal circumstances, you would never use a prescription medication for someone other than the person to whom it is prescribed, but, in this case, you are taking unusual steps in hopes of saving a life.

In case of cardiac arrest, cardiopulmonary resuscitation should be initiated, although artificial ventilations may be difficult until the bronchospasm relaxes.

Evacuation Guidelines

Any patient who has been treated for anaphylaxis should be evacuated from the wilderness as soon as safely possible to be evaluated by a physician. During the evacuation the patient should be kept on an oral antihistamine, such as diphenhydramine. Follow the recommended dose for over-the-counter diphenhydramine (25 to 50 milligrams every 4 to 6 hours).

Conclusion

In the meadow near Crested Butte, you think back to your patient's medical form, recalling she reported once having a "bad reaction" to a bee sting. Recognizing an anaphylactic reaction in your budding naturalist, you grab the medical kit from the top of your pack. Your own labored breathing includes sighs of hopeful relief based on the foresight that allowed you to acquire a prescription for injectable epinephrine and written protocols for its use from your physician. Uncapping the Ana-Guard's syringe, you carefully plunge the needle into your patient's upper arm. After injecting the drug, you remove the needle, protecting the syringe in case you need to make a second injection. Within 30 seconds, your patient begins to breathe more easily, returning slowly to a normal level of consciousness. When she is able to accept a water bottle and drink, you give her 50 milligrams of Benadryl in 2 25-milligram tablets. You begin plans for an immediate evacuation.

William W. Forgey, MD, contributed his expertise to this chapter.

CHAPTER TWENTY-NINE

Abdominal Illnesses

You Should Be Able To:

1. Demonstrate an assessment of an acute abdomen.
2. Describe the treatment for a patient with acute abdominal pain.
3. List the guidelines for deciding when to evacuate a patient with acute abdominal pain.

It Could Happen to You

Jake begins to complain of abdominal discomfort about nine o'clock on a sunny morning during a backpacking trip in the Superstition Mountains of Arizona. You give him a quick assessment that reveals the discomfort is general to his mid-abdominal area, nothing of significance, and he completes the day's hike. By evening, however, his discomfort has increased, and he complains of nausea. Before dinner he vomits, and reports some relief of pain.

Jake declines food, saying he isn't hungry, and soon complains the pain is back and worse than before. The pain now centers in his lower right quadrant, pain he describes as a "dull ache with pressure, a sharp pain when I breathe deep, and a sharp pain when you push on it." He calls the pain a 5 on a scale of 1 to 10 in general, but it increases to a 7 when palpated.

By nine o'clock in the evening, as the rest of the group begins to think about sacking out, Jake is found lying on the ground in a fetal position, groaning in pain, unable to straighten out because, he grunts, "It hurts too bad."

The abdomen contains the liver and spleen, the digestive, urinary, and reproductive systems, and the major blood vessels supplying the lower extremities—to name most of its contents. For anatomical reference all of these body parts lie in one of four *abdominal quadrants* (see Chapter 11: Abdominal Injuries). Numerous medical problems can develop in this region of the body, but, for convenience, they can be divided into four categories:

1. *Hemorrhage*: An ectopic pregnancy, for example, may cause life-threatening bleeding if the ectopic pregnancy ruptures (see Chapter 32: Gender-Specific Emergencies). Puncture wounds and blunt trauma are covered in Chapter 11: Abdominal Injuries.
2. *Perforation*: An ulcer, for example, may "eat" a hole through the stomach or intestinal wall, allowing contents to spill into the abdomen.
3. *Infection*: Any organ of the abdomen may get inflamed or infected, appendicitis being a common example. The lining of the abdomen itself may suffer inflammation, a problem called *peritonitis*.
4. *Obstruction*: The tubelike pathways of the abdomen may become blocked; kidney stones, gallstones, fecal impactions, and twisting of the intestines are the most common blockages.

You may have a "good guess" about the problem existing in the abdomen of the patient, but you will not know for sure until a physician has examined the patient, and even then you may not know. Some abdominal pain goes away without leaving substantial clues to its cause. In all cases of acute abdominal pain, the primary responsibilities of the Wilderness First Responder are (1) assess if the condition is serious, calling for an evacuation of the patient; and (2) properly care for the patient up to and perhaps during the evacuation.

General Abdominal Illness Assessment

1. Observe the patient's body position when you first approach. A patient who lies still with legs drawn up into a fetal position is often assuming a posture that minimizes serious pain.
2. When assessing abdominal pain, inspect the abdomen. Make the patient as comfortable as possible, warm and insulated

from the ground, and remove enough clothing to see the entire abdomen.

3. Look at the abdomen with the patient lying in a supine position (face up), if possible. A normal abdomen is gently rounded and symmetrical. Look for distension and/or an irregularly shaped abdomen, both of which may indicate a serious illness.

4. Palpate the abdomen with flat fingers. Press gently in all four quadrants of the abdomen. Watch for a pain response. Normal abdomens are soft and not tender to palpation. Feel for rigid muscles, lumps, and pain specific to a local spot, and guarding by the patient, all of which may be signs of serious illness. Check for rebound pain—pain that increases when you release the pressure of palpation, another indication the illness could be serious. Your palpation of the abdomen is a key assessment tool.

5. Ask the patient about his or her condition. Has this ever happened before? The OPQRST questions, in relation to pain, may be very helpful:

    O (for onset): Did the pain come on suddenly or gradually? Some problems bring sudden pain and some gradual pain; knowing which can help you guess what the problem may be.

    P (for provokes or palliates): Does anything make the pain worse or better? Pain increased by movement is not always but may be more serious.

    Q (for quality): How does the patient describe the pain? Although descriptions can be misleading, many patients describe nonserious pain as dull, cramping pain.

    R (for radiate, refers, region): Where is the pain, and does it radiate or refer to another region? Specific pain can help you guess what part of the body has a problem.

    S (for severity): On a scale of 1 to 10, how does the patient rate the pain? The worse the patient considers the pain, the higher your level of concern.

    T (for time): How long has the pain been there? Prolonged pain, greater than 12-24 hours, is a key sign of serious illness.

6. Listen with an ear or stethoscope pressed against each of the four quadrants. This takes time. In 2 to 3 minutes, within each quadrant, bowel sounds (gurgling noises) should be heard. Absence of noise means something is not working right.

7. Ask about nausea and vomiting. Although nausea and vomiting are common (see Chapter 31: Common Simple Medical Problems), pain associated with prolonged nausea and vomiting could indicate a serious problem.

8. Ask about diarrhea and constipation. When was the patient's last bowel movement? Diarrhea and constipation are seldom indications of a serious abdominal illness, but both can become serious problems if they continue for several days (see Chapter 31: Common Simple Medical Problems).

9. Ask about blood that may appear in the urine, stool, or vomit. Blood leaking from infections of the urinary tract may turn

urine pink. Mild *gastritis* (inflammation of the stomach) may produce a bit of blood in vomit, and anal fissures or hemorrhoids may cause some bright or brown blood in stools. This type of bleeding is typically not serious, but the patient should be monitored closely. Dark blood in vomit, appearing somewhat like coffee grounds, and dark blood in stools, appearing something like tar, and frank bleeding from the mouth or rectum are serious signs.

10. Check for a fever. A fever of 102° F (35° C) or higher, especially if it comes on fast, could indicate a serious infection.

11. Monitor vital signs for indications of shock.

Some Types of Abdominal Illnesses

Appendicitis

The *appendix* is a pouch of the intestine located in the right lower quadrant (RLQ). *Appendicitis* is an inflammation of the appendix with numerous causes. Due to an obstruction, mucus builds within the appendix, causing pressure, swelling, and infection. The highest incidence of appendicitis occurs between the ages of 10 and 30.

The problem typically begins as mid-abdomen discomfort that grows worse over 6 to 24 hours, localizing in the RLQ. Rebound pain usually develops in the RLQ. Movement usually aggravates the pain once it is localized. Loss of appetite is very common. Nausea, vomiting, and a low-grade fever (less than 102° F/35° C) are also common. If the appendix ruptures, peritonitis will develop, which may lead to septic shock.

Appendicitis is a serious illness and often difficult to diagnose. You do not want to miss at least recognizing the possibility that the patient has a serious illness. The patient should be rapidly but gently evacuated in a position of comfort. Food should be avoided, but small sips of fluid may be given to avoid dehydration.

Fecal Impaction

A *fecal impaction* results when hardened feces form a blockage in the descending colon, preventing the passage of fecal material. The patient usually reports a history of constipation. In the wilderness constipation is often caused by a lack of ease defecating outdoors and/or dehydration. Gradually increasing pain in the left lower quadrant (LLQ) builds to severe cramping. You might be able to palpate a mass in the LLQ, and the lower abdomen may be distended. Nausea and vomiting often occur. If untreated, the condition can lead to bowel obstruction, dead bowel, a perforation, and septic shock. A high level of hydration is an important aspect of treatment. Sometimes a bowel movement can be stimulated by giving the patient a large drink of cold water, followed immediately by a cup of hot liquid, such as coffee or tea. Sometimes a caffeinated drink such as coffee will do the job. In the case of a true impaction, you have to resort to manual disimpaction, going up the rectum with a well-gloved finger to dislodge and remove the mass.

Food Poisoning

Gastrointestinal distress caused by ingesting bacterially contaminated food is termed *food poisoning*. Leftovers served later are common wilderness sources. Abdominal cramps with diarrhea, nausea, and vomiting are common signs that appear 1 to 12 hours after ingesting the germs. The problem usually affects more than one person in a group of wilderness travelers, and the problem is typically self-limiting. Patients should be kept well hydrated and monitored for changes. Anti-emetic (antivomit) and antidiarrheal drugs may be considered. Bloody diarrhea, a fever above 102° F (35° C), and the signs and symptoms of shock are reasons for evacuation.

Gallstones

Gallstones form in the gallbladder most often when bile contains more cholesterol than can be kept in solution. Pain usually comes on gradually in the right upper quadrant (RUQ) when the gallbladder becomes inflamed or when a stone tries to escape through the cystic duct, the duct carrying bile to the small intestine. The pain may radiate into the right shoulder and sometimes into the back. Nausea and vomiting are common. The pain often subsides in as little as a few hours, but the patient should be evacuated if the RUQ pain persists and either a fever and/or jaundice (yellowness of skin and of the whites of the eyes) develops. Strong painkillers and adequate hydration are recommended.

Gastroenteritis

Gastroenteritis, an inflammation of the gastrointestinal tract, is a common wilderness medical condition, most often caused by poor camp hygiene (see Chapter 30: Communicable Diseases and Camp Hygiene). The patient complains of gradually increasing discomfort, usually diffuse in the abdomen, often worse in the lower quadrants. Cramps come and go, and diarrhea is common. Nausea and vomiting may be present, sometimes with a low-grade fever. Malaise, a general feeling of discomfort or indisposition, is a common complaint. Seldom a serious problem, gastroenteritis typically resolves in 24 to 48 hours, less often in as long as 3 days. Gastroenteritis, however, is a diagnosis

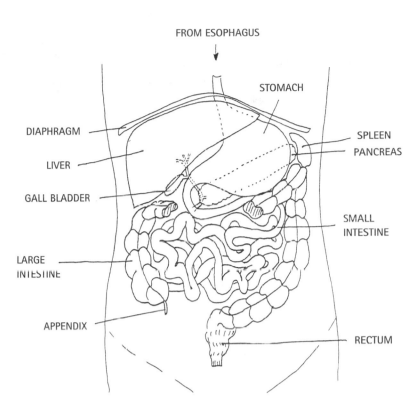

Figure 29-1: *The digestive system.*

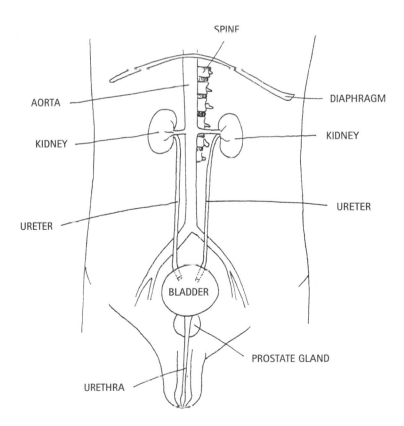

Figure 29-2: *The urinary system.*

of exclusion, and the most common misdiagnosis in patients with serious abdominal illness. Hydration is critical. Antidiarrheal and/or anti-emetic drugs may be considered. Watch the patient closely for signs of serious illness: fever rising above 102° F (35° C), blood and mucus in stools, and the signs and symptoms of shock.

Kidney Stones

Kidney stones are formed from minerals in urine. They can grow to 1 inch (2.5 centimeters) or more and cause sudden, sharp, stabbing flank pain that radiates down into the lower abdomen and/or groin as the stones pass from the kidney and down the ureter. The pain tends to wash over the patient in excruciating waves, and no position of comfort can be found. Blood may appear in the urine, and nausea and vomiting are possible. The pain will last for 24 hours or more, and it will suddenly subside if the patient is able to pass the stone from his or her ureter into the bladder or out of the body through the urethra. A patient with a kidney stone appreciates strong pain killers. Adequate hydration is of great importance, but overhydration should be avoided. Any patient with a fever and a suspected kidney stone must be evacuated immediately.

Ulcers

An *ulcer* is an open sore or lesion developing, in this case, in the lining of the stomach or intestine. In the stomach an ulcer hurts more when the patient eats and the production of stomach acid is stimulated. In the duodenum, the first section of the small intestine, the pain of an ulcer may be reduced when the patient eats, especially bland foods, but the burning discomfort returns 1 to 4 hours after eating. Pain is typically described as "burning" in the mid-epigastric region (upper middle abdomen). Alcohol, caffeine, nicotine, aspirin, and nonsteroidal anti-inflammatory drugs worsen the condition. A history of similar pain can often be described by the patient, and the pain usually resolves spontaneously in a few days. Watch the patient for signs and symptoms of a serious ulcer: unremitting pain; vomiting "coffee grounds"; dark, foul-smelling, tarry stools; weakness; fainting; and shock. For patients with mild ulcers, antacids may provide relief of symptoms. Monitor the patient in the wilderness, and evacuate if serious signs and symptoms appear.

General Treatment for Abdominal Illnesses

Assist the patient with severe abdominal pain in maintaining a position that provides the most comfort. If an evacuation is planned and expected to be short, give nothing by mouth. In extended-care situations give the patient clear fluids if they are well tolerated. Cool fluids are often best tolerated.

Evacuation Guidelines

Determining the exact cause and even the severity of abdominal pain in the wilderness is a baffling undertaking. For that reason it is recommended to evacuate any patient with abdominal pain if any of the following applies:

1. The pain is associated with the signs and symptoms of shock.
2. The pain persists for longer than 12 to 24 hours.
3. The pain localizes, and especially if the pain involves guarding, tenderness, or abdominal rigidity.
4. Blood appears in the vomit, feces, or urine.
5. Nausea, vomiting, or diarrhea persist for longer than 24 to 72 hours.
6. The pain is associated with a fever above 102° F (35° C).
7. The pain is associated with signs and symptoms of pregnancy (see Chapter 33: Obstetrical Emergencies).

Conclusion

In the Superstition Mountains, you recognize Jake's signs and symptoms, especially the classic progression from generalized mid-abdomen pain to localized RLQ pain, as suggestive of appendicitis. The group, a strong and mature party, meets and decides a team will be formed to hike out with a SOAP note (see Chapter 3: Patient Assessment) and an appeal for help, but, for safety reasons, the team will not leave until first light. You make Jake as comfortable as possible, staying by his side during the night, offering sips of cool water frequently.

With dawn almost more of a suggestion than a reality, a team of three leaves for the roadhead. By early afternoon, the *thwop-thwop* of a helicopter brings you to your feet. Later that day, Jake's appendix lies in a stainless steel pan in a hospital's operating room.

CHAPTER THIRTY

Communicable Diseases and Camp Hygiene

You Should Be Able To:

1. Describe the ways germs are communicated to humans.
2. Describe ways to prevent the spread of communicable diseases.
3. Describe the basic principles of camp hygiene.

It Could Happen to You

The tortured turns and twists of the Badlands of South Dakota have always held a mystical attraction for you, and near one of the area's trails you're setting camp. The group of eight you lead is young. For many, these past five days have been an initiation into laboring under a backpack and sleeping on the ground. You've been as busy as a homeless beaver teaching these kids to pitch tents, cook meals, and Leave No Trace. Although your stomach feels a little upset, tonight, with the group better trained, you're hoping for a more relaxing evening.

Your hopes are dashed when two 13-year-olds come to you with complaints of bad stomachaches. Your investigation reveals they both have been suffering from diarrhea. As you question them, a third student, overhearing the conversation, admits to the same problem: diarrhea and "bad cramps." You are encouraging hydration and rest when, with a sigh, you realize you've got to excuse yourself for a trip to the bushes. Into your cat hole goes the first of what will be numerous explosive bowel movements.

As an outdoors enthusiast you might think the great majority of *germs*—a scientific term representing all the microscopic organisms that might infect a human and cause disease—lurk in the wilderness waiting for a suitable host to pass near enough for an attack. Not so. In fact, generally speaking, the contrary is true: Most germs hitch a free ride into the wilderness with you or some other unwary bipedal primate. As more and more *Homo sapiens* show up more and more often in wilderness areas, the presence of humans, even if only for a short while, builds a community of disease possibilities.

Germs are *pathogens,* microorganisms that cause at least some *pathos* (suffering) as part of their nature. Some pathogens are more pathological than others, and some hosts are more susceptible. Hence the degree of pathos is a product of the attributes of the "bug" and the host. Some humans, you probably have noticed, tend to be continuously ill while others seem to be superhumanly immune. Despite your personal level of immunity, however, you should be determined to prevent the spread of pathogens at all times.

Agents of Infection

Unseen and ubiquitous, microorganisms live among us and on us and in us. Some are beneficial. Pathogens, not of benefit, are divided, generally, into four classifications: *viruses, bacteria, fungi,* and *parasites.*

Viruses cannot exist for long outside of living tissue and must penetrate human cell walls to multiply and cause disease. They are unimaginably tiny and account for the respiratory infections that are responsible for approximately half of all acute illnesses. Major viruses include influenza, the common cold, mumps, and measles. Some viruses set up housekeeping in the central nervous system and cause forms of meningitis and encephalitis. Herpes simplex virus type 1 causes cold sores, and herpes simplex virus type 2 causes genital lesions. The Epstein-Barr virus produces infectious mononucleosis. The varicella virus causes chicken pox, while the herpes zoster virus causes a related disease, shingles. Viruses also cause hepatitis and AIDS (acquired immunodeficiency syndrome).

Bacteria grow independently, without need for a host cell. *Staphylococcus aureus* lives on the surface of human skin and is responsible for wound infections, abscesses, bacterial pneumonia, and some food poisonings. Some bacteria grow in chains, called *streptococci,* and cause problems such as strep throat, scarlet fever, rheumatic fever, and acute sinusitis. *Clostridium tetani* lives in soil and causes tetanus. Many bacteria live helpfully in the human gastrointestinal tract, but this group also includes *Escherichia coli,* which causes urinary tract infections, and *salmonella,* which causes typhoid fever. Both E. coli and salmonella produce diarrheal illness. *Shigella* causes dysentery, and another group, the *mycobacteria,* cause tuberculosis and leprosy.

Fungi are primitive life forms that feed on living plants, decaying organic matter, and animal tissue. Fungal infections are usually bothersome but relatively mild, such as athlete's foot (see Chapter 31: Common Simple Wilderness Medical Problems). In the immunosuppressed, however, a fungal infection can be overwhelming and fatal.

Parasites that cause disease in humans are protozoa that include the mosquito-borne *plasmodia* of malaria, and the waterborne *Giardia lamblia* of giardiasis and *Cryptosporidium* of cryptosporidiosis.

Communication of Disease

Germs can be transmitted in a variety of ways, but all communication falls into one of two broad classifications: *direct* and *indirect.* Direct transmission means the person carrying the germs passes them directly to another person. Contact with blood and other body fluids may pass germs directly. Particulates expelled by coughs and sneezes and inhaled by a nearby person are another example of direct transmission. Sexual activity offers an opportunity for direct contact. Germs also may be picked up directly through cuts or other open wounds. Indirect transmission methods include those in which germs are passed without direct contact with another person. Eating contaminated food and drinking contaminated water are examples of indirect transmission. Bites from infected insects may pass pathogens indirectly. Sharing a water bottle, a towel, or an eating utensil provides a means of indirect contact with germs.

Some Specific Communicable Diseases

Norwalk Virus

Norwalk virus, named for Norwalk, Ohio, where it was first isolated, makes more people sick than any other food-related virus. It's passed easily from one sufferer to another by hand and mouth, and shifts into high gear 24 to 48 hours after contact has been made. Although it lasts about a week, the problems of vomiting and diarrhea are relatively mild, rarely requiring a doctor's care.

Hepatitis

Hepatitis A virus can be ingested with some fecal-contaminated foods and water. Undercooked shellfish from water polluted with human wastes has been a common source of hepatitis A. It can spread through sharing water bottles and utensils, improperly washing hands (fecal-oral route), and intimate contact between people. Stomach pain, nausea, vomiting, fatigue, and loss of appetite show up 15 to 50 days after ingestion. Severe cases may cause jaundice (yellowing of the skin and the whites of the eyes) and dark urine. Hepatitis B produces similar but more severe symptoms, but it's transmitted primarily by blood contact.

AIDS

AIDS—acquired immunodeficiency syndrome—is a sincerely life-threatening infection that destroys the body's ability to fight off other types of infections. AIDS is the final, completely fatal stage of a continuum of problems caused by the *human immunodeficiency virus* (HIV). Once the HIV is in human blood, antibodies to the virus develop; those antibodies show up in a blood test. Seropositivity (testing positive for a certain antibody) usually shows up 4 to 6 weeks after HIV infection. HIV is communicated by blood-and-body fluid transmission. Saliva, tears, sweat, urine, semen, vaginal secretions, and stool can all carry HIV, but only blood and semen have been known to transmit the virus.

Patients who have been infected but have not yet developed the antibodies are considered to be in the first stage of the continuum. Stage 2 patients have the antibodies but no symptoms of the disease. Once signs and symptoms appear, the patient may still not be technically classified as an AIDS sufferer. Regardless of the stage of the disease, the HIV-infected patient can pass the virus to others.

Prevention of Blood/Body Fluid-Borne Diseases

1. Wear protective gloves when handling all patients.
2. Wear protective glasses when splashing/coughing of body fluids occurs.
3. Use a rescue mask with a one-way valve when performing rescue breathing.
4. Wash your hands after contact with blood/body fluids even if you were wearing gloves.
5. Double-bag in plastic all soiled bandages and dressings and dispose of them properly after leaving the wilderness.
6. Keep your vaccinations up to date.

Bacillus Cereus

Bacillus cereus lives as a bacterial spore (a dry, seedlike structure) in grains and spices; it germinates when the food is moist and when contaminated cooked food is improperly stored. Stomach pain, nausea, vomiting, and sometimes mild diarrhea usually occur within 8 to 16 hours of ingestion. The problem almost always runs its course in less than 24 hours.

Staphylococcus Aureus

Staphylococcus aureus may drop off contaminated hands into breakfast, lunch, and dinner. It multiplies with great speed in protein-rich foods at warm temperatures. Rather than an infectious disease, the bacteria produce a toxin. The reaction that erupts suddenly 30 minutes to 6 hours after eating produces cramps, vomiting, diarrhea, headache, sweats, and chills. Although the problem may last 1 to 2 days, medical treatment is seldom required unless the patient gets seriously dehydrated.

Shigella

Shigella most often attacks through food and water contaminated with fecal matter, usually from the hands of the person who last handled the food and water. Shigellosis causes dysentery (bloody, mucus-ridden diarrhea), fever, bad stomach cramps, and a search for a doctor. Illness usually appears less than 4 days after ingestion, but some cases have shown up 7 days later.

Salmonella

Salmonella are bacteria common in eggs and poorly processed dairy products. Within an average of 12 to 24 hours, sometimes faster, symptoms appear: stomach pain, diarrhea, nausea, vomiting, headache, chills, weakness, and thirst. Fever may be present. Although cases have been known to become severe, most people recover by drinking lots of fluids and waiting in distress.

Campylobacter Jejuni

Campylobacter jejuni contaminates meats primarily, especially chicken, although some types of campylobacter are common in backcountry water and fecal matter. The likelihood of contacting the bacteria increases if you handle raw meat or eat undercooked flesh. An average of 4 to 7 days passes after ingestion before the onset of stomach pain and bloody diarrhea. The problem may last 2 to 7 days. A visit to a doctor is recommended.

Clostridium Perfringens

Clostridium perfringens are bacteria found in meat usually stored at too high a temperature before serving. Clostridium causes abdominal cramps, nausea, and diarrhea 8 to 22 hours after ingestion. Vomiting, headache, fever, and chills are rare with clostridium. Symptoms usually go away within 36 hours.

Giardia

Giardia lamblia parasites swim or float around as cysts in many wilderness water sources and spread through fecal contamination by humans and other animals. Giardiasis is a common waterborne illness in the United States. Unpleasant, but typically benign, the illness usually causes more than a week of diarrhea with bloating, flatulence, and stomach cramps. Longer lasting illness requires the use of drugs available by prescription. Symptoms take about 10 days to show up, but the parasites may hang around for weeks before symptoms show up. Some patients never develop the typical signs and symptoms of giardiasis. They have periodic mild cramping and bloating, but they never explode with diarrhea. Some people can carry the *Giardia lamblia* bacteria without symptoms.

Cryptosporidium

Cryptosporidium protozoa, although transmittable via food and body contact, are primarily waterborne parasites. They get in the water from feces of infected animals, including humans. Explosive diarrhea and stomach cramps appear after an incubation period of 4 to 14 days, and usually go away after 5 to 11 miserable days.

Camp Hygiene

Many wilderness-related diseases are preventable by following acceptable camp hygiene practices.

Hand Washing

Skin, the outer layer, is an overlapping armor of dead cells that protect the living cells beneath. Under a microscope, this outer layer looks like the surface of the Colorado Plateau from 30,000 feet: canyons, mesas, cracks, and fissures. Resident microbes are wedged firmly into the low spots. Some of these microbes are friendly, serving to keep skin slightly acid and resistant to other microbial life forms, like fungi. Others, such as *Staphylococcus aureus*, can make you severely sick. In addition to the residents, transient germs come and go as fortune dictates. They can accumulate rapidly after bowel movements, and they congregate most thickly under fingernails and in the deeper fissures of fingertips. Hands are the single most important "tool" in communicating illness. **Hand washing is the single most important method of preventing the spread of disease.**

Hand washing, even with detergents, does not remove all the organisms living on hands, but it does significantly reduce the chance of contamination. The following eight-step hand washing technique is recommended for maximum cleanliness:

1. Wet hands with hot flowing water (100 to 120° F/38 to 49° C).
2. Soap up until a good lather is attained.
3. Work the lather all over the surface of the hands, concentrating on fingernails and tips.
4. Clean under fingernails.
5. Rinse thoroughly with hot water (very important).
6. Resoap and relather.
7. Rerinse.
8. Dry (very important).

In most cases, hot water is a rare wilderness commodity. You can still get clean hands with this modified wilderness technique, which substitutes germicidal soap for hot water. In tests adequate hand sanitation was achieved with as little as 0.5 liter of water.

1. Wet hands thoroughly.
2. Add a small amount of germicidal soap, such as Betadine Scrub or Hibiclens.
3. Work lather up, especially with the fingertips.
4. Clean under fingernails (and keep nails trimmed).
5. Rinse thoroughly.
6. Repeat soap, lather, and rinse.
7. Dry.

In a rush? Join the crowd, but not the sick crowd. Some products—hand sanitizers—provide quick hand sanitization with a fast but thorough rub and a let dry. They work, especially if they utilize alcohol. With constant use, however, alcohol breaks down tissue, leaving hands less healthy. Choose a hand sanitizer with an added moisturizer to help counter the drying of the germ-killing agent. Hand sanitizers do not remove grit and grime, and a hand washing with soap is still recommended at least once a day.

Food for Thought

Leftovers result, most often, from cooking more than you can eat. Storage of cooked-but-uneaten food in the wilderness poses an almost insurmountable problem. Bacteria grow optimally at temperatures ranging from 45 to 140° F (7 to 60° C), and unhealthy populations of bacteria can be reached in a brief period of time. Reheating cooked food may kill bacteria, but it often leaves dangerous toxins produced by the bacteria at sickening levels. Your safest bet is to get rid of leftovers—with a couple of exceptions. In temperatures below 45° F (7° C), food allowed to cool rapidly usually remains safe to eat. Out of necessity you may choose to eat some leftovers. If you do, add water and raise the temperature of the food until the water boils.

Wilderness food usually is stored in plastic bags, and food

contamination can be further reduced by pouring the food out instead of reaching in for it.

Another major source of food contamination in the wilderness is dirty cookware. Cooking and eating utensils should be boiled daily or rinsed in water disinfected with chlorine. If the water smells strongly of chlorine, it is usually adequately chlorinated for cookware disinfection.

Fecal Contamination

Even if you're washing your hands after a bowel movement, fecal contamination of the environment remains a primary source of disease communication. You can, with an adequate plan, reduce the risk of fecal contamination to an absolute minimum. Transmission of fecal-borne pathogens occurs in four ways: direct contact with the feces, indirect contact with hands that have directly contacted the feces, contact with insects that have contacted the feces, and drinking contaminated water.

Human waste products break down to a harmless state as a result of two mechanisms: (1) bacterial action in the presence of oxygen, moisture, and warmth; and (2) sterilization from direct ultraviolet radiation. Deposition of solid body wastes should include placement to maximize decomposition, to minimize the chance of something or someone finding it, and to minimize the chance of water contamination. Latrines are out, except in established spots. They concentrate too much human waste in one place. They carry a high risk of water pollution. They invite insect and mammal investigation. They are unsightly, and they stink. If you are ever required to dig a latrine, make it at least 1 foot deep, add soil after each deposit, and fill it in when the total excreta lies several inches below the surface.

Cat holes are the best receptacle for your fecal matter. Cat holes should be at least 200 feet, or approximately 70 adult paces, from water and placed where little chance of discovery exists, preferably in a level spot. A cat hole should be dug several inches into an organic layer of soil, where decomposing microorganisms live most abundantly. After you've defecated, stir the waste into the soil to speed decomposition, then cover it with a couple of inches of soil and disguise the spot to hide it from later passersby.

■ SPECIFIC FECAL MATTER MANAGEMENT GUIDELINES

Wilderness areas are not created equal. Some are especially wet, some dry, some cold, and some hot. Special sanitation considerations may be required in special environments to stop the spread of disease.

Lakes and rivers: Moving well away from bodies of water and carefully selecting your site will eliminate most of the health risks associated with water contamination. In some places, however, such as deep dry-country canyons, moving well away from water isn't possible. In those spots the only safe alternative is packing

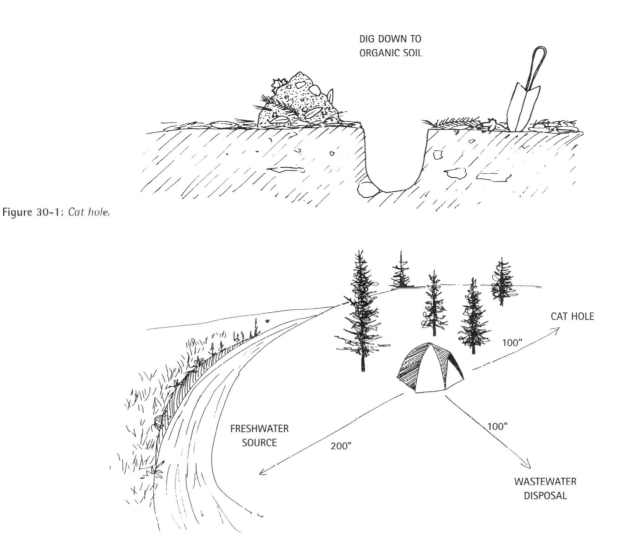

DIG DOWN TO
ORGANIC SOIL

Figure 30-1: *Cat hole.*

CAT HOLE
100"

FRESHWATER
SOURCE
200"

100"

WASTEWATER
DISPOSAL

Figure 30-2: *Camping healthy.*

out the waste. The most acceptable means to do this requires a portable toilet (commercially available) or a sturdy sealable can and several heavy-duty garbage bags. Line the can, such as a large ammo box, with a couple of garbage bags folded out over the rim. Before and after each use, throw in some chemicals to reduce the smell and slow decomposition. (Rapid decomposition inside a plastic bag may produce a disgusting explosion.) Bleach or quicklime will do. Toilet paper goes into the bag, too, but urine should be squirted elsewhere. Urine dilutes the added chemicals and greatly increases bag weight. Before packing the bag for the day's travel, squeeze out the air and tie it firmly closed.

Deserts: Deeply buried human excrement won't decompose in sandy, predominantly inorganic desert soil. Instead, it filters down through the ground. For this reason, deposits should be made far from water sources, out of gullies and other obvious drainages, and away from slickrock. The best choice in most areas is shallow burial. High near-surface temperatures cook pathogens to death in short order.

Above timberline: In the frozen north and in the fragile oft-frozen high country, decomposition goes slowly due foremost to the cold. Fecal monuments may stand for ages. If you cannot dig a cat hole, choose a secluded spot well away from water sources.

Snow: Stools deposited in snow, no matter how far they're buried, will appear on the surface come springtime. For that reason, proper choice of burial sites remains of paramount importance. Select a spot that is well away from water sources that will eventually thaw, trails, and camp sites.

Urine

Although urine is often considered a sterile waste product, it can carry germs. To stay on the safe side, urinate on rocks or in non-vegetated spots far from water sources whenever possible. On some wilderness waterways, travelers are encouraged to urinate directly into the water. In some areas, this practice is discouraged. Follow local recommendations.

Water Disinfection

Long gone are the days when you could drop your exhausted body to the ground beside a sparkling flow of wilderness water and plunge your face into the cold rush for a drink. Pathogens inhabit

Communicable Diseases and Camp Hygiene 197

most of the world's water to some degree, and, unless you're willing to risk gut-ripping misery, it is of critical importance to carry some means of water disinfection on wilderness trips.

There are three proven ways to guarantee your wilderness water is safely disinfected:

1. **Boiling.** The rule is very simple: Once the water is hot enough to produce one rolling bubble, it is free of organisms that will cause illness worldwide and up to at least 19,000 feet above sea level. The reason: All of the time it takes to bring water to a boil works toward the death of organisms in the water. By the time water *reaches* the boiling point, it's safe. *Giardia lamblia* cysts, for instance, die at approximately 122° F (50° C). Boiling is cheap—the only cost is fuel—and effective, but it consumes time, and it's inconvenient if you run out of water on the trail.

2. **Halogenation.** As for chemicals that kill waterborne pathogens, both chlorine and iodine have been proven relatively effective in most cases, given enough of the chemical and enough time. Halogenation is affected by water's temperature, pH level, and turbidity. Halogens are generally more convenient and faster than boiling the water (when you consider lighting the stove or building the fire), but they cost a bit more and can't be guaranteed to work as well as other means of water disinfection. Chlorine and iodine, for instance, have not been proven fatal to *Cryptosporidium.* Halogens also tend to leave the water tasting bad, a phenomenon reversible by adding flavoring, such as powdered drink mixes, *after* the disinfection process has been completed. If you add anything to the water prior to complete disinfection, the added substances may disrupt the disinfection process. A simple rule is: Follow the directions on the label.

3. **Filtration.** Water filters physically strain out some of the organisms and contaminants in water that could cause disease. Structurally, there are two basic kinds of filters. *Surface* or *membrane filters* are thin perforated sheets that block impurities. *Depth filters* are made of thick and porous materials that trap impurities as the water is forced through. The effectiveness of filters varies greatly, from one that removes only relatively large particles, such as *Giardia lamblia,* to one that removes virtually everything removable. Viruses are too small to be filtered out, but some filters kill most viruses with iodine from resins on the filter as the water passes through. Mechanically, once again, there are two basic types: pump-feed filters, which require manual force to push the water through the filter; and gravity-feed filters, which just hang there while water drips via gravity through the system. Filtered water looks clean, but the purity of the water depends on the specific filter. Read the claims of a filter carefully before your purchase. They are available in a wide variety of costs, shapes, and sizes. Filtration, in general, costs more but offers the quickest route to safe water.

Principles of Camp Hygiene

1. Wash your hands after a bowel movement and before food preparation.
2. Do not share handkerchiefs, toothbrushes, lip balm, water bottles, eating utensils, or other such items. If you can't finish your candy bar or your lunch, dispose of the leftovers properly instead of passing your germs to someone else.
3. Keep your hands out of food bags and your personal utensils out of group cooking pots and pans.
4. Keep all sick people away from food preparation areas.
5. Wash all community and personal kitchen gear daily.
6. Do not eat leftovers unless they have cooled off quickly and remained cold, as in a winter environment. If you do eat leftovers, add water and heat until the water boils.
7. Disinfect all drinking water.

Conclusion

By morning in the Badlands, you notice a little blood in your diarrhea. Questioning the group uncovers reports of blood in the diarrhea of several other members of the party. There is nothing to do—for the sick or yourself—except to stay hydrated and walk out.

A drive to the nearest hospital and a series of lab tests later, and a doctor tells you there is campylobacter in stool samples from the sick. Someone or several someones failed to properly disinfect drinking water. How it got passed around to so many members of the group will remain a mystery.

The doctor says the problems may resolve eventually, but she recommends an antibiotic and writes prescriptions. You carry the prescriptions to the parents, and you will, you promise yourself, pay closer attention to camp hygiene in the future.

Section Six

Special
Emergencies

CHAPTER THIRTY-ONE

Common Simple Medical Problems

You Should Be Able To:

1. Describe the most common simple wilderness medical problems.
2. Describe the proper wilderness care of the most common simple problems.

It Could Happen to You

With clear skies and no wind, the weather couldn't be better for your ski trip into the Rockies of northern Colorado. Last week's storm dropped another foot of white stuff, a foot of snow that has settled and firmed up to the perfect consistency for strong kicks and long glides. But the third morning dawns with a problem. The problem is Phil's right heel.

Inside new boots, the blister that Phil kept secret blossomed early on the second morning, and ruptured on the second afternoon. This morning Phil's groan from where he sits pulling on his right boot draws your attention. An inspection of his heel reveals a "volcanic" hole, and an inquiry from you stimulates Phil's announcement: "I just can't go on. It hurts too much."

When a Wilderness First Responder reaches for her or his first-aid kit, the injury or illness to be treated, most of the time, is a minor and simple one—a small wound, a headache, a stomachache, sunburn. It is important for the WFR to be well acquainted with common simple problems because (1) proper management can ease suffering and speed healing, (2) proper management often keeps a simple problem from becoming more complex, and (3) recognizing when a simple problem is no longer simple allows the WFR to better care for and to arrange an evacuation for a patient who requires definitive medical treatment. Let's review some of the most common problems you are likely to face.

Headaches

Headache is a generic term meaning anything from mild discomfort to agonizing debility. The message carried by the pain could be as insignificant as "slow down and rest" or as critical as "you will soon be dropping pack forever in that Great Campsite in the Sky."

Long a medical mystery (and still to some degree today), headaches have been blamed on such factors as wimpishness, psychological disorders, repressed emotions, demons, mothers-in-law, red wine, old cheese, and bad karma. Some of that blame may be valid. More than two dozen types of headaches have been identified by experts, including pain related to diet, stress, heredity, and personality traits. What a WFR should know for sure is (1) what to do for the pain, (2) how to prevent the pain, and (3) when to consult a physician.

Head pain usually can be traced to one of three sources, or a combination of two of the three: dehydration, tension, and vascular problems.

In the wild outdoors *dehydration headaches* are probably the most common. Very few people give their bodies and their brains enough water to function at maximum efficiency (see Chapter 17: Heat-Induced Emergencies). The first symptom of a drop in optimal internal water level is usually a headache, which is the brain's way of complaining of the inadequacy.

The problem rates initially as merely a nuisance. Prolonged dehydration, however, eventually leads to the breakdown of important body parts. If a dehydration headache strikes, an hour of rest during which the patient slowly sips a quart of water should put him or her back in the pain-free zone.

Tension headaches usually arise as a dull ache in the forehead or sides of the head, often associated with muscular tightness in the back of the neck. They are usually the result of exhaustion coupled with physical and/or emotional stress. Muscles in the head, and muscles in the neck and shoulders that attach to those reaching up over the head, squeeze down tight, putting grief on nerves and blood vessels. Massaging the muscles provides some relief. An over-the-counter painkiller, with lots of water and rest, and the pain typically dissipates.

Tension headaches are nothing more than a nuisance. The chance of one can be reduced by maintaining a fair to middling level of physical fitness, making sure packs fit well and ride comfortably, and talking a companion into carrying the heaviest gear. Tension headaches also can be caused by sleeping uncomfortably; adequate sleeping pads and a flat tent site can be of benefit in prevention.

Vascular headaches can be brought on by a variety of causes, any of which produce unusual dilation or constriction of blood vessels in the brain. Blood pounding through the vessels sometimes creates the throbbing often associated with a vascular headache. Hangover and hunger headaches are of the vascular type. Certain foods cause vascular headaches in susceptible people, including alcoholic beverages (especially red wine), bananas, caffeine-rich drinks, chocolate, citrus fruits, onions, peanut butter, and ripe cheese. Shifts in hormones can trigger these headaches; boys and girls are affected about equally, but adult women suffer more than twice as often as adult men.

Migraine, from a combination of French and Greek meaning "half a head," describes more vascular headaches than any other term. Hour after hour of severe pain usually dominates one side of the brain, with unwelcome accompaniment that may include blinding light in one eye, blurred vision, nausea, and numbness or tingling in arms or legs. Migraine headaches have numerous causes, often unknown, and they are announced in one out of five patients by an aura such as flashing lights or prepain numbness.

Patients with migraine headaches sometimes find relief by pouring ice-cold water over their heads, which apparently constricts swollen blood vessels in the brain. Strong prescription drugs typically provide the only reliable source of pain mitigation.

Even worse than migraines are *cluster headaches,* which most often occur in clusters of up to three headaches a day over a period of several days. The pain may feel as if someone has pounded a tent stake through your eye directly into your brain. Fortunately, the agony lasts an average of no more than 45 minutes, but 45 minutes can be a long time to have an imaginary stake in your brain. The cause of this particular head pain remains unknown, but, in the plus column, four out of five patients report the pain eases off in minutes after they start breathing 100 percent supplemental oxygen.

If headaches come on strongly and suddenly, like nothing the patient has ever felt before, persist for 24 hours unrelieved by rest and medication, and cause weakness, dizziness or other neurological manifestations, evacuation to a doctor as soon as possible is the best choice. Other indications of a serious headache include the following:

1. The patient also has a high fever.
2. The patient also has an unusually stiff neck.
3. The patient experiences a series of headaches growing steadily worse each time.
4. The patient's arms and legs are tingly, weak, numb, or paralyzed.
5. The patient has an altered mental status.

Eye Injuries

The *eyeball* is truly round, but only a small slice shows to the outside world. A jelly-like fluid called *vitreous humor* fills the ball. A bulge, the *cornea,* sits on the front of the eyeball and is filled with salty fluid called *aqueous humor.* Between the bulge and the ball lies the *lens.* Over the lens is a special muscle, the *iris,* which has an adjustable opening, the *pupil.* Working together, the lens, iris, and pupil focus images on a layer of sensitive cells, the *retina,* covering the back of the eye. Translated into electrical signals and carried by the optic nerves from each eye to the back of the brain, the images are decoded into thought. Since the images are translated from the back of the brain, sometimes a blow to the back of the head causes visual disturbances.

Nature has determined to protect our eyes in several ways. The outside of the eyeball, except over the cornea, is covered with a tough membrane called the *sclera,* the "white" of the eye. A thin, mucous membrane, the *conjunctiva,* covers the exposed part of the eye and the inner side of the eyelid. Together these see-through "skins" are much hardier than the skin that covers the rest of our bodies. *Lacrimal* (tear) glands keep the eye moist, washing out most of the dust and debris. The eyelids add protection and help keep the surface of the eye clean when you blink. A bony socket, the *orbit,* surrounds everything and forms a protective shield. It is lined with fatty "shock absorbers," a final barrier against physical abuse.

Still, it is possible to injure the eye. The most common problem is when something is in your eye that doesn't belong, such as debris, hair, insects. The discomfort can be enormous, but usually the patient's own tearing mechanism washes the eye clean before any serious damage occurs. Healthy first aid is simply to avoid rubbing the eye and immediately washing it out with a lot of clean water. Lie the patient down while a steady stream of water is poured on the bridge of the nose. Rapid blinking encourages the flushing process.

There is nothing harmful about removing large chunks of

debris from an eye, as long as force is never used. If matter on the surface of the eye does not come loose with the aid of a soft tissue or additional rinsing, leave it alone, cover the eye with a folded gauze pad or sterile eye patch, and get to a doctor. But **do not try to remove anything stuck to the surface of the cornea.**

Occasionally the eye will feel irritated after the matter has been removed, making it painful to blink. Most likely, the eye has suffered an abrasion. Rinse the eye again, and if still uncomfortable, patch it shut. Administer an anti-inflammatory analgesic, such as ibuprofen. Ophthalmic ointment or drops often provide relief. For irritation that is unrelieved in 24 hours, head for a doctor.

The classic "black eye" is a result of rapid swelling and discoloration following a blow. The whites may also turn an alarming red. But don't worry. A cold compress reduces the pain and swelling. Within 12 to 24 hours the eye returns to a relatively normal size. It can take 7 to 10 days for the black to fade. Reasons to see a doctor after a punch in the eye include persistent blurred vision, double vision, extraordinary sensitivity to light, and discharge of something other than tears.

An eyeball impaled with a sharp object is a serious injury, not a simple problem. If the object is still there, do not try to remove it. Keep the patient sitting at an angle of approximately 45 degrees—a trade-off between lying down, which causes pressure to increase within the eye, and sitting up, which would encourage gravity to pull the critically important vitreous humor out. Stabilizing what has pierced the eye is necessary. One way to do this is to make a "donut" out of a rolled handkerchief or triangular bandage. Place it gently around the eye, adding a cup over the donut so nothing can catch or jar the object. Tape it all securely in place, and patch the other eye shut as well. If the good eye looks around nervously, the bad eye will try to follow

it, possibly causing more damage. The patient should be evacuated to a hospital, with a strong recommendation to maintain the 45-degree angle.

A cut in the lid produces a lot of blood. Relax, and check the eyeball for damage. If it has been cut, the patient must be kept still, reclining at 45 degrees, with both eyes patched. Without damage to the eye, cover the wound with a light, sterile dressing. Although the slice may be quite small, a scar on the eyelid can be a lifelong discomfort. Doctors can stitch the wound neatly with little or no noticeable scarring.

Ear Injuries

That flap of flesh, the external ear, directs sound down a canal to the *eardrum* (the *tympanic membrane*), where the vibrations are transmitted to the three little bones of the middle ear. The *eustachian tube* leads from the middle ear to the nose, balancing the pressure and allowing you the interesting pastime of pinching your nose and blowing to make your ears pop as pressure changes. The little bones of the middle ear pass the vibrations on to the snail-shaped *cochlea* of the inner ear, which breaks the sound up before sending it along nerve pathways to the brain for decoding.

Common ear injuries include external damage that is obvious and treated as any soft-tissue injury is handled (see Chapter 15: Wilderness Wound Management), ruptured eardrums, or foreign objects stuck in the ear canal. Damage to the eardrum can be caused by a blow to the side of the head, a change in pressure while diving underwater (see Chapter 22: Diving Emergencies), or a nearby explosion (possibly in a lightning strike). The patient complains of pain, ringing or whistling in the ear, loss of hearing, and maybe a loss of coordination (an

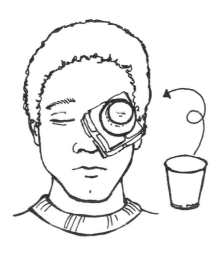

4 X 4 STERILE GAUZE PADS

Figure 31-1: *Stabilization of object in eye.*

upsetting phenomenon called *vertigo*). Keep the ear covered to reduce the chance of infection. Evacuation depends on the level of pain and discomfort the patient is experiencing.

A foreign object lodged in the ear can be carefully removed with forceps if the object is visible. Otherwise objects lodged in the ear should be removed by a physician, unless it is an insect. Insects can be drowned with water, alcohol, or vegetable or mineral oils. Once dead, the insect will ooze out with the oil or drop out later. The insect can be encouraged to fall out by irrigating the ear canal with warm water from an irrigation syringe.

Ear infections can develop in the outer, middle, or inner ear. An outer ear infection, called *otitis externa* or "swimmer's ear," typically is caused by swimming in contaminated water and/or by periods of high humidity that encourage bacterial or fungal growth in the ear canal. The patient complains of ear pain that undergoes a marked increase when the external ear is tugged. The ear canal should be flushed with a dilute solution of alcohol or vinegar, approximately 4 parts water to 1 part alcohol or vinegar, and the ear should be kept as dry as possible. If the condition worsens, it may be *otitis media,* middle ear infection. Middle and inner ear infections should be treated with appropriate antibiotics, which means an evacuation is needed to provide the necessary care.

Nose Injuries

A blow to the nose usually produces one or two manifestations: pain and blood. Keep the patient sitting up and leaning slightly forward, if possible, to prevent blood from entering the airway. If blood is visibly running out, pinch the fleshy part of the nostrils closed, using direct pressure to promote clotting. Try to prevent the patient from moving around for about a half hour after the bleeding stops, to give clotting a chance to complete. Instruct the patient to restrain from nose picking, sneezing, and bending over and straining for the next couple of days.

Epistaxis (a nosebleed) without trauma is almost always anterior, meaning the blood runs out of the nose. A few, especially in people with chronic high blood pressure, may be posterior, meaning the blood runs down the throat. A patient with a posterior nosebleed may require a doctor's care to stop the bleeding. Decongestant nasal sprays, such as Afrin, may help reduce bleeding.

Susceptible people are prone to have nosebleeds in dry weather. Applications of an ointment inside the nose can reduce the dryness and prevent bleeds. Any topical substance used for dry lips can be gently applied to the inside of the nose once or twice a day.

To avoid the possibility of serious damage, an object securely wedged inside a nose should be left in place until a physician can remove it.

Rapid swelling often makes it difficult to make an assessment of a broken nose. If you aren't sure, give the patient painkillers and apply cold packs for 20 to 30 minutes, 3 to 4 times a day. If the patient's level of discomfort is acceptable, it's OK to stay in the wilderness. When the nose is obviously broken, the same treatment is advisable. If you act quickly, before gross swelling takes over, you may be able to realign an obviously misaligned broken nose. But consider evacuation to a medical facility as well. Your patient may wish to have a bit of surgery done so the nose eventually looks the same as before. There is no rush. The surgery will be just as successful if a week goes by first. In fact, some surgeons like to have the patient wait until the swelling goes down because it's easier to tell where to realign the nose.

Skin Damage

Fungal Infections of the Skin

The fungi that cause skin infections are known as *dermatophytes* ("skin plants"), and they live on *keratin,* a hard protein in skin, hair, and nails. *Ringworm,* a popular name for several fungal infections, takes its name from the leading edge of the inflammation that is wavy and wormlike, sometimes circular. A fungal infection of the foot bears the name of *tinea pedis* or athlete's foot; of the groin, *tinea curis* or jock itch; of the skin in general, *tinea corporis;* of the scalp, *tinea capitis;* of the bearded area in men, *tinea barbae.* Most cases are passed human to human or furry animal to human. Dermatophytes like it warm, wet, and dark and are encouraged by poor skin hygiene, skin chafed by tight clothing, and skin puffy from long exposure to moisture.

Foot fungus is highly contagious, as most fungal infections are. The best place to pick up a foot fungus is off the floor of a locker room or public shower after a workout, which is how the name "athlete's foot" was derived. But you don't have to be athletic—you don't even have to pass a gym on the way to the trailhead—to get athlete's foot. The fungi can sometimes be picked up from the soil outside your tent. For some unknown reason, foot fungus grows on men twice as often as women. The fungus prefers the bottom of feet and between the toes. Starting with mild scaling, athlete's foot progresses to burning pain and irresistible itching. Untreated, the skin cracks, blisters, and smells. The inflammation can spread up the sides and onto the top of the foot.

Almost half of athlete's foot sufferers claim the problem comes and goes for years. The reason for recurring *tinea pedis—* and worsening *tinea pedis—*is failure to treat it correctly. Proper treatment involves the following steps:

1. Start treatment as soon as the signs and symptoms appear.
2. Keep feet as clean and dry as possible.
3. Apply an antifungal cream, lotion, or spray—such as Tinactin Antifungal or Lotrimin AF—twice a day.

4. Maintain antifungal treatment for 4 weeks, even if it appears that the infection has subsided.

Giving feet some direct sunlight every day may also help curtail the growth of fungus. If the itch and smell persist after more than 4 weeks of treatment, a physician should be consulted.

Most foot fungi can be stopped before they blossom. Feet should be washed and dried well every day, if possible, especially in warm, humid climates. Socks should be thoroughly dry before being put on. Boots should be allowed to dry out at night. If a patient seems prone to athlete's foot, suggest the use of athlete's foot powder or antifungal deodorant spray every 24 hours.

The treatment of any fungal infection of the skin is basically the same: Change the moist, dark environment that bred the infection. Stay as clean and dry as possible. Wear loose-fitting clothing. Apply antifungal lotions or sprays. Apply a thin layer of 0.1 percent hydrocortisone cream a couple of times a day for intense itching. Expose the infected skin to fresh air and sunshine a couple of times a day. Prescription drugs are available for severe fungal infections.

Poison Ivy, Oak, and Sumac

These poisonous plants grow in every state except Alaska and Hawaii and can be a ground cover or a small shrub as well as a woody, ivy-like vine, with several different varieties east of the Rockies bearing the name *poison ivy*. West of the Rockies you're more apt to encounter *poison oak*. Wet areas of the southeastern United States provide habitat for *poison sumac*. All three—ivy, oak, sumac—fall into the genus *Toxicodendron*. Although the leaves may be smooth-edged, or sawtooth-edged, or lobed, the leaves do indeed *almost* always grow in threes, with the middle leaf extending farther than the other two. The exception is poison sumac, which grows in a complex leaf of seven to thirteen paired and pointed leaflets.

Learn to identify the members of this genus that grow in your area because what all these plants have in common, without any exceptions, is *urushiol*, a usually colorless (sometimes light yellow) oil. Just 2.0 to 2.5 micrograms of usushiol can trigger a reaction in the very sensitive, and about 50 percent of all adults in the United States are very sensitive. Another 35 percent will have a reaction to higher concentrations of urushiol. The rest do not react, even to extremely high concentrations. No one knows why some people are tolerant of urushiol—perhaps it's an acquired immunity or a genetic blessing—but, without a doubt, sensitivity to the oil is the single most common source of allergic skin reactions in the United States and perhaps the whole world.

When urushiol soaks into human skin, an allergic reaction takes place. Not everyone reacts exactly the same, but most people first develop redness where they contacted the oil. The redness often appears in streaks where the plant brushed the skin. There may be swelling. Blisters—sometimes large, sometimes small—erupt later and discharge the fluid that fills them. The discharge eventually crusts over. The entire area itches with an intense ferocity.

After contact, it takes varying amounts of time for the reaction to show up. On thicker areas of the skin, such as on hands and feet, the oil soaks in more slowly than on thinner areas. Still, those sensitive to usushiol can see the reaction begin to happen within 2 to 6 hours. Those of low sensitivity may not develop signs or symptoms for days to as long as 2 weeks. For most people, 12 to 48 hours brings assurance that the next 10 days to 2 weeks or more will be cursed with the misery of the reaction. Adding agony to misery, parts of the body that have reacted to usushiol in the past could become reirritated when a new body part reacts to a fresh contact.

Unpredictability in the way the rash emerges has created the myth that scratching the blisters open spreads the poison. Not so: The fluid in the blisters is harmless, but you can spread the oil easily when it gets on your hands, causing the rash to show up in places you *know* never touched a plant.

For stability urushiol has few equals—it has been found active in dried plants that date back more than 100 years. Hike through it, store your boots for a couple of years, slip back into your boots, and you could suddenly react to poison ivy. You can pick up the oil from your clothing, the bottom of your tent, your hiking staff, or the hair of your dog or cat. Neither dogs nor cats, by the way—nor any other critter—seems to react to urushiol only humans.

The good news is that the oil is contained within the plants, not on the surface, which means casually brushing against an intact plant will not spread urushiol on your skin. The bad news is that the plant may not be intact, even though it appears so, for many reasons, such as a previous hiker stomping the plants, a bug chewing the plants, or a strong wind tearing the leaves.

Speaking of wind, it will not carry the oil, so standing downwind of one of these poisonous plants is safe. When plants burn, however, smoke particles *will* take the oil aloft, allowing you the opportunity to break out all over—even in your airway if you breathe in the smoke.

An allergen for all seasons, urushiol remains potent throughout the year. Beware, therefore, even the brown stems and roots of winter and the red leaves of fall if they're still attached to the stems. Plant juices return to stems and roots in the fall, leaving dead poison ivy leaves littering the ground with virtually no urushiol in them.

Nothing cures a rash caused by urushiol, but there are several methods of relieving the itch. Nonsteroidal anti-inflammatory drugs (such as ibuprofen) have no effect, but an oral antihistamine may relieve some of the itching. Topical lotions, creams, or sprays that contain antihistamines or anesthetics (words that usually end in -*caine*) should be avoided because the additives have a tendency to make things worse. Topical corticosteroids sold over the counter are too weak to work well. A physician may be able to prescribe a strong topical steroid that works, if use is started before the reaction has turned to blisters. Topical applications of calamine lotion (because it dries the area) or soaking in a tepid bath with one cup Aveeno oatmeal or two cups of linnet starch added can reduce the itch. A cold, wet compress sometimes brings relief, but, on the other end of the heat spectrum, many sufferers report substantial relief from standing in a hot shower for several minutes, then gently patting the water off the itchy area. Simply soaking the affected body parts in water as hot as tolerable often provides anti-itch relief for hours.

Home remedies abound, with numerous recipes for folk medications against the itch. If you've tried one, and it works for you, you should use it.

Unproven but often recommended—and, therefore, worthy of a few words—are two plants whose juices may ease the torment. Plantain, common and buckhorn, has leaves that release a pale, green sap when crushed. Dabbed on the rash, plantain sap reportedly stops the itch for 24 to 48 hours. The second plant, jewelweed, has as many supporters as plantain. Once again, the juice from crushed plants is applied to the rash. Jewelweed supporters also claim relief from soaking in a bathtub of water to which the juice of approximately 1 pound of the plant has been added.

Forego self-treatment and get to a doctor with any patient who shows significant swelling if the airway, face, or genitals are involved, or if any reaction seems serious to you.

In addition to recognizing and avoiding the enemy, several actions may prevent the reaction. Of prime importance is washing as soon as possible after contact with one of these poisonous plants is suspected. Lightning-fast reflexes are not required since even the extremely sensitive have an estimated 5 to 10 minutes to wash off the urushiol before it soaks in enough to cause trouble. Those of low sensitivity may have 2 hours. Cold water, lots and lots of it, inactivates urushiol, so plunging into a nearby stream or lake, depending on the person's ability to swim, would be a reasonable act. If plenty of cold water is available, especially within the first 3 minutes of contact, soap does not help, according to experts. Avoid hot water, which may spread the oil around more than off and/or open the pores so the oil soaks in faster.

Recommendations to wash with soap and water after exposure to urushiol go back to at least the 1930s, and soap is still recommended by many dermatologists after the initial 3-minute period ends. As to what kind of soap, the experts vary in exact recommendations, but detergents—such as dish and laundry—seem to work as well as anything. Technu is a soap marketed especially for removal of urushiol. Of greater importance than

what kind of soap is used is how it is used. Dermatologists recommend: repeated rubbings, not scrubbings, with sufficient rinses in between.

Organic solvents such as alcohol and gasoline work even better than water or soap and water. The preferred method is to dab repeatedly with several pieces of solvent-soaked cotton to pick up urushiol from the skin before giving your skin a good rub with fresh solvent-soaked cotton. Once the solvent dries, urushiol caught in the cotton will redeposit on your skin, so don't use the same piece of cotton for more than a few moments. The National Safely Council, opting on the side of extreme caution, recommends washing 5 to 6 times, followed by a wash of rubbing alcohol, followed by a clear water rinse. Solvents are especially useful for removing urushiol from gear.

Clothes that may be contaminated with usushiol should be washed. Clothing may hold the oil, protecting the skin at first, but the oil remains active for a long, long time.

Two products carry U.S. Food and Drug Administration approval as skin-protectant barriers against urushiol: IvyBlock and Work Shield. Applied prior to exposure, these barrier creams prevent urushiol from contacting skin, which means the cream and the oil can be washed off together after exposure. Work Shield may also be used as a cleansing cream after your skin has contacted urushiol.

Figure 31-2: *Poison ivy.*

Blisters

The fluid-filled bubble of a *blister* is a mild partial-thickness burn caused by friction. The friction produces a separation of the tough outer layer of skin from the sensitive inner layer. Only where skin is hardened is it thick enough for blisters to happen, such as at the heels, soles, and palms. Loose skin just wears away with friction, leaving an abrasion.

The space between the outer layer of the blister (the roof) and the inner layer (the base) fills with fluid drawn from the circulatory system to protect the damaged area while it heals. Gravity encourages this to happen, causing foot blisters to swell rapidly. Wet skin blisters much more quickly than dry skin, and warm skin more quickly than cool skin—and what skin is more moist and hot than feet in heavy boots after a long hike?

Blisters would probably heal best—or at least most safely—if patients sat with their feet propped up for a few days, but that doesn't happen. They keep moving. You don't want the blisters to pop inside a dirty sock inside a dirty boot, so the best wilderness medicine is to drain the blister in a controlled setting. Besides, the patient feels better when the bubble is deflated.

Here's the procedure for draining a blister:

1. Wash the site thoroughly.

2. Sterilize the tip of a knife in a flame, or use a sterile scalpel, or wipe a sharp point with alcohol.

3. Carefully slice the blister open and let it drain until the fluid is gone.

Leaving the roof intact will let it feel better and mend faster. If the roof has already been rubbed away when you discover the injury, treat the wound initially as you would any other. Clean it and keep it clean to prevent infection.

After draining, you want to reduce the friction on that area as much as possible while keeping the foot in action. Many techniques and products are available for treating a deflated blister. Some products are specifically made and marketed for blister treatment—and some of them work well. A simple and proven "improvised" technique involves creating a moleskin or molefoam "donut" to surround the blister site, then filling the hole of the "donut" with a glob of ointment, creating a sort of jelly-filled donut. Any ointment works. A liberal application of tincture of benzoin compound on the skin before applying the moleskin greatly increases its adhesiveness. A second patch of moleskin or a strip of tape over the filled donut keeps the ointment in place. A product called 2nd Skin works as well or better to reduce friction over a deflated blister. 2nd Skin can be used to fill a "donut," or it can be placed over the deflation site and held in place with moleskin or tape.

Precautions should be taken to prevent blisters from forming. Of critical importance is the fit of boots. It doesn't matter how expensively feet are shod if the fit is poor. Fit boots with the

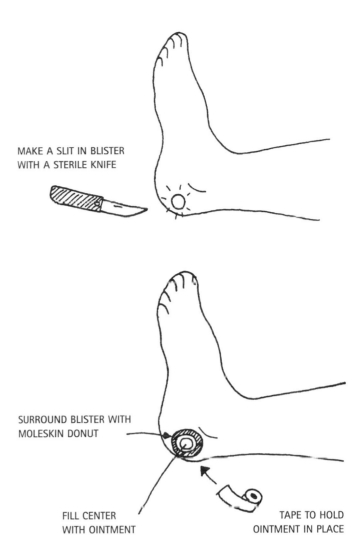

MAKE A SLIT IN BLISTER
WITH A STERILE KNIFE

SURROUND BLISTER WITH
MOLESKIN DONUT

FILL CENTER
WITH OINTMENT

TAPE TO HOLD
OINTMENT IN PLACE

Figure 31-3. *Blister treatment.*

socks that will be worn in them. Boots that are the right size and are well broken in go a long way toward preventing blisters.

Keeping feet cool and dry is also important. Hikers should take frequent breaks with boots off. A thin liner sock can wick moisture away from feet and into a thicker outer sock. Some folks report success from applications of antiperspirants to their feet. Most people think it makes their feet sticky, which increases friction. Some foot powders help reduce moisture and friction, but they also tend to cake up and require frequent applications.

If a "hot spot"—a sore, red spot—develops at a prime blister site, stop and cover the area with tape, moleskin, or 2nd Skin *before* the blister has a chance to form. Moleskin conforms to the shape of the foot better if it is cut into strips or ovals first instead of using a wide piece that inevitably refuses to go flat. Once again, tincture of benzoin compound applied first helps keep the tape or moleskin from peeling off when feet start sweating. Without tape or moleskin, benzoin alone can be applied to the skin to prevent blisters. Benzoin hardens protectively over the outer layer of skin.

Fishhook Removal

There are three ways to remove an embedded fishhook: the good, the bad, and the ugly.

The good way is the *string jerk technique:* Loop a length of string, at least 1 foot long, around the curve of the hook, and wrap the ends around your index finger. Push down on the eye and shank of the hook with your free hand to disengage the barb. Align the string with the long axis of the fishhook, then give it a jerk. The hook will come out easily and almost always without pain.

The bad way is the *push-through and snip-off technique:* Wash the skin around the wound with antiseptic solution, then numb it with ice or another source of local anesthesia. With pliers, grasp the fishhook and push the point through the skin. Snip off the barb, then back the hook out. "Bad" is a relative term; this technique does work and is practiced in some emergency rooms.

The *ugly technique* involves gently slicing through the skin with a sterile scalpel until the barb is released. This technique is not recommended except in extreme situations.

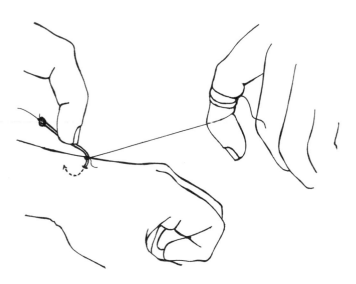

Figure 31–4: *Fishhook removal.*

Splinters

Forceps (tweezers) are one of the handiest tools to keep in your first-aid kit, and splinter removal is one of their best uses. Splinters and other small embedded objects, such as cactus spines, should be removed as soon as possible. In addition to being irritating, an organic splinter left in place can lead to an infection. If the end of the splinter is visible, grasp it with the tweezers and pull it gently out. If the end is buried, probe with your fingers until you find the orientation of the splinter, and push it toward the opening of the wound until the end is graspable. With deeply buried splinters, you may need to cut superficially, preferably with a sterile scalpel from a well-equipped first-aid kit, to expose the embedded object. Clean and dress any wounds left after removal (see Chapter 15: Wilderness Wound Management).

Subungual Hematoma

Smashing a finger or toe may cause an accumulation of blood under the nail, an often painful accumulation called a *subungual hematoma.* You can relieve the pressure—and the pain—by drilling a tiny hole in the nail to allow the blood to escape. Heat a sharp point, such as a needle, safety pin, paper clip, or scalpel, to red hotness and gently drill through the nail into the accumulation of blood. Or simply slowly and gently drill through the nail with a sharp point.

Gastrointestinal Problems

Constipation

Constipation is infrequent and/or difficult movement of the bowels usually created by either a fluid level too low to lubricate the tract, a fiber level too low to keep things rolling along, and/or a reluctance to defecate in wilderness settings. Constipation can be encouraged by poor exercise habits.

Treatment should start with forcing fluids. Encourage the patient to eat lots of whole grains, fruits (dried is OK), and vegetables. Peanut butter, cheese, and high-fat foods should be avoided. For patients who are prone to the problem, add a stool softener—preferably a suppository—to your first-aid kit.

Stimulating the gastrocolic reflex sometimes works to relieve constipation. Have the patient drink cold water, at least a half liter, followed by a cup of hot coffee or tea.

Someone who hasn't had a bowel movement in 3 days typically feels uncomfortable. A patient without a movement in 5 days could be developing serious problems from the toxins building up in the intestinal tract. Check for a fecal impaction (see Chapter 29: Abdominal Illnesses). It may be necessary to go in with a lubricated, gloved index finger to break up the impaction and pull it out. Disgusting as it may sound, it might prove life-saving.

Prevention of constipation is a combination, essentially, of drinking plenty of water (try for at least 3 liters per day) and eating healthy. A wilderness leader can encourage those who have psychological trouble going without a porcelain toilet by creating a relaxed attitude toward "pooping in the woods." You may suggest the patient pick a secluded spot with a scenic view and a minimum of tickling grasses—and practice.

Diarrhea

Diarrhea falls into two broad types: invasive and noninvasive. From bacterial sources, *invasive diarrhea*, sometimes called *dysentery*, attacks the lower intestinal wall causing inflammation, abscesses, and ulcers that may lead to mucus and blood (often "black blood" from the action of digestive juices) in the stool, high fever, stomach cramps, and significant amounts of body fluid rushing from the patient's nether region. Serious debilitation, even death, can occur from the resulting dehydration and from the spread of the bacteria to other parts of the body. *Noninvasive diarrhea* grows from colonies of microscopic organisms that set up housekeeping on, but without invading, intestinal walls. Toxins released by the colonies cause cramps, nausea, vomiting, and massive gushes of fluid from the patient's lower intestinal tract. Noninvasive diarrhea also carries a high risk for dehydration.

Diarrheal illnesses come and go with the vagaries of what one voluntarily and involuntarily ingests; erupt from numerous sources that include bacteria, viruses, and protozoas (see Chapter 30: Communicable Diseases and Camp Hygiene); and last as briefly as a few hours or as long as 3 weeks or more. *Traveler's diarrhea* is not a specific disease but a syndrome. Although *E. coli* bacteria get the nod as the cause of the largest number of traveler's diarrheas, many other waterborne or food-borne germs may be the source.

Whatever the causative agent, a diarrheal illness can be mild, moderate, or severe, depending on the frequency of discharge, the pain of cramping, the wateriness of the bowel movement, and the vileness of the gas, the latter being often a matter of personal opinion. All cases, however, have in common the departure of water from the anus—as much as 25 liters in 24 hours, in the most severe cases. And it's not just water your body spills onto the ground. An impressive amount of electrolytes can be lost during an episode of diarrhea.

Initially, the field management of all diarrheal illnesses looks the same: Replace the lost water. Clear liquids are the best choice, including plain water, broths, herbal teas, and fruit juices you can see through. If the illness continues and dehydration threatens, the patient grows weaker, with bouts of light-headedness and dizziness, and he or she requires additional electrolytes. You can pack oral rehydration salts in your first-aid kit or whip up a mixture in your water bottle. To 1 liter of water add 1 teaspoon of salt and 8 teaspoons of sugar. If you have got baking soda, throw in a pinch, but you can get by without it. Mix well. Approximately one-third of the solution should be taken every hour along with all the plain water the patient can manage to get down. Look for clear urine, the most reliable field sign of a well-hydrated person.

Pepto-Bismol not only relieves some of the torture of diarrhea but also, according to controlled studies, may provide reasonable protection from contacting some forms of diarrhea. Stronger drugs should be withheld for 24 hours—since the patient's body obviously wants to get the bad bugs out—unless travel via vehicle is necessary and/or the patient is dehydrating rapidly. Imodium, a stronger over-the-counter drug, reduces the cramps of diarrhea and the frequency and volume of stools. The prescription drug Lomotil probably ranks as most seen at the scene of diarrhea. If you've been giving antidiarrheal drugs for 24 to 72 hours and the diarrhea persists, you should stop the drugs and start considering an evacuation of the patient.

Antidiarrheal drugs should *not* be used if dysentery is suspected. Severe diarrhea, bloody stools, high fever, and tenacious vomiting are indications that the body eagerly wants to get rid of something. In case of dysentery, you should not be stopping the flow, and you should be looking for a physician.

In the best interest of the patient, stick to liquids for persistent and voluminous diarrhea. If and when the problem subsides in the field, provide bland foods, such as bread, crackers, cereals, rice, potatoes, lentils, pasta, and bananas. Avoid alcohol, caffeine, spices, fruits, hard cheeses, and other fat-laden foods.

Nausea and Vomiting

Nausea and vomiting, a pair of ugly cousins, can have many causes. Infections (usually viral) somewhere in the gastrointestinal tract, lumped under the name *gastroenteritis*, are a prevalent source. Other causes are contaminated food, motion sickness, and altitude illness (see Chapter 18: Altitude Illnesses). Usually nausea and vomiting are just the body's way of dealing with something that is upsetting the normal balance.

Protracted vomiting can lead to dehydration and electrolyte imbalances. You can give the patient an anti-emetic, such as prochlorperazine (Compazine) or promethazine (Phenergan) suppositories every 4 to 6 hours as needed. Once the vomiting has stopped, ask the patient to keep drinking plenty of clear liquids.

The patient should be evacuated if either (1) you cannot keep him or her hydrated or (2) blood appears in the vomit.

Flu-Like Illnesses

Colds and Coughs

Rhino comes from the Greek word for "nose," and *rhinoviruses* are a group of viruses, of which there are probably more than 100, that cause the common cold, an illness that seems to affect the nose as much or more than anything else. Although you can only grow ill from a specific rhinovirus once, there are enough such viruses to offer numerous chances to get sick every year. To make matters worse, viruses other than rhinoviruses can cause a cold.

A cold begins with sneezing and clear mucus running from the nose, followed by congestion (the stuffy nose) and perhaps a bit of scratchiness in the throat. The patient may complain of a mild headache and fatigue, but, overall, there isn't too much suffering. Symptoms usually last 4 to 9 days, fading away without complications.

The patient needs rest and plenty of water. Pseudoephedrine (Sudafed), a decongestant, often alleviates some of the symptoms. Acetaminophen or aspirin may also help the patient feel better.

Droplets carrying the virus are easily passed by coughing, sneezing, and hand-to-hand contact, allowing a cold to sweep through a group that doesn't practice good camp hygiene (see Chapter 30: Communicable Diseases and Camp Hygiene).

If the patient develops a profuse green or yellow nasal discharge, he or she may have *sinusitis* (a sinus infection). Rest and hydration are important, and warm compresses on the face may help, but a broad spectrum antibiotic such as amoxicillin usually offers the patient the greatest benefit.

Most wilderness coughs are the result of mechanical irritation of the throat caused by rapid breathing, such as panting in high, dry air. Hydration and rest usually ease the cough, and sucking on a throat lozenge or a piece of hard candy tends to help. A patient coughing up green, yellow, or rust-colored phlegm may have bronchitis or pneumonia (see Chapter 24: Respiratory Emergencies). Both may require treatment with antibiotics. Coughing is best controlled by increasing oral fluids and a cough suppressant, such as Robitussin DM. If the patient has a high fever, severe shortness of breath, vomiting, and/or if the cough lasts more than a week, an evacuation should be organized.

Flu

Some viruses cause different types of *influenza*. It's often difficult to distinguish between the flu and a cold, especially because they both may cause nasal congestion, coughing, and malaise (general discomfort and uneasiness). Colds, however, seldom cause much fever, and the flu does. In addition, the flu commonly causes headaches and *myalgia* (muscle pain and tenderness). Although colds typically last longer, the flu tends to be much more debilitating.

In general, treat the flu as you would a cold—with rest, hydration, and ibuprofen or acetaminophen for fever, aches, and pains. Decongestants may be given for congestion and antihistamines for a really runny nose. The patient may feel too debilitated to continue the wilderness trip, in which case an unhurried evacuation is recommended.

Fever

The need for an elevated core temperature is recognized by the *hypothalamus,* a small chunk of tissue near the bottom and slightly toward the back of the brain. The hypothalamus is like a thermostat. Poisons circulating in blood, say the virus of a flu, stimulate the hypothalamus to increase metabolism, thus increasing core temperature. This initial rise in temperature provides an advantage over the invading pathogen. The body's ability to defeat the invaders is enhanced.

Paradoxically, this rise in temperature is most often accompanied by abrupt chills. The patient with a high fever may tremble as if exposed to a raging snowstorm. The chilling process encourages the patient to wrap up in blankets, throw more logs on the fire, and snuggle up to hot water bottles. This is the first phase of a fever.

The brain withdraws from normal activity. It wants to be left alone, and it doesn't want the body that houses it to move around. Warmth and rest is the febrile brain's greatest good. This marked reduction in activity is the second phase of a fever.

When the fever reaches a satisfactory level, the brain seems to relax. Chills tend to go away, aches and pains ease off, a sense of relief washes slowly over the patient. This is a second paradox: a feeling of comfort in the midst of a high fever that has reached its maximum. This third phase of a fever may be very long and filled with unseen changes taking place on the inside of the patient.

A high fever in humans is accompanied by rises in heart rate, respiratory rate, and blood pressure. The gastrointestinal

Flu vs. Common Cold	
Signs and symptoms of influenza	**Signs and symptoms of a cold**
Mild to severe fatigue	Mild fatigue, if any
100 to 104° F fever (37 to 40° C)	No fever to very low fever
Muscle and joint pain	No muscle and joint pain
Headache and cough	Cough and nasal congestion and drainage
Duration 2 to 5 days	Duration 4 to 9 days

tract may temporarily cease to function as a digesting system. In a prolonged high fever, growth stops and aging accelerates.

Appropriate treatment for fevers involves four areas: (1) establishing an healthful external environment, (2) facilitating the increased metabolic state, (3) protecting the vulnerable organs at risk from the fever, and (4) administering drugs that specifically fight the infection.

To ignore the early stage of a fever is undoubtedly a poor choice. Encourage the patient to rest. Avoid exposure to cold. Keep the patient near a source of warmth, wrapped in protective clothing and other insulating materials. Give warm drinks. Mild fevers—less than 102° F (38° C)—may be left to run their course.

As a fever rises, adequate hydration is critical to the patient's well-being. Be aware that thirst will not increase to match the patient's need for water. Since digestion begins to fail, there is no point in forcing the patient to eat, but if hunger is present, feed it. An *antipyretic* (antifever) drug such as aspirin, acetaminophen, or ibuprofen should be administered. Sponge the patient's skin, especially the face and head, with tepid water.

Children and young adults can tolerate fevers of 104° F (40° C) for several days without permanent damage. Older adults, especially the elderly, may be susceptible to damage almost immediately at that temperature. Anyone who persists at a fever greater than 102° F (38° C) for more than 72 hours should be evacuated. Anyone who reaches 104° F should be considered an emergency evacuation.

Because fevers are difficult to measure in the wilderness, patients with obviously hot and flushed skin, a glassy look in the eye, and an altered level of consciousness should be evacuated as soon as possible to a medical facility.

Sore Throat

Sore throats of a mild variety are not uncommon complaints with viral infections, especially if the patient has been coughing. Sore throats may also develop from simple dryness due to altitude, dehydration, and/or prolonged panting during exercise. Simple sore throats can be treated with hydration, decongestants, and lozenges that can be sucked on to coat the throat. A severe sore throat, however—a throat so sore it hurts to swallow and associated with a fever and a beefy red pharynx, perhaps splotched with patches of white—could be strep throat, a serious bacterial infection requiring antibiotic therapy. Strep throat is not associated with coughs and stuffy noses. Patients with suspected strep throat should be evacuated for evaluation and, perhaps, antibiotic therapy.

Solar Radiation Problems

Sunburn

Sunshine, essential for life, strikes the Earth in rays of varying wavelengths. Long rays (infrared radiation) are unseen but felt as heat. Intermediate-length rays are visible as light. Shorter rays (ultraviolet radiation, or UVR) are also invisible and are further divided into three groups. *Ultraviolet A* (UVA) is beneficial in low doses but increases the risk of cancer in high doses. *Ultraviolet B* (UVB) is primarily responsible for sunburn and cancer. *Ultraviolet C* (UVC), the shortest and most dangerous rays, are reflected or absorbed by the atmosphere's ozone layer. UV rays contain enough energy to damage DNA in living skin and eye cells. DNA controls the ability of cells to heal and reproduce.

More Facts and Protection from UVR

1. UVA rays bombard the Earth at an almost constant rate throughout the day, but approximately 80 percent of UVB rays strike from 9:00 a.m. to 3:00 p.m. Plan to be in the sun early and late in the day for minimum exposure.
2. Smoke absorbs UVR, but clouds do not. Cool, overcast days in the summer are dangerous, because UVR penetrates the densest cloud cover, while heat-carrying infrared waves are filtered out. People feel cool, fail to take precautions, and often get severe sunburns.
3. Activities associated with water provide an excellent opportunity for serious sunburn because sunlight bounces off the water's surface to attack exposed skin. The more directly overhead the sun is, the more the reflectivity, and rough, choppy water is much more reflective than calm water. Snow is also highly reflective—as much as 85 percent as reflective as water.
4. For every 1,000 feet of elevation gain above sea level, UVR increases around 5 to 6 percent.
5. Some medications, combined with sunshine, decrease the time it takes for UV light to damage skin. These medications include tetracyclines, antihistamines, sulfa drugs, diuretics, and some oral contraceptives.
6. UV light damages eyes as well as skin. The conjunctiva can swell from UV exposure, sun-induced cataracts can form from repeated exposure, and direct UVR can burn the retina. Wear sunglasses that absorb or reflect 100 percent of UV light, and that wrap around or have side shields to protect from reflected UVR.
7. According to the Skin Cancer Foundation, 80 percent of lifetime exposure to skin-damaging UV light occurs during the first 18 to 20 years of life. Protect the children.

The earliest sign of skin damage is *sunburn:* red, painful, slightly swollen, perhaps blistered skin. Treatment should include avoidance of the damaging rays of the sun. Cool compresses provide relief. Moisturizing lotions and creams, including aloe, help, as does ibuprofen for pain and swelling. Keep the patient well hydrated.

Sunburns are preventable. Prevention can be worn in the form of tightly woven clothing that blocks a large amount of UVR, especially if the clothing stays reasonably dry. A full-brimmed hat can shade face and neck, and a floppy brim breaks up scattered UV better than a rigid brim. A bandanna hanging down behind a baseball cap protects the neck.

Sunscreens reduce the chance of skin injury. Most experts agree that screens with a sun protection factor (SPF) of 15 sufficiently protect most skin. Higher SPFs give more protection, but not at a rate comparable to the numbers. SPF 30, for instance, does not offer twice the protection of SPF 15. There is probably no reason to choose an SPF higher than 30, but make sure the sunscreen you choose protects against both UVA and UVB rays. Sunscreens are maximally effective if smeared on when skin is warm and allowed to soak in for about 30 minutes before extreme exposure. People with very susceptible skin types may do better to completely block UVR on exposed skin with an opaque substance, such as zinc oxide.

A glorious tan, by the way, does not promote skin health. All tanning should be considered visible evidence of toxic injury.

Windburn

Wind dries skin, removing the natural skin protection of urocanic acid. Wind does not truly burn skin, but it makes skin more susceptible to sunburn and irritates already sunburned skin.

Sun Bumps

Some people who are sensitive to sunlight develop small pimples or blisters—called *sun bumps*—after exposure to UVR. The skin often swells and reddens, and it may itch. The hands and face, typically getting the most UVR, are the places most likely to develop the bumps. These bumps usually appear 1 to 4 days after initial exposure, and they usually take about 2 weeks to go away. The best guess by experts is that sun bumps indicate an allergic reaction to UVA. Protect the patient, and prevent the problem, as you would for sunburn. Antihistamines, in addition, may provide some relief.

Snow Blindness

Six to 12 hours after overexposure to the sun's ultraviolet radiation, to which eyes are especially susceptible as it bounces off snow, the eyes may feel painful, like an "eye full of sand." The patient often complains of blurred vision, and the eyes may appear red and swollen. The cornea of the eye has been sunburned, and it is very sensitive to light. In this condition, called *snow blindness,* the patient is not truly "blind" but practically so, because it hurts so much to open up the eyes. Cool, wet compresses may be applied for pain. A small amount of an ophthalmic (eye) ointment may be applied several times a day for 2 to 3 days. Pull down the bottom lid of the eye and apply a thin line of ointment, and ask the patient to blink a few times and then to keep still with eyes shut until the ointment melts. If possible, remain in camp, allowing the patient to rest his or her eyes for 24 hours, during which time exposure to UV light should be avoided. Snow blindness almost always resolves harmlessly in 24 to 48 hours. If the problem doesn't resolve, an evacuation is recommended.

Problems can be prevented by wearing sunglasses that block all UV light. On snow or water, sunglasses should fit well and have side shields or a wraparound fit to block reflected light.

Motion Sickness

Three systems process information about body movement and position in space, providing a sense of body orientation: the visual system (the eyes), the vestibular system (three semicircular canals and two small bones in the inner ear), and a system of specialized nerve cells in the skin, muscles, and joints that relay data on body position. Normally, these three systems work together, sending information to the brain to maintain a sense of equilibrium. However, when you're on a moving platform on another moving body, such as a boat, you get mixed messages from these systems. Your eyes may say you're sitting still while the specialized nerve cells feel your tension and motion as you try to stay balanced. At the same time, the vestibular system reports movement side to side, back and forth, and up and down. The confused brain sends out "sick" messages, and the stomach may respond with vomit.

A plethora of home remedies recommended for motion sickness have passed down many throats, remedies that include dill pickles and saltine crackers, stewed tomatoes and saltine crackers, horseradish with or without the crackers, and crackers all by themselves. They universally have had the same effect on most people: none. Some experts claim a stomach containing only liquids deals with motion easier than one full of solids. Others claim the opposite. If you've tried something and it works, keep using it. If motion sickness seems to persist, perhaps one of the following remedies will work.

Ask the patient to focus his or her eyes, whenever possible, on a distant stationary point. Something on the horizon would be appropriate, such as a mountain peak or a tall tree. A fixed point of reference gives the brain help in sorting through the mess of messages. If the patient can go below on board a ship, don't allow it. The lack of a point of reference tends to aggravate a condition of motion sickness.

Since motion is the culprit, attempt to avoid as much of it as possible. On a large boat have the patient sit or stand near the center, with head still and eyes straight ahead. On an airplane suggest that the patient sit near the wings, where the ride rocks the least.

By lying semireclined, with eyes closed, taking deep calming breaths, and reducing the flow of messages to the brain, the patient may be able to put a stop to the ailment.

Acupressure on the *Nei-kuan* point may control nausea. The magic spot lies on the inside of both wrists, between the two tendons on the thumb side of the arm, just below the bones of the wrist itself. Pressure can be applied here with one finger for at least a full minute each on both wrists, or the patient can wear Sea-Bands, a wristband available in many pharmacies that presses constantly on the *Nei-kuan* point.

Two capsules of powdered gingerroot (500 milligrams each), taken 20 to 30 minutes before motion, may prevent sickness for numerous sufferers. With raw gingerroot, grate 1 teaspoon, add it to 4 ounces of water, and steep it for 10 minutes before drinking for the same result. A cup of ginger tea must be downed approximately every 30 minutes to maintain a nonnauseous state. Other sources of ginger may also work.

Antihistamines are medications that prevent motion sickness apparently by interrupting the nerve impulses between the vestibular system and the brain. Over-the-counter antihistamines that have proven to be particularly effective for motion sickness include meclizine (such as Antivert or Bonine), a drug that should be taken once every 24 hours starting at least an hour *before* the rocking and rolling starts. The recommended dose is 25 to 50 milligrams. Meclizine may cause mild to moderate drowsiness and a dry mouth. Another antihistamine, one of the oldest and least expensive, is dimenhydrinate (Dramamine), which must be taken every 4 to 6 hours starting at least an hour before the trip. Dimenhydrinate tends to cause extreme drowsiness, but, in the plus column, it comes in liquid form suitable for small children. Available with a prescription only, promethazine (such as Phenergan) works to prevent and (for some people) treat motion sickness, a bonus if the patient forgot to take a pill before she or he started feeling nauseous. It is recommended, however, to take 25 milligrams prior to motion. Promethazine also causes severe drowsiness in some people. Another prescription drug, Transderm-Scop (scopolamine), a patch placed behind the ear, releases small amounts of the drug over a period of 3 days right through the skin to prevent motion sickness. People report successes and failures with these drugs, and the only sure-fire way to know if one of the antihistamines will work is to try it.

Dental Problems

The heart of a tooth is the soft, inner *pulp* containing blood vessels and nerves that support the outer, hard *dentin* and *enamel*. The part of the tooth you see rising above the *gingiva* (gum) is called the *crown*. The part you don't see that fits into the socket in the jaw is called the *root*.

Dental problems are a common source of minor wilderness emergencies. A lost filling or artificial crown often creates discomfort, mostly when cold, sweets, or the tongue hits where the filling or crown has fallen out. If the pulp is exposed and becomes inflamed (*pulpitis*) the pain can be much more than a discomfort. Gently clean the area by rinsing and brushing to remove any food that may be trapped in the tooth. You can dip a cotton pellet in *eugenol* (oil of cloves) and swab out the vacancy to help relieve pain. Carry a tube of Cavit in your first-aid kit. A squeeze of the tube sends a bit of premixed paste onto your finger. Roll it into a ball and place it where the filling or crown used to be. Have the patient bite the Cavit gently into shape. The Cavit will harden, after saliva mixes in, into an excellent temporary filling that may be periodically replaced, if needed. Sugarless gum (chewed to softness), ski wax, or candle wax can sometimes be successfully used to plug the hole where a filling or crown fell out. All of these temporary fillings will soon wash out, requiring you to monitor each regularly.

If a crown is found, clean it and coat it with a little oil of cloves. Place it back on the tooth to see if it can be fitted properly. Remove it again, and put a dab of Cavit on the bottom of the crown. Set it immediately back in place, and have the patient bite it gently into position. It will probably stay put.

A broken tooth that doesn't expose the pulp is merely a nuisance that can wait until the wilderness venture is over, but a broken tooth that exposes the pulp can be a source of infinite pain. A tiny piece of a crushed aspirin placed directly on the pulp will cauterize the pulp, a technique that produces fierce pain followed by relief. Do *not* put an aspirin on the gum next to an aching tooth. This will cause an acid burn of the gum that can be severe. The void left by a broken tooth may possibly be filled with Cavit or another temporary filling. When the pulp is exposed, the patient should see a dentist as soon as possible, preferably within 48 hours.

For a tooth that has been knocked out, pick it up by the top, not the root. Rinse it off, but do not scrub it. Your best chance of saving the tooth is by pushing it gently back into the hole from whence it came. If this procedure is going to be successful, it must be attempted soon after the tooth is knocked out, within about 10 to 15 minutes, after which swelling will prevent the effort from working. If you get the tooth back in the socket, you can "splint" it with Cavit or a sticky substance such as sugarless gum or wax. If it hurts the patient too much, or refuses to go back in, don't force it. Wrap it in sterile gauze, and have the patient hold it in his or her mouth, if possible—otherwise wrap it in moist sterile gauze. If evacuation is very fast, less than an hour, the tooth might possibly be replaced by a dentist. With longer evacuations, a dental surgeon will be required to either reattach the tooth surgically or, more commonly, replace the missing tooth with an artificial one. If a blow to the mouth has simply loosened a tooth, don't wiggle it. It may firm up on its own, or a dentist might be needed. Eugenol applied to the gum might ease the pain, but stronger painkillers may be preferred by the patient.

Infection is indicated by swelling of the gum and cheek around an aching tooth and may indicate a periodontal abscess. Gas and pus can be trapped inside the gum, causing extreme pain. If the swelling from infection reaches a major stage, the gum may form an elevated bulge that can turn blue. The patient needs antibiotics and a dentist. Ice packs on the jaw can help alleviate pain. If evacuation is not possible, you might have to consider lancing the abscess. Have the patient brush, floss, and rinse out her or his mouth. A gentle and careful slice horizontally through the abscess with a sterile point will allow the pressure to release, after which the patient needs to repeatedly rinse out his or her mouth, flushing disinfected water through the open abscess. The rinsing process should be repeated several times a day until healing appears complete and/or the patient is evacuated. You may also be able to stuff a small piece of sterile gauze into the lanced abscess to keep it draining. Lancing is dangerous and not advisable except in radical circumstances. Likewise, attempting to pull a tooth out of an infected socket is dangerous. Both can cause significant bleeding. Ask your physician for information concerning antibiotic treatment for a dental infection.

A friendly dentist, outdoors enthusiast or not, should be willing to help you in finding and learning how to use the first-aid items specific to dental problems. Some excellent commercial wilderness dental kits are available.

Other Mouth Problems

Any bleeding inside the mouth, for any reason, can be treated with direct pressure. You can do this with a gauze pad that the patient bites to hold in place. A moistened tea bag (nonherbal) can be used instead of gauze, and may work better. The tannic acid in tea initiates the formation of clotting. Avoid irritating the wound, which can renew the bleeding. Irritants include smoke, extremely hot food, chewing on the "bad" side, and sucking on the wound site.

Mouth lacerations tend to look horrible for the first couple of days, but they also tend to heal nicely with little care. The patient should rinse his or her mouth with warm water every 2 to 3 hours and after every meal until the wound is completely closed.

If the mouth wound gapes open significantly, the patient should be evacuated from the wilderness for stitches, especially if the lips are involved. Gaping wounds to the mouth and lips that heal puckered are unsightly, in the way, and difficult to repair once they heal that way.

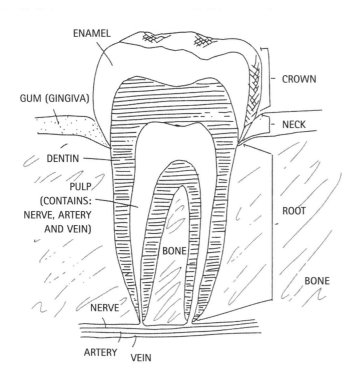

Figure 31-5: *Basic anatomy of the tooth.*

Conclusion

After a thorough cleaning of Phil's right heel, a process through which you wish Phil had suffered wordlessly, you cut a "donut" from the molefoam in your first-aid kit. Applying tincture of benzoin compound around the hole in his heel, you press the donut into place and fill it with antibiotic ointment. The strip of tape you apply over the donut promises to hold the ointment in place. But once in his boot, Phil says it still hurts too much to continue.

Challenged to create a better donut, you slice a corner from Phil's foamlite sleeping pad. Cutting the foamlite into a super donut, you remove the molefoam. With more tincture of benzoin compound smeared onto his heel and the foamlite, you "glue" your new donut in place. More ointment fills the hole. You now add longer strips of tape, secure with more benzoin. The final product at last produces enough comfort for Phil to keep on skiing. You can't resist extracting a promise from him that he will never attempt a long trip in new boots again.

CHAPTER THIRTY-TWO

Gender-Specific Emergencies

You Should Be Able To:

1. Describe the basic anatomy of male and female genitalia.
2. Describe medical problems specific to males and specific to females.
3. Explain the proper wilderness treatment of these specific problems.

It Could Happen to You

One week into your sea kayaking journey along the Maine Island Trail you have found a landing more lovely than you thought possible. The five boats that belong to your kayaking group have been pulled up onto a wide sandy beach. Just above the high tide line, tall trees shade flat campsites from the summer sun. Nearby a spring brings fresh water to a tiny pond. You're settling down for a quiet evening when Bill rushes up to say Ted is "totally in pain."

Ted lies on his back, gently supporting his scrotum with both his hands. Embarrassed, he hesitantly, between gasps of pain, relates his problem. His left gonad is swollen, and it hurts so much he has trouble moving. With Bill witnessing the assessment, you inspect the affected area. Ted's scrotum is indeed alarmingly red, and the left gonad appears more than twice the size of the right one.

Injuries involving the genitalia can be embarrassing, frightening, and, from time to time, life threatening. A variety of illnesses also can affect the reproductive systems of males and females. These emergencies range from epididymitis, urinary tract infections, and vaginal infections to the more serious testicular torsion and pelvic inflammatory disease. These problems have occurred on wilderness trips. The Wilderness First Responder should be able to assess, treat, and know when to evacuate a patient with genitalia injury or illness.

General Assessment Guidelines for Gender-Specific Emergencies

Both you and the patient will appreciate a private place to talk. The patient will benefit if you can maintain eye contact while being straightforward, respectful, and nonjudgmental. Use proper medical terminology and/or terms that you both understand. Avoid jokes or slang. A member of the patient's sex should be present before and during any physical exam—especially when minors are involved—unless it is impossible. You may choose to explain to a competent companion with less medical training, someone of the same sex as the patient, what you require in a physical examination, and ask that person to perform the exam.

Basic Anatomy of the Male Genitalia

The *penis* and *scrotum* comprise the visible external male genitalia. The penis contains a canal, the *urethra,* that provides a route for urine expelled from the bladder and sperm expelled from the *testes.* The scrotum is a pouch-like structure located to either side of and beneath the penis. Two roundish glands called testes lie within the scrotum and are the site of sperm and testosterone production.

Sperm travels out of the testes via the *epididymis,* a comma-shaped organ that lies behind the testes. The epididymis is composed of approximately 20 feet of ducts. From the epididymis, sperm travel through the *ductus deferens,* a tube approximately 18 inches long, which loops into the pelvic cavity on its journey to the ejaculatory duct.

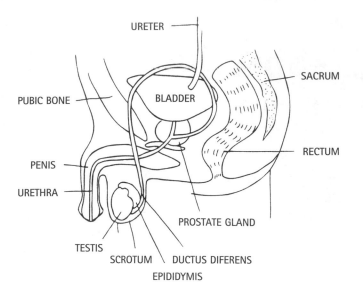

URETER
SACRUM
PUBIC BONE
BLADDER
RECTUM
PENIS
URETHRA
PROSTATE GLAND
TESTIS
SCROTUM
DUCTUS DIFERENS
EPIDIDYMIS

Figure 32-1: *Male-specific anatomy.*

Male-Specific Emergencies

Inguinal Hernia

An *inguinal hernia* occurs when part of the intestine protrudes into the groin or scrotum. Weak abdominal muscles due to congenital malformations, trauma, aging, or any activity that increases intra-abdominal pressure such as coughing, straining, lifting, or vigorous exertion may cause a hernia. Inguinal hernias are possible in women but far more common in men.

The patient will complain of a lump or swelling in the groin and a sharp, steady pain. If the hernia becomes *incarcerated* (can't be reduced), the intestine may become blocked. The portion may then become *strangulated,* the blood supply cut off, with death of the bowel resulting. If the intestine is blocked, the lump in the groin will become more swollen and more tender, the patient may vomit and complain of wavelike abdominal cramps, the abdomen will become swollen, and the patient will not have any stools. If you monitor bowel sounds by periodically placing your ear on the abdomen, you will hear an increase in bowel sounds, then a decrease, and finally an absence of bowel sounds.

Attempt to reduce the hernia by lying the patient on his back with the head and chest lower than the abdomen. Apply moderate, steady, upward pressure on the hernia to reduce it. It may take 10 minutes or longer to reduce the hernia. Monitor the patient for the next 24 hours for signs of intestinal obstruction.

If you are unable to reduce the hernia or if the hernia reappears after reduction, the patient needs to be evacuated. If signs and symptoms of intestinal blockage or strangulation are present, evacuate the patient immediately. Do not give the patient anything to eat or drink unless dehydration becomes a problem during a long evacuation, in which case small amounts of water at regular intervals may be given.

Epididymitis

Epididymitis is an inflammation of the epididymis, a problem that can be caused by gonorrhea, syphilis, tuberculosis, mumps, prostatitis (inflammation of the prostate), or urethritis (inflammation of the urethra). Epididymitis is not caused by traumatic injury to the scrotum.

The patient suffers from pain in the scrotum, possibly accompanied by fever. The scrotum may be red and swollen. Epididymitis tends to come on slowly, perhaps over several days, unlike torsion of the testis, which sometimes comes on rapidly.

The immediate treatment is bed rest and support of the scrotum with a jock strap or whatever can be improvised to create support for the testicles. Antibiotics, however, are necessary, so the patient must be evacuated when epididymitis is a possibility. A nonsteroidal anti-inflammatory drug (NSAID) such as ibuprofen may decrease the fever and pain. It may be difficult to differentiate epididymitis from torsion of the testis.

Torsion of the Testis

Torsion of the testis is a twisting of the testis within the scrotum. Mechanisms for testicular torsion could be as dramatic as violent physical activity or as simple as rolling over in a sleeping bag. The ductus deferens and its accompanying blood vessels become twisted, decreasing the blood supply to the testis. If the blood supply is totally cut off, the testis dies. After 24 hours without blood supply, little hope of saving the testicle remains.

With many patients, the scrotum is suddenly and intensely full of pain, sometimes rendering the patient unable to move. The scrotum grows red and swollen, and the testis may appear slightly elevated on the affected side. The pain, however, can come on slowly, and the other signs and symptoms may be absent.

Cool compresses and pain medication can provide some relief. An improvised jock strap—made, for example, from a triangular bandage—elevates the scrotum and may increase blood flow to the testis. The patient who is unable to walk must be evacuated for treatment immediately.

If evacuation is delayed, attempt to rotate the affected testicle back into position. Since most testicles rotate "inward," a gentle lifting and a gentle rotation "outward" may give immediate relief. The patient may wish to make the rotation himself. If it doesn't work, perhaps the testicle rotated in the opposite direction, so rotate the testicle two turns in the opposite direction. If you fail, the patient is no worse off than before the attempt was made. In either case, the patient suspected of suffering torsion of the testis should be evacuated as soon as possible. With torsion of the testis (unreduced) and with epididymitis, an evacuation that takes more than 24 hours may not be fast enough to save the testicle.

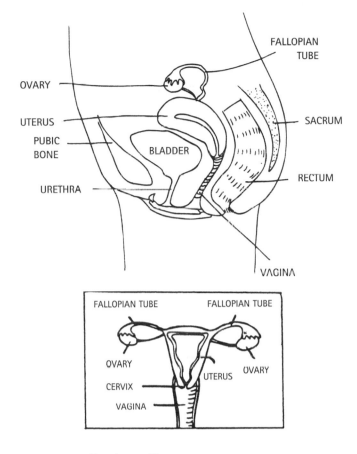

Figure 32-2: *Female-specific anatomy.*

Basic Anatomy of the Female Genitalia

The female reproductive organs lie within the pelvic cavity. The *vagina*, or birth canal, is approximately 3 to 4 inches long. The vagina is continuously moistened by secretions that keep it clean and slightly acidic. At the top of the vagina is the *cervix*, a circle of tissue pierced by a small hole that opens into the uterus. The cervix thins and opens during labor to allow the baby to be expelled.

The *uterus* is about the size of a fist and is located between the bladder and rectum. Pregnancy begins when a fertilized egg implants in the tissue of the uterus. The uterus is an elastic organ that expands with the growing fetus.

On either side of the uterus are the *ovaries*, which lie approximately 4 to 5 inches below the waist. The ovaries produce eggs and the female sex hormones estrogen and progesterone. Each month an *ovum* (egg) is released from one of the ovaries and travels down the fallopian tube to the uterus. The *fallopian tubes* are approximately 4 inches long and wrap around the ovaries but are not directly connected to them. When the egg is released from the ovary, the *fimbria* (fingerlike structures at the end of the fallopian tube) make sweeping motions across the ovary, sucking the egg into the fallopian tube.

The Menstrual Cycle

At approximately 12 years of age, a woman starts her *menstrual cycle*, the monthly release of ova. Hormones regulate the cycle, which continues until menopause at approximately 50 years of age.

The *endometrial tissue* that lines the uterus undergoes hormone-regulated changes each month during menstruation. The average menstrual cycle is 28 days long, counting from the first day of the menstrual period. For the first 5 days the endometrial tissue sloughs off from the uterus and is expelled through the vagina. The usual discharge is 4 to 6 tablespoons of blood, tissue, and mucus. During days 6 to 16, the endometrial tissue regrows in preparation for implantation of an ovum, becoming thick and full of small blood vessels.

Ovulation occurs on day 14. The egg takes approximately 6½ days to reach the uterus. During days 17 to 26 the endometrium secretes substances to nourish the embryo. If conception has not occurred, the hormones progesterone and estrogen decrease, causing the uterine blood supply to decrease and the lining of the uterus to be shed (see Chapter 33: Obstetrical Emergencies).

Assessment Guidelines Specific to Women

Many emergencies specific to females arise because they have menstrual cycles and sometimes become pregnant. For these reasons, it might be necessary to gather information about the patient's menstrual and sexual history. When was her last menstrual period? How long is her cycle? Does she use contraception? Has she had sexual intercourse in the past month, and what is normal for her? If she has had the problem before, how did she treat it? Ask open-ended questions, such as, "Tell me about your normal menstrual cycle." This type of questioning usually allows you to gather more information.

Female-Specific Emergencies

Abnormal Vaginal Bleeding

The most common female reproductive system–related problem is heavy menstrual bleeding. This is usually due to an imbalance in the amount of estrogen and progesterone that are produced by the ovaries, and commonly occurs at the times in a woman's life when her ovaries are either gaining or losing their function: when she first starts having a period and prior to menopause. It also may occur under times of stress, when ovulation does not occur—an *anovulatory* cycle. When ovulation does not occur, the ovaries do not produce progesterone, and the uterine lining does not undergo its usual maturation phase. The endometrium is shed in an abnormal fashion, and the menstrual flow is abnormally heavy and irregular.

An anovulatory cycle, with its heavy, prolonged bleeding, is usually not associated with any of the usual cramping that occurs with the menstrual period, and this may be a clue as to the cause of the abnormal bleeding.

There is not much that can be done in the field except reassurance for the patient that the problem will likely be of limited duration. It is usually more of a nuisance than a serious medical problem, though it may be alarming for a young woman who is just starting to menstruate. She should seek medical attention if the problem persists or is unusually heavy. If there is any chance the patient could be pregnant, suspect ectopic pregnancy and evacuate immediately.

Mittelschmerz

Some women experience cramping in the lower abdomen on the right or left side or in the back during the middle of the menstrual cycle, when the ovary releases an egg. The pain is sometimes accompanied by bloody vaginal discharge. This is called *mittelschmerz*, from the German *mittel* for "middle" and *schmerz* for "pain."

The pain may be sudden, sharp, and severe enough to be confused with appendicitis or ectopic pregnancy. Ask the patient where she is in her menstrual cycle. Has she ever had this pain before? Typically, a woman will have had similar cramping in the past. Any light bleeding or pain should cease within 36 hours. Her abdomen will be soft. Women taking birth-control pills do not ovulate, so they cannot have mittelschmerz.

Managing the symptoms with over-the-counter pain medications and general support of the patient is all that is usually required unless there is reason to suspect ectopic pregnancy. Monitor the patient closely for signs of a serious abdominal emergency (see Chapter 29: Abdominal Illnesses).

Dysmenorrhea

Dysmenorrhea is pain in association with menstruation. Possible causes include *prostaglandin,* a fatty acid that can cause the uterus to cramp; *endometritis,* inflammation of the endometrium; *pelvic inflammatory disease;* or anatomic anomalies, such as a displaced uterus.

Drugs such as ibuprofen reduce the pain, as do relaxation exercises such as yoga and massaging the lower back or abdomen and applying heat to the abdomen or lower back. A change in diet may help. Decreasing the amount of salt, caffeine, and alcohol in the diet while increasing the B vitamins—especially B6, found in brewer's yeast, peanuts, rice, sunflower seeds, and whole grains—can offer some relief during the acute phase of the cramps. Because exercise causes *endorphins* (natural opiates) to be released by the brain, many women find that cramps diminish when they participate in strenuous exercise.

Secondary Amenorrhea

Secondary amenorrhea is the absence of menstrual periods after a woman has had at least one period. Causes of secondary amenorrhea include pregnancy, ovarian tumors, intense athletic training, altitude, and stress (physical and emotional). In the past it was thought that excessive weight loss in female athletes, resulting in low body fat, was the cause of amenorrhea. It is now believed that stress (physical and emotional) may cause hormonal changes resulting in amenorrhea. Changes in the menstrual cycle are common on wilderness expeditions and may be normal adjustments to unfamiliar stresses. But if pregnancy cannot be ruled out, the patient should be evacuated.

Ectopic Pregnancy

When the developing embryo implants in the fallopian tube rather than into the uterine cavity wall, it begins invading into the wall of the tube, and the blood vessels in the tube enlarge dramatically. The patient has an *ectopic pregnancy,* a life-threatening emergency. (An ectopic pregnancy can occur in other pelvic sites, but the fallopian site is the most common.) Shortly after she first notices some of the symptoms of early pregnancy, as soon as a couple of weeks, the woman may notice some mild cramping and vaginal bleeding. The pain is usually on one side or the other, but it may be midline. The pain and bleeding increases as the placenta continues to eat its way into the fallopian tube. Eventually it can destroy the enlarged tubal vessels, and the woman can begin to hemorrhage into her abdominal cavity. Emergency surgery is required to stop the bleeding and repair the damage.

An ectopic pregnancy is obviously not a problem that can be dealt with in a wilderness situation. The real problem is determining whether a pregnant woman's cramps and bleeding are due to an impending miscarriage or an ectopic pregnancy. Therefore, to be safe, a woman with symptoms of early pregnancy and with lower abdominal pain, with or without vaginal bleeding, should be considered as having an ectopic pregnancy. She should be treated for shock, if the signs and symptoms develop, and evacuated as rapidly as possible.

Premenstrual Syndrome

Premenstrual syndrome (PMS) is a cluster of symptoms that occur prior to menstruation. PMS typically starts when a woman reaches her middle to late twenties and disappears in the late thirties to early forties. The most common symptoms of PMS are depression, anxiety, breast tenderness, and food cravings. Other symptoms include anger, anxiety, irritability, bloating, edema, headache, fatigue, and acne. Most researchers believe PMS is a physiological phenomenon, but the exact cause is still unknown.

Treatment includes decreasing stress, such as through yoga, meditation, and deep-breathing exercises. A high-carbohydrate diet, low in caffeine and fat, decreases breast tenderness. Reducing salt consumption helps control edema; reducing alcohol intake helps alleviate depression; reducing nicotine and caffeine lessens anxiety. A physician may prescribe diuretics to decrease edema. Vitamin B6 supplements (25 to 50 milligrams to start) may help to decrease the bloating, moodiness, and depression that come with PMS. The dose should not exceed 200 milligrams a day. The body better absorbs B6 when taken along with or as part of a B-complex vitamin.

Oral progesterone has been prescribed by physicians to treat PMS, but studies of its effectiveness are inconclusive. Daily aerobic exercise such as running or hiking apparently helps ease all symptoms.

Vaginal Infections

The normal pH of the vagina is slightly acidic. An alteration in the pH may cause a vaginal infection to develop. Infections usually result from lowered body resistance due to stress (physical and emotional), a diet high in sugar, or taking birth-control pills and/or antibiotics. A diabetic or prediabetic condition also increases the risk of infection. Cuts and abrasions from intercourse or tampons, not cleaning the *perineal* area (the area between the vagina and anus), or not changing underwear can lead to an infection.

There are three major types of vaginal infections: *yeast* (fungus), *bacterial vaginosis* (bacteria), and *trichomoniasis* (a parasitic protozoan). For the purposes of field diagnosis, the symptoms are similar, and initial treatment is the same.

Signals of vaginal infection include redness, soreness, or itching in the vaginal area. There may be an excessive or malodorous discharge from the vagina. There may also be a burning sensation during urination.

Women with a history of vaginal yeast infections may want to take an over-the-counter medication, such as Gyne-Lotrimin or Monistat, with them on extended expeditions. These are the most common medications used to treat yeast infections. Fluconazole (Diflucan), a prescription drug, cures most yeast infections and should be considered for long wilderness trips. Women with bacterial vaginosis or trichomoniasis require antibiotic treatment. Acetaminophen and warm, moist compresses should provide symptomatic relief from the pain. Itching, which may be severe, often responds to cool compresses and over-the-counter 0.5 percent hydrocortisone cream, such as CortAid. If these treatments don't provide relief within 48 hours, the patient should be evacuated. An untreated infection can develop into pelvic inflammatory disease.

The best prevention for vaginal infections is education. To help prevent vaginal infections, women should take care to (1) clean the perineal area with plain water or a mild soap daily and (2) wear cotton underpants and loose outer pants. Unlike cotton, nylon doesn't allow air to circulate, thus giving bacteria a moist place to grow. Women should avoid coffee, alcohol, and sugar—all of which can change the pH of the vagina—when the signs and symptoms of infection exist.

Urinary Tract Infection

Urinary tract infection (UTI) is relatively rare in men but common in women due to the relatively short length of the urethra, through which the pathogens causing the problem are introduced. The infection can affect the urethra, bladder, ureters, even the kidneys.

Urinary tract infection causes increased frequency or urgency of urination, with decreased urine output and/or a burning sensation during urination. The patient usually complains of pain above the pubic bone and a heavy urine odor with the morning urination. Blood and/or pus may be present in the urine. UTI can progress to kidney infections. If the kidneys are infected, the patient usually complains of pain in the small of the back, where the ribs join the backbone, and usually complains of tenderness when that area is palpated. The patient may have a fever.

UTI may be treated early and successfully by having the patient drink as much water as possible. Repeated emptying of the bladder may "flush out" the offending pathogens. A patient with a UTI should get as much rest as possible. The perineal area should be cleaned with water or mild soap daily.

On extended expeditions consider carrying antibiotics, such as Bactrim or Keflex, to treat UTI. Consult your physician for antibiotic specifics. Pyridium, available over the counter, can help relieve the pain of urination and should also be considered for first-aid kits on extended wilderness trips. UTI can be managed in the wilderness, but not kidney infections—therefore, if an infection persists for more than 48 hours despite the use of antibiotics, the patient should be evacuated.

Pelvic Inflammatory Disease

Pelvic inflammatory disease (PID), another common source of abdominal pain in women of child-bearing years, is an inflammation of the fallopian tubes, peritoneum, ovaries, and/or uterus. It is primarily caused by gonorrhea, chlamydia, or enteric bacteria. PID is most often seen in women who are sexually active, who use an intrauterine device, and/or who have decreased immune defenses.

Because PID affects the reproductive organs bilaterally, the patient usually complains of diffuse pain in the middle of the lower abdomen. The pain begins gradually and develops into a constant ache. She may also complain of pain in the right upper quadrant due to a bacterial irritation of the tissues surrounding the liver. Lower back or leg pain may also occur.

Signs and symptoms include a fever, nausea, vomiting,

anorexia, and abdominal bloating. Ask about a pus-filled, foul-smelling discharge from the vagina. The patient may complain of irregular bleeding, an increase in menstrual cramps, and pain or bleeding during or after intercourse. Some women develop acne-like rashes on the back, chest, neck, or face. Usually these signs and symptoms start within a week following the menstrual period.

The treatment for PID is evacuation to a physician for antibiotic therapy. Untreated PID can lead to peritonitis (inflammation of the lining of the abdomen), scarring of the fallopian tubes, and sterility.

Toxic Shock Syndrome

Toxic shock syndrome (TSS) is an infection caused by the bacterium *Staphylococcus aureus*. If the patient uses a tampon, the fibers in the tampon may abrade the vaginal lining, allowing the bacteria to enter the vaginal tissues, causing an infection. This particular bacterium is capable of producing a chemical, or toxin, that can result in death for the patient. TSS also occurs with the use of highly absorbent tampons left in place for prolonged periods of time—more than eight hours—which happens sometimes on long trips when supplies are short.

The onset of toxic shock syndrome is abrupt. The patient has a high fever, chills, muscle aches, a sunburn-like rash, abdominal pain, sore throat, vomiting, diarrhea, fatigue, dizziness, and/or fainting. In some people the onset may be gradual and the characteristic rash does not appear for 1 to 2 days. Mucous membranes are beet red. The sunburn-like rash appears on the palms or all over the body. It typically peels, just like sunburn, 1 to 2 weeks later.

Any woman who is menstruating and has the above symptoms should remove her tampon, drink lots of liquids to avoid dehydration, and seek medical attention as soon as possible. If shock is indicated, the patient should be treated for shock. Interestingly, although antibiotics have not been definitely proven to speed recovery, they are still considered a mandatory part of treatment.

To decrease the risk of TSS, women should change their tampons frequently and use pads at night and on light flow days. Pads and tampons, by the way, should be carried out of the mountains or burned in a very hot fire. The staphylococcus organism is frequently found on the hands; adequate hand washing prior to inserting a tampon is a must. If the patient has had TSS previously, there is a 30 percent chance of recurrence.

Pelvic Pain

Pelvic pain, a common complaint of women, is usually due to some abnormality in the reproductive organs and may produce bleeding or a vaginal discharge. There is little to be done in the wilderness except for over-the-counter pain medication. Acetaminophen is a good choice because aspirin and drugs like ibuprofen can increase the blood loss with some of the more serious conditions, such as ectopic pregnancy. Since some of these conditions can have severe consequences and may be difficult to differentiate from some of the lesser problems, a woman with increasing pelvic pain, especially with symptoms of early pregnancy, should seek medical attention as soon as possible (see Chapter 33: Obstetrical Emergencies).

Conclusion

Beyond the shade of the trees, the strength of the sun has just begun to fade from the small island off the coast of Maine. Ted, 14 years old, teeth clenched in pain, denies any and all sexual activity. That information, in addition to the sudden onset of his distress, leads you to make an assessment of torsion of the testis. Your first choice right now would be to have an emergency room on the other side of the island. But it's a long day of paddling to the next town on the mainland.

Your second choice is to attempt a reduction of the twisted testicle. Your foresight has provided you with a release signed by Ted's father, written permission to care for Ted should the need arise. You relay your opinion to Ted.

Ted, not exactly thrilled by your description of the maneuver, agrees to let you try. Anything that might relieve the pain.

Taking a deep breath, you gently lift the left gonad, rotating it slowly outward. Your reward is a great sigh of relief from Ted. Although the swelling does not immediately resolve, Ted's face tells you he is now on the trail to recovery.

Linda Lindsey, RN, contributed her expertise to this chapter.

CHAPTER THIRTY-THREE

Obstetrical Emergencies

You Should Be Able To:

1. Describe the basic anatomy and physiology of pregnancy.
2. Describe a normal pregnancy.
3. Describe complications of pregnancy and treatment for those complications.
4. Describe the process of normal childbirth.
5. Describe the assistance a mother might need with her childbirth and with the proper care of her newborn infant.
6. Describe the treatment for women with common complications of labor and delivery.

It Could Happen to You

At this spot on the Noatak River of far northern Alaska, the river lies wide from bank to bank. Camping is possible on both sides, and across the river another party has already set up tents and stretched a tarp over a cooking area. As you're unloading your canoe, a cry carries from the opposite bank: "Is anyone in your group a doctor? We need a doctor!"

A long way from a physician, both educationally and geographically, you are at least confident of your WFR skills. You paddle across to find a pregnant woman, her face tight with pain.

What were these people thinking? What is this woman doing here?

A man who identifies himself as the husband answers your unspoken questions: "She's not quite eight months along. We thought we had plenty of time."

The woman cries, "I need to push!"

To begin with a word of caution: All wilderness travel should be avoided during the last 4 to 6 weeks of pregnancy. Some women, however, leave on wilderness journeys not knowing they are pregnant—not a problem unless they experience an early and potentially life-threatening complication and some women in otherwise healthy pregnancies develop a later complication prior to the last 4 to 6 weeks of pregnancy.

Some women, furthermore, may go into labor early and find themselves a little farther from civilization than they had planned when the birth process starts. Most of these women deliver normally with or without assistance. Occasionally, delivery will be complex and life-threatening to mother and baby, made so by one of several possible complications.

Although medical problems associated with pregnancy and birth are rare in wilderness situations, the Wilderness First Responder should be able to manage the emergency.

Basic Anatomy and Physiology of Pregnancy

At *menarche*, the initial menstrual period, the hypothalamus begins stimulating the pituitary gland to release *follicle stimulating hormone* (FSH) and *luteinizing hormone* (LH), which, when produced in exactly the right quantity and timing, cause the ovaries to produce estrogen and to *ovulate*, release one egg each month. The estrogen that is released during the first half of the cycle causes the *endometrium*, the lining of the uterus, to thicken. Following ovulation the ovary that released the egg begins producing progesterone, which causes the endometrium to mature, so as to be ready to receive a fertilized egg should intercourse occur at the proper time. A single sperm fertilizes the egg well up in the fallopian tube, and the developing embryo travels down to the uterus, where it sticks to the mature endometrial lining and begins developing. If a woman does not become pregnant during a menstrual cycle, the falling levels of ovarian hormones causes the built-up uterine lining to be shed during menstruation, and the cycle starts all over again.

If conception does occur and the embryo begins developing, there is still a high probability, about 25 percent, that the pregnancy will not continue. If that particular egg or sperm were nearing their normal life span of 2 to 3 days at the time of conception, then there is a strong likelihood that their genetic material may not divide and reproduce properly and, therefore, the fetus will not develop properly. In such cases the fetus will fail to develop beyond a few weeks, and a miscarriage will occur.

Normal Pregnancy

The *placenta* is a fairly complex organ. It may be thought of as a tissue that is similar to a tumor, one that invades into the wall of the mother's uterus to provide the blood supply from her side. It is constructed something like a lung, in that the mother's blood is brought in close contact with the baby's blood, with only a thin membrane in between. Oxygen, carbon dioxide, water, nutrients, and waste products cross over this membrane much in the same way that the same molecules cross the alveolus in the lung. The fetal heart provides all the blood flow to the fetus through the placenta, which is connected to the fetus by the *umbilical cord,* which contains two arteries and one large vein. A thin membrane, the *amniotic sac,* covers the developing fetus, placenta, and *amniotic fluid,* which is essentially fetal urine made from the excess fluid that crosses the placenta from the mother's circulation.

The fetus continues to develop normally through the usual 280 days (40 weeks) of pregnancy. The length of pregnancy is measured from the beginning of the woman's last, normal menstrual period, with conception occurring approximately 2 weeks later. A woman usually feels her baby move within the uterus at about 20 weeks for the first pregnancy, or a couple of weeks earlier for a second or subsequent pregnancy. At term, the fetus normally weighs from 5 to 8 pounds, lies head down within the uterus, and is surrounded with about 1 liter of amniotic fluid. The ready-to-be-born infant, placenta, and amniotic fluid take up most of the space within the mother's abdomen.

Signs and Symptoms of Pregnancy

The first sign of pregnancy and the most common reason a woman thinks she might be pregnant is the absence of an anticipated menstrual period. If a period is 2 weeks late and the woman has been sexually active, she may suspect that she is pregnant. Additional early signals of pregnancy may include breast tenderness and enlargement, nausea that may be associated with vomiting, unusual fatigue (probably the most common symptom), and frequent urination. Some women will note abdominal enlargement unusually early in a pregnancy.

Obstetrical Emergencies

The following problems relating to pregnancy are discussed in the order during the pregnancy in which they may be encountered. This may help you determine the cause of the particular problem depending on how far along the woman is in her pregnancy.

Miscarriage

Miscarriage, or *spontaneous abortion,* is a common occurrence during the first few weeks of a pregnancy. There may be none of the usual symptoms of early pregnancy. Many times a woman may not even know she is pregnant. In these instances the woman may have a slightly delayed, heavy period.

A miscarriage is a natural process that can usually take place without medical intervention even if the pregnancy is more advanced, although a physician may be able to shorten the process and reduce the amount of discomfort involved. The woman may begin to have midline menstrual-like cramps that occur in a regular pattern. These pains increase in frequency and severity, and vaginal bleeding begins. The bleeding increases and the contractions intensify until, after 1 or 2 hours of hard cramps and heavier-than-menstrual bleeding, the woman passes fetal tissue and the "hamburger-like" placenta. Prior to passing the placenta she may pass some smooth, dark-colored blood clots. Within minutes after passing the placenta, the cramping diminishes markedly and the bleeding decreases to a very light flow.

Some women may have the placenta fail to pass within a reasonable length of time, and the blood loss may become excessive. In these cases definitive medical intervention may be necessary to terminate the process and prevent further blood loss. Much of the blood lost during a miscarriage is new blood that was made by the mother during the first few weeks of the pregnancy. If the woman's uterus can be made to contract forcefully, the miscarriage process may be accelerated. This is commonly done with a medication by her physician when the miscarriage process has begun. It is possible to accomplish the same effect in the wilderness by simulating breast-feeding. Having the woman lightly stroke her nipples for 1 minute out of every 3 or 4 minutes will cause her pituitary gland to release the hormone pitocin, which will augment her contractions, shorten the miscarriage process, and diminish the amount of blood lost during the process. She can stop the stimulation about 30 minutes after the tissue passes. The woman's uterus will be back to normal within 2 weeks following the miscarriage, and she could reattempt pregnancy after her next normal menstrual period. This will usually occur about 6 weeks after the miscarriage.

Ectopic Pregnancy

With an ectopic pregnancy, a pregnancy in an abnormal position, most commonly a fallopian tube, the patient has a truly life-threatening emergency due to the probablility of dramatic blood loss (see Chapter 32: Gender-Specific Emergencies).

Placenta Previa

Bleeding during the last 3 months of pregnancy may signal a potentially serious problem with the placenta. If the placenta should implant right over the cervix, it may separate off the uterus when the cervix starts to thin and dilate prior to labor. This condition is known as *placenta previa,* and it is potentially fatal unless an emergency *cesarean section* is performed promptly. Bleeding from a placenta previa is usually painless but may be profuse. A bleeding placenta previa is one of the causes of the rare *audible hemorrhage,* bleeding so rapid that it can be heard, like a running faucet. Again, placenta previa is not a problem easily dealt with in the wilderness. Fortunately, most women now have an ultrasound scan early in their pregnancy and know if they have a placenta previa before they make their wilderness plans. These women are advised to minimize their activities during the last 2 months of their pregnancy and will have a cesarean section delivery as soon as their baby's lungs are mature. If they start to bleed heavily they will have to have an emergency C-section.

Placental Abruption

Another problem that may be signaled by bleeding during the last few months of pregnancy is a separation of the placenta off the uterine wall, known as a *placental abruption.* The separation may be partial or complete, the latter being catastrophic for the developing infant and potentially lethal for the mother. Partial abruptions may result in only a small amount of blood loss, or they may cause massive hemorrhage if they extend to involve more of the implantation site. Placental abruptions are more common when the woman has high blood pressure complicating her pregnancy, may be caused by trauma, and also may be caused by cocaine use during pregnancy. When the placenta separates from the uterus it causes uterine irritability with contractions that increase in frequency and intensity. Within a fairly short time the uterus may contract continuously, called *tetanic contraction,* and cause constant, severe pain. If the placental implantation site is high in the uterus, or if the blood ruptures through the membranes into the amniotic sac, there may be no visible bleeding.

Note: Women suspected of either having a placenta previa or a placental abruption must be quickly transported to medical care, preferably where a surgical team and operating room are standing by. Transport should be as rapid and as smooth as possible. Any bumping or jostling may aggravate the bleeding.

Premature Labor

Since most pregnant women usually do not plan wilderness trips during the latter weeks of their pregnancy, there is a good chance that, if you are asked to assist with a delivery in a remote area, the woman will be in *premature labor.* When a woman goes into labor more than 3 weeks early (37 weeks or less into the pregnancy), the labor is considered to be premature. If the delivered infant weighs less than 5 pounds, then the infant is considered to be premature. Most of the causes of premature labor remain unknown. Most women just go into labor early, for no obvious reason. Premature rupture of the membranes, cervical and uterine infections, incomplete prenatal care, and trauma to the uterus are several of the known, but less common, causes of premature labor.

Premature labor, in itself, causes no problems for the woman. The infant, however, is at risk of several problems by being born early. First, and most significantly, the baby's lungs may not be fully formed and he or she may have varying degrees of respiratory distress, depending on the degree of lung immaturity. The infant may also have difficulty maintaining its temperature and meeting its requirements for food and water due to its immature organ systems. Premature infants are somewhat more fragile than are infants who develop to full term and do not have their reserves of lung function and fat. A markedly premature infant, one born more than 2 months early, may not tolerate the stress of labor, suffering some degree of damage that would not have occurred to a full-term baby.

Except for paying particular attention to maintaining the premature infant's temperature, the method of delivery is the same as for the delivery of a full-term infant. The other complications of prematurity usually do not develop for several hours to days after birth.

Normal Childbirth

Any woman who suspects that she is going into labor should be encouraged to leave the wilderness and reach a hospital as soon as possible, where her condition can be evaluated and her delivery conducted in as safe an environment as possible. Although most deliveries occur without incident, a delivery away from facilities where the capability for an emergency cesarean section delivery, intravenous blood and fluid administration, and the administration of oxygen and medications to the mother and/or child unnecessarily jeopardizes the health and possible survival of either or both the mother and child.

Assessment of the Pregnant Patient

How do you decide that a mother cannot possibly reach medical attention in time for her labor and delivery? You must estimate how long it will take to get her to a hospital or birthing center by whatever means available, versus how long you have before she delivers. Here are a few tips to estimate the time that might be available:

1. The more prior deliveries a woman has had, the faster the labor will be. First pregnancies usually—but not always—have at least 4 to 6 hours of hard labor, while the labor from a fifth and subsequent pregnancy may last only 1 hour. Women who have had short labors tend to have shorter subsequent labors.

Ask the patient how long her last labor was and subtract 1 hour as an estimate.

2. The smaller the infant, the faster the labor will be. A premature labor that starts 6 weeks early may be several hours faster than a labor at full term for the same pregnancy.

3. Ask the patient if she had a "gush" of clear fluid and has been leaking fluid since. If her membranes have ruptured, her labor will be about 25 percent faster than if her membranes were intact.

4. A woman who is having regular contractions every 2 to 3 minutes, lasting 45 seconds or more, and has to stop walking during her contractions is likely to have her baby within 1 to 2 hours.

5. If the patient has the urge to push with the contractions, then delivery is imminent.

Three Stages of Labor

Labor is divided into three stages. During the first stage, which lasts 8 to 12 hours for the first pregnancy and progressively less in subsequent ones, the cervix gradually *effaces* (thins) and dilates with each contraction of the uterus to allow the infant's head to pass through. The second stage, which takes from 30 minutes to 2 hours, commences as the woman starts to forcefully push the baby down the birth canal. It starts with the complete dilation of the cervix, and ends with the delivery of the baby. The third stage, which usually lasts 5 to 10 minutes, involves the delivery of the placenta.

Delivery of the Baby

When you have decided that it will be impossible to reach a hospital, then your focus should switch from transporting the mother to preparing for the delivery. The best position for a mother in labor is lying on her side, and the left side is preferable. This takes the weight of the uterus, which weighs 10 to 12 pounds with the baby, placenta, and fluid, off the mother's *vena cava*, the large vein in the abdomen that is bringing blood back to her heart. Compression of the vena cava acts like a loose tourniquet around the middle of the mother's body. Blood continues to flow through her aorta to her legs and uterus, but it cannot return to her heart to complete the circuit. This blood that is "trapped" in the lower half of her body causes a progressive decrease in the patient's cardiac output, leading to falling blood pressure and decreased blood flow to the placenta, the lifeline for the fetus. Many women who lie on their backs during the last months of pregnancy feel light-headed for this reason. Turning the mother on her side quickly solves the problem. The lateral recumbent position (side position) can also be used during delivery and is a preferred position when the mother is lying on a flat surface, as opposed to a birthing bed or delivery table with her bottom at the foot of the bed. When the baby is delivered there will be space between its head and the "bed" that will keep the infant's face out of the normal fluids that accumulate there during a delivery.

When planning on where to deliver the baby, you should consider several factors. If possible, the mother should be elevated off the ground to make your task of attending the delivery easier. A cot, bed, picnic table, or even a pile of brush covered with a tarp can serve to raise her body a little. If the weather is cold or raining, it is best to fashion some sort of shelter for the delivery. Have the patient remove her undergarments and place some clean clothing beneath her buttocks to soak up the amniotic fluid and blood that will be passed during her labor and delivery. You will be placing the newborn baby on the mother's bare chest and upper abdomen following the delivery, so she should be wearing loose clothes on her upper body and no brassiere. If possible, heat some clean water and wash your hands and the mother's hands and her perineal area (vagina and rectal area) with soap and water. Then cover her lower body with sufficient insulation to keep her warm during the rest of her labor. If the mother prefers to walk about during labor that is fine, but have her lie down when she starts to feel the urge to push during her contractions. The delivery is near.

When the patient has the urge to push, you should begin inspecting her perineum to watch for the top of the baby's head to appear. If you gently stretch the opening of her vagina, between contractions, when the top of the baby's head becomes visible, you can reduce the amount of tearing that will occur with the delivery somewhat. Insert a couple of fingers along side the baby's head and gently run your fingers around the baby's head, stretching her tissues away from the head. The baby's head descends a little with each contraction until, finally, the top of its head, down to its ears, is visible outside the mother's vagina. From this point on, due to the fact that the rest of the baby's head will be tapering down to its neck and getting smaller, it will be easier to expel.

At this point the mother will have a tremendous urge to push the baby out as rapidly as possible. Here your assistance can prevent damage to the baby. If the baby's head is allowed to "pop" out rapidly, the sudden change in pressure may cause the rupture of some small blood vessels in the baby's brain. You can prevent this by simply placing the palm of your hand on the top of the baby's head to *gently* slow the delivery of the head. Tell the mother to stop pushing—you may have to be quite loud and forceful to overcome her tremendous urge to do so. Then have her bear down gently to complete the delivery of the head. Allow the head to come out at the same speed that it had been descending up to that point: *slowly*.

The baby's head usually comes out face down, or more rarely, face up. Have the mother stop pushing so that you can clean some of the amniotic fluid and mucous from the baby's mouth. If the baby's face is down, gently rotate the head in either direction until the head is facing sideways. It will rotate in one direction quite easily as its shoulders align with the birth canal. When it is facing to the side, gently wipe the mucous and fluid from the baby's nose and mouth. Now ask the mother to push with the next contraction, and gently press down on the upper side of the baby's head and neck so that the front-most shoulder of the baby passes beneath the front of the mother's pelvic bone. As the shoulder and chest start to come out, gently lift the baby's head and allow the rear shoulder to be delivered. Have the mother continue to bear down gently—the largest part is now out and you do not want her to force the baby out so rapidly that you cannot hang on to it—and finish the delivery. If the mother looks down between her legs and watches the baby's body being delivered, she can control her pushing so that the delivery is slow and smooth.

As the body comes out, you need to pay attention to how you are going to hold the slippery little body that has begun to squirm around. You want to hold the baby with its head below the rest of its body so that the amniotic fluid within its airway can drain out of its nose and mouth during the first few breaths. The easiest way to do this is to place your index finger and thumb around the baby's neck, with the rest of your hand supporting the baby's head, and then let the baby's body rest on your arm. This leaves your other hand free. Support the baby with its body inclined towards its head, and rotate the head toward the side so that secretions can run out of its nose and mouth. Now have the mother lift up her shirt or jacket and place the baby against the skin of her chest and upper abdomen, keeping the baby warm and keeping baby's head lower than the rest of its body.

Wipe the baby's skin dry with the softest, cleanest clothing you have, paying particular attention to getting his or her hair and scalp dry. This drying process not only serves to prevent critical heat loss but also stimulates the infant sufficiently to ensure that it is breathing adequately. You should be vigorous in the drying process, especially if the baby seems sleepy, limp, or is not breathing and crying loudly. Rubbing the baby's back is a particularly safe and effective way of stimulating it to breathe. As you finish the drying, you should feel the umbilical cord between your fingers and check the baby's pulse, which will be very easy to feel through the two umbilical arteries in the cord. It should be about 140 beats per minute, roughly twice as fast as yours would be if you were not just delivering a baby.

If the baby's pulse is less than 100, and it is still limp and not breathing vigorously, you should perform mouth-to-nose-and-mouth ventilation until the baby responds (see Appendix C: Basic Life Support for Healthcare Providers). Do this about 60 times per minute, or once each second, until the baby's pulse has increased to over 100, it has begun to breathe, and it begins to

flex and squirm. As the baby's brain begins to receive an adequate amount of oxygen, the baby starts to wake up and fuss. The fussing and crying serves to continue to provide adequate oxygen, which causes further awakening and the cycle continues upward.

Carefully tie the umbilical cord with a couple of pieces of wide string, such as clean shoelaces, several inches from the baby, and about 1 inch apart. Now cut the cord between the strings. Check for and immediately stop any bleeding that may occur from the umbilical cord. The baby has little blood to lose.

Delivery of the Placenta

Several minutes will have passed by now, and the placenta should be separating off from the wall of the uterus. This is usually heralded by a gush of blood. Applying a little gentle traction on the umbilical cord causes the placenta to be delivered. Again, do this slowly so that all of the placenta comes out, and so that the membranes do not tear off, leaving some inside to cause complications later. Grasp the delivering placenta and twist it continuously as it comes out. This causes the membranes, which are trailing behind the placenta, to form a stronger ropelike strand as they are delivered and prevent them from tearing.

As soon as the placenta is delivered, you should begin massaging the mother's uterus to make it contract. The blood vessels in the uterus that supply the site where the placenta was embedded will bleed profusely until the uterus contracts. They run through the various layers of the uterine muscle, and when the uterus contracts they are pinched off. The bleeding will go from a heavy stream to just a trickle when the uterus contracts. Place the flat of your hand on the mother's lower abdomen, just above her pubic hair, and massage the uterus firmly—hard enough so that it is uncomfortable for the mother. The uterus should contract down to the size of a small grapefruit, and the bleeding will slow to a minimal flow. Continue massaging the uterus for 10 minutes or so, and then repeat the process anytime the mother starts bleeding heavily again. Another method that may be used in addition to uterine massage is to have the baby begin nursing at the mother's breast. The act of nursing causes the release of pitocin from the mother's pituitary gland. This hormone is necessary for the milk letdown reflex but also causes uterine contractions. Newborns do not need any liquids at all for 24 hours, but they may try and nurse for the first hour or so after birth. The nursing also helps start the mother-child bonding process.

Care of Mother and Child

If the mother and child are doing well, your wilderness living situation is comfortable and warm, and you have sufficient food and supplies, then you should defer trying to travel any long distances for a couple of days so that the mother can have some time to recover from the delivery. She should nurse the baby every 3 hours, drink lots of liquids, and enjoy some well-deserved rest. If the above conditions are not met then you should probably attempt to transport the mother and child to a more tolerable environment several hours after her delivery. The infant can stay warm if the mother carries him or her against her chest with some covering clothes for insulation.

Complications of Delivery

Umbilical Cord Around the Neck

If you see the umbilical cord wrapped around the baby's neck, gently attempt to slip the cord over the head. Usually the cord will be loose enough to slip over easily. If the cord is wrapped too tightly to move, it must be tied off securely, with two pieces of string a couple of inches apart, and cut between the ties. Delivery from that point should not be delayed.

Prolapsed Cord

When the umbilical cord slips down into the vagina before the baby, a complication called a *prolapsed cord* may occur. Once this happens, the descending baby may compress the cord against the bones of the mother's pelvis, cutting off all or part of the baby's blood supply.

If you can see the cord presenting before the baby, elevate the mother's hips or place her in a knee-to-chest position, and attempt to lift the presenting part of the baby off the baby's life-sustaining cord. Instruct the mother to stop pushing. In such cases, however, rapid transportation to a facility capable of performing a cesarean section is all that saves mother and child.

If transport is delayed, attempts should be made to reduce the prolapsed cord. Use the techniques described above to relieve compression of the cord. Then attempt to push the cord above the presenting part of the baby. Keep cord manipulation to a minimum. Deliver the baby as rapidly as possible and be prepared to resuscitate the newborn.

Breech Presentation

Should you encounter a *breech presentation*, in which the baby's feet or buttocks are coming out first, you should initiate transport to a hospital as rapidly as possible. If transport is delayed, do not do anything until the mother has pushed the baby out beyond the point where the baby's belly button and umbilical cord are showing. At that point you should grasp the baby by the hips and pull gently until its armpit is visible. Put your finger in the vagina, behind the baby's back, and try to pull the baby's arm forward and down in front of its chest. Then rotate the baby one half turn and repeat the process for the other arm. If you cannot get the arms to come out, then rotate the baby until it is face down and back up. Put your index finger in the mother's vagina beneath the baby's head and place the tip of your finger in the baby's mouth. Have another person or the mother push down firmly on her lower abdomen while she bears down, and while you lift upward on the baby's body and pull down (outward) with your finger in the baby's mouth. The idea is to "rock" the baby's head out of the vagina and keep the baby's neck flexed, with its chin on its chest. It is very important that the baby be delivered within a couple of minutes of when its belly button is visible. From this point on the baby no longer receives any blood or oxygen from the mother because the umbilical cord is pinched off. If you have difficulty delivering the head, which is the largest part and is coming last, then you will have to push harder from above and pull harder from below. Just remember to keep the baby's neck flexed, using your finger in the baby's mouth to keep the chin down.

Note: It is beyond the scope of this chapter to describe all of the complicated positions the baby can assume during childbirth, and it would be extremely difficult to remember the ways of managing these abnormal fetal presentations. Fortunately, more than 98 percent of babies are born head first.

Evacuation Guidelines

It is recommended to evacuate for a check up any woman assessed as possibly being pregnant if the woman did not know she was pregnant prior to the wilderness trip. All pregnant women should be evacuated if an assessment reveals the possibility of complications with the pregnancy. Should a baby be delivered in the wilderness, mom and child should be evacuated, speed being not especially relevant if both patients are doing well.

Conclusion

Inside the tent near the Noatak River, the pregnant woman cries again. In the next few tense minutes, you learn a couple of important facts. The plane that will pick up this group for the flight out to Bettles, Alaska, is scheduled to land in 4 days and will land more than 50 miles downriver. Your pick-up is even farther away. The woman's contractions are about a minute apart and a minute long.

Explaining your level of training, and lack of experience in this area of medicine, the man and woman agree to accept your assistance. Washing your hands, you direct the husband to wash his, and his wife's, and his wife's perineal region. Making the woman as comfortable as possible, you perform your initial inspection and see, during the next contraction, the top of the baby's head. Delivery is indeed imminent.

The following two hours, however, pass in a blur. During an especially long contraction and an urgent push by the woman, the head of baby begins to emerge. Even though you gently press your palm against the baby's head to slow the delivery, the tiny newborn rapidly but safely slips out to be bundled quickly in the cleanest clothing available. The baby boy is so very small. Will he survive? You will do your best—but only time will tell.

Tying and cutting the cord, you prepare to assist in the delivery of the placenta.

Richard Sugden, MD, contributed his expertise to this chapter.

CHAPTER THIRTY-FOUR

Psychological and Behavioral Emergencies

You Should Be Able To:

1. Describe the difference between crisis stress and critical-incident stress and the management of each.
2. Define the basic psychological impairments, including depression, mania, grief, anxiety and panic, suicidal behavior, and assaultive behavior, and describe the management of each.

It Could Happen to You

Of all the members of the group you're leading, Harold has been the most reticent, the least talkative, the one you find day after day hiking alone, sitting alone, eating alone. He sometimes refuses to answer even a direct question, as if he didn't hear you. True, you anticipated spring in the Jarbidge Wilderness of Nevada would bring warmth and sun, but snow has fallen in thin flurries 5 of the last 7 days. Still, that doesn't fully explain Harold's remoteness.

Then, at tonight's camp, Harold's tentmate brings you startling news: "Harold told me he's going to kill himself!"

If it were a perfect world, every wilderness experience would proceed happily and without conflict. Indeed, many an account of expeditions paints a rosy picture of lifelong friendships forged on the way to a mountaintop. It's inevitable, however, that stress in the wild outdoors sometimes produces behavior that is less than impeccable. It's reasonable to call these types of situations psychological and/or behavioral emergencies—emergencies in which normal interpersonal interaction is negatively interrupted.

This chapter addresses psychological and behavioral emergencies from three points of view:

1. *The responses of healthy people to a crisis.* Everyone in a group may experience certain reactions to a crisis, or unusual stress, which are perfectly normal responses to abnormal events, such as a serious injury to a member of the group. A Wilderness First Responder should be able to recognize and deal with these responses.
2. *The responses of healthy people to a critical incident.* Everyone in a group may react in certain ways to a critical incident, or an overwhelmingly stressful event, such as the death of a group member. A WFR should be able to recognize the need for critical-incident stress management.
3. *Basic psychological impairment.* Less likely to occur in the wilderness, mental disorders, difficult to assess even by trained mental health workers, are especially difficult to assess by the average WFR. One important principle to remember is that behavioral changes are sometimes due to physical disease, and a WFR should be able to recognize at least the possibility of needing to evacuate the patient for professional help.

Normal Responses to Crisis

A *crisis* is an unstable period when people have to respond and adapt to unusual and possibly critical changes in the state of affairs. The changes could be as mild as running out of food or as serious as a group member with a broken leg. People generally have strong emotions and high expectations when their worlds are altered by unexpected events. Plans are disrupted, sometimes enough to seriously or permanently alter the expedition. An activity previously designed as enjoyable is transformed into one requiring hard work to resolve. People are likely to harbor some intense and often negative feelings about this because many people do not know how to behave when faced with a crisis. They may have preconceived notions of what to do, but sometimes these are based on the unreality of television and cinema, not on good judgment and common sense.

People are initially affected, sometimes drastically, by the impact of adrenaline as it engages the "fight or flight" response to anything they perceive as a threat. They may get very excited, angry, even explosive, all solely due to this powerful hormone. The initial effects usually fade within a few minutes as the group adjusts to the idea that an emergency has occurred. Levels of adrenaline, however, may remain high enough after the initial surge to sustain everyone for quite some time, but eventually individual coping mechanisms do engage.

Wilderness leaders need to understand that there are varying normal responses to a crisis. Until there is time to regroup, behaviors may seem unusual when, in truth, they should be expected. Some behaviors that may emerge in the face of a crisis include:

1. **Regression.** Many grown people revert to an earlier stage of development. The theory is that, since their parents used to care for them as children, someone else may care for them now if they behave in a childlike manner. In particular, tantrums used to be very effective. Tantrum-like or very dependent behavior is not unusual.

2. **Depression.** Closing into one's inner world is another common response to crisis. This is where some people find the sources of strength to cope with an emergency. This is characterized as a shutdown effect: fetal positioning, slumped shoulders, downcast eyes, arms crossed over the chest, and unwillingness or difficulty in communicating.

3. **Aggression.** Some people lash out, physically or emotionally, at threats, including the vague threat of an emergency. High adrenaline levels may intensify the response, and so may the feelings of frustration, anger, and fear that commonly surround unexpected circumstances. This response is characterized by explosive body language, including swinging fists and jumping up and down.

What one should do about the various behaviors that surface during a crisis depends somewhat on the individual circumstances. As a general rule, open communication, acknowledgment of the emotional impact of the event, and a healthy dose of patience and tolerance can go far during resolution of the situation. Some basic procedures to consider in crisis management might include the following:

1. *Engage the patient in a calm, rational discussion.* You can start the patient down the trail that leads through the crisis.

2. *Identify the specific concerns about which the patient is stressed.* You both need to be talking about the same problems.

3. *Provide realistic and optimistic feedback.* You can help the patient return to objective thinking.

4. *Involve the patient in solving the problem.* You can help the patient and/or the patient can help you choose and implement a plan of action.

Someone who completely loses control needs time to settle down to become an asset to the situation. Breaking through to someone who has lost control can be a challenge. Try *repetitive persistence,* a technique developed for telephone interrogation by emergency services dispatchers. Remain calm, but firm. Choose a positive statement that includes the person's name, such as, "Todd, we can help once you calm down." (An example of a negative statement would be, "Todd, we can't help unless you settle down.") Persistently repeat the statement with the *same* words in the *same* tone of voice. The irresistible force (you) will eventually overwhelm the immovable object (the out-of-control person). Surprisingly few repetitions are usually needed to get through to the patient, as long as the tone of voice remains calm. Letting frustration or other emotions creep into the tone of voice, or changing the message, can ruin the entire effort. Over time, the overwhelming responses that generated the reaction may occasionally resurface. This is normal. Without being judgmental or impatient, regain control through repetitive persistence.

A crisis may bring out a humorous side (sometimes appropriately, sometimes not) among the group. When you wish to release the intensity surrounding a situation of crisis, *appropriate* laughter is one of the best methods. It should also be noted that many people cope just fine with emergency situations and unexpected circumstances. They are a source of strength and an example of model behavior for the others.

Critical-Incident Stress

There is an expectation that a trip into the wilderness—even just for the weekend—entails certain risks not found in daily life. A good trip entails a lot of physical effort and teamwork. People expect to be able to cope with the usual demands of the wilderness, and, thus, they develop unusual coping mechanisms. Sometimes, however, for some or all of the people on the trip, events surpass standard coping mechanisms. Then a wilderness-style *critical incident* has occurred.

A critical incident is almost any incident in which the circumstances are so unusual or the sights and sounds so distressing as to produce a high level of immediate or delayed emotional reaction that surpasses an individual's normal coping mechanisms. Critical incidents are events that cause predictable signs and symptoms of exceptional stress in *normal* people who are having *normal* reactions to something *abnormal* that has happened to them. A critical incident from a wilderness perspective may be caused by such events as the sudden death or serious injury of a member of the group, a multiple-death accident, or any event involving a prolonged expenditure of physical and emotional energy.

People respond to critical incidents differently. Sometimes the stress is too much right away, and signs and symptoms appear while the event is still happening. This is *acute stress;* this member of the group is rendered nonfunctional by the situation and needs care. More often the signs and symptoms of stress come later, once the pressing needs of the situation have been addressed. This is *delayed stress.* A third sort of stress, common to us all, is *cumulative stress.* In the context of the wilderness, cumulative stress might arise if multiple, serial disasters strike the same wilderness party.

The course of symptom development when a person is going from the normal stresses of day-to-day living into distress (where life becomes uncomfortable) is like a downward spiral. People are not hit with the entire continuum of signs and symptoms at once. However, after a critical incident, a person may be affected by a large number of signs and symptoms within a short time frame, usually 24 to 48 hours.

The degree of impairment an event causes an individual depends on several factors. Each person has life lessons that can help, or sometimes hinder, the ability to cope. Factors affecting the degree of impact an event has on the individual include the following:

1. **Age.** People who are older tend to have had more life lessons to develop good coping mechanisms.
2. **Degree of education.**
3. **Duration of the event,** as well as its suddenness and degree of intensity.
4. **Resources available for help.** These may be internal (a personal belief system) or external (a trained, local critical-incident stress debriefing team).
5. **Level of loss.** One death may be easier than several, although the nature of a relationship (marriage partners or siblings, for example) would affect this factor.

Signs and symptoms of stress manifest in three ways: physical, emotional, and cognitive. Stress manifests differently from one person to the next. Signs and symptoms that occur in one person may not occur in another, who has responses of his or her own.

Signs and Symptoms of Critical-Incident Stress

When an experience is an unusually powerful emotional event, there may be a series of reactions. These are both common and normal. Signs and symptoms of critical-incident stress include the following:

1. **Physical**—enduring fatigue, sleep dysfunction (either needing too much or insomnia), change of appetite (eating too much or too little), gastrointestinal upset, headache, back-

ache, chills, nausea, muscular twitches or tremors, shock-like symptoms (especially in acute stress), hyperactivity, or its opposite, underactivity.
2. **Emotional**—anger, irritability, fear, grief, anxiety, guilt, depression, feeling overwhelmed, identification with the patient(s) in a rescue, emotional numbness, feelings of helplessness or hopelessness.
3. **Cognitive**—memory loss, especially *anomia* (the inability to remember names); inability to attach importance to things other than the incident; concentration problems; loss of attention span; difficulties with calculations, decision-making, and problem-solving; flashbacks; nightmares (especially recurrent ones); amnesia for the event; violent fantasies; confusing the importance of trivial and major tasks.

Critical-Incident Stress Management

The sooner the event is defused or debriefed, the faster the reactions will ease or disappear. Denial prolongs the pain and can keep the event freshly in mind far longer than necessary. Once a situation has been identified as a critical incident, there are several options for managing the group's response. During a critical incident, watch for acute stress symptoms. Someone allowed to continue functioning when suffering acute stress can cause additional, if inadvertent, rescue burdens to arise.

Soon after the event, within a few hours, a *defusing* is likely to help the group. Everyone is brought together and the event is discussed informally. This is *not* a critique of how the event was handled. A defusing is a time for examining how people are responding to the situation emotionally, physically, and cognitively. It is an acknowledgment that something unusual happened and that unusual responses may be occurring because of it. Defusing these intense reactions allows healing to begin.

As a WFR, you may be called upon to manage a defusing. It is generally best to form the group into a circle with no one hanging back "in the shadows." Establish guidelines for the defusing. Encourage everyone to speak, but do not allow anyone to cast blame or dwell on things he or she thinks were done wrong. Let no one interrupt while another is speaking. Ask each person to relate (1) his or her role during the incident, (2) how he or she felt and now feels, and (3) what he or she thought and now thinks.

A formal critical-incident stress *debriefing* requires the assistance of a trained group. Many critical incident stress management (CISM) or critical incident stress debriefing (CISD) teams exist. You may wish to check for local availability even before leaving the trailhead.

A formal debriefing is conducted by a group composed of both peer counselors (in this case, the ideal would be wilderness-oriented peers) and mental health workers who have been specially trained in CISM. Only those who were involved are invited. The process usually takes 2 to 4 hours.

The relief of a properly debriefed group is palpable. The ability for an untrained, or well intentioned but naive, group to cause permanent damage to participants is also very real. Call in *only* an established, trained CISD group.

Basic Psychological Impairment

There is a wide range of potential for psychological impairment stemming from various causes. Although not common to wilderness expeditions, psychiatric problems (or behavioral disturbances of various sorts) may be encountered. This may be especially true for groups that use the wilderness as an experiential education forum for students with health or social challenges, including alcoholics, status offenders, and others.

It is important to recognize that apparent psychological emergencies may be due to organic causes, which means the root of behaviors that seem bizarre or deranged is not psychological at all, but physical. Medical conditions, such as meningitis or encephalitis, can cause significant behavioral changes, sometimes in the course of just a few days. Hypoxia, hypoglycemia, hyperglycemia, and hypothermia can all cause an alteration in a person's behavior. Trauma, such as a subdural hematoma, can do likewise. Substance abuse may be the culprit.

Deciding whether someone in the wilderness has a psychiatric illness is not the domain of untrained people, regardless of how well intentioned. The goal in the field is to protect the patient and those around him or her from potentially dangerous behavior while access to proper help is arranged. In particular, dangerous behavior may occur as a result of depression, mania, grief, extreme anxiety or panic, suicidal feelings, and loss of control for some reason that leads to assaultive or menacing behavior.

Depression

Everyone gets "the blues" now and then. *Clinical depression* is different because it is enduring. Depression is known as an affective disorder. This means it is outwardly manifested by mood, feelings, or tone, in this case characterized by lowered or diminished actions and feelings. Two criteria can be used to identify clinical depression: (1) a mood characterized by feeling blue, irritable, hopeless, depressed, or sad; and (2) at least four of the following symptoms present almost continuously *for at least 2 weeks*:

1. Sleep disturbance (too much or too little).
2. Eating disorder with significant weight loss or gain.
3. Psychomotor agitation or retardation.
4. Loss of interest or pleasure in usual activities.
5. Fatigue and loss of energy.
6. Feelings of worthlessness and guilt.

7. Difficulty concentrating or paying attention.
8. Preoccupying thoughts of death or suicidal feelings.

Unless someone heads out on a shorter wilderness trip already feeling depressed, it would require a lengthy trip to encounter clinical depression in the backcountry. If the condition does present, however, it is easy to see that depression can have serious impact on the group's safety and enjoyment of the expedition. Furthermore, there is little room for preoccupation and problems with concentration or for fatigue and poor self-care in terms of getting adequate rest and nutrition. There are as many as 74 medical conditions that can be indicated by a depressed mood; it remains essential to seek proper diagnosis of clinical depression when the signs and symptoms of it appear.

Note: People with a known history of depression may already have medication that masks its signs and symptoms. Although it can take up to 3 weeks for the effects of the medication to start, the medication allows a person to function in a relatively normal way.

Mania

Mania is a mood disorder characterized by excessive elation, agitation, accelerated speaking, and hyperactivity. Ideas may flood out so fast that one seems unconnected to the next, a phenomenon known as a "flight of ideas." Mania can be dangerous if the hyperactivity, which is hard to control, occurs in dangerous locations, such as precipitous terrain and whitewater.

Sometimes mania and depression combine in a cyclical pattern known as *bipolar disease*. This patient experiences extreme and uncontrollable mood swings from the depressed end of the spectrum to the manic. Such people can be difficult to be around, generating frustration and anger within the group, especially if the nature of the disease is not understood.

Grief

Grief is the emotional response to loss. Even something as simple as permanently losing a favorite pocketknife in the grass can generate a small degree of grief. Sorrow and regret are sharp and painful. There may be times on a wilderness trip that unusual behavior, such as crying easily or unexplained sadness, arises in a person who is in mourning. Identifying the cause of the behavior might be a relief for everyone concerned.

If news must be given to the group that someone has just died, be prepared for any behavior. Some people lash out, others draw within, just as in any other personal crisis. The group leader must assess the group and decide the best means for breaking the news. There may be times to tell everyone together, and times to tell group members individually. Use the words "dead" and "died" so there is no chance of misinterpretation.

The first response to news of a death is a *grief spike*, a totally consuming period of time, usually 5 to 15 minutes, in which the world shrinks down to that single acute pain. Give the news, but do not expect anyone to hear anything else until after the grief spike passes. Expect both anger and guilt feelings to be prominent, depending on the circumstances surrounding the death. Grief dissipates over time through a course of mourning, which for major losses normally takes about 1 year.

Anxiety and Panic

Acute anxiety or panic attacks can be very disruptive, especially if they occur with poor timing, such as above treeline with an electrical storm fast approaching. These are characterized by a discrete but unpredictable onset of symptoms, often without warning or an apparent precipitant. Physical symptoms include rapid heartbeat or palpitations, light-headedness, sweating and shaking, and sometimes fainting. There are feelings of terror or apprehension and fear of dying, losing control, or "going crazy."

If someone with such symptoms has good contact with reality, the cause is likely to be a panic disorder—if not, there may be an affective disorder or organic cause. Other conditions associated with anxiety or panic include phobias, hyperventilation syndrome, and generalized anxiety disorder (characterized by persistent anxiety of at least a month-long duration).

Field care begins with close, careful assessment of the person's behavior and identification of the apparent cause. Many people who have not had a panic attack believe they are dying of a heart attack. Anything that can help the person regain a sense of control can help. Because anxiety can be infectious, watch the rest of the group and communicate in a supportive, calm fashion with everyone. If calm, patient support does not work after a while, it may be necessary to be firmer and set limits, especially when the safety of the group is at stake.

Suicidal Behavior

Suicide was recently the ninth overall cause of death in the United States. About two-thirds of suicidal gestures or attempts are made by females, although males are successful twice as often, since they tend to use more definitive means, such as gunshots and hangings. Teenage males are a high-risk population, as are men over 45 and women over 55. Anyone who speaks of suicide or makes suicidal gestures should be taken seriously. It may or may not happen, but the group that ignores the suggestion may be unpleasantly surprised. Suicidal behavior, by the way, is *not* inherited.

It may seem difficult to broach the subject with someone who is causing concern, but if suicidal thoughts are suspected, inquire directly. There are numerous myths about suicidal behavior. One myth is that talking with someone about suicide will drive him or her to do it. After developing good trust and rap-

port, ask, "Are you thinking of suicide?" If the answer is "Yes," ask what method he or she is thinking about. The more concrete the plans, the more likely suicide is to happen. If a person who was previously seriously considering suicide seems suddenly much better emotionally and behaviorally, the "miraculous" recovery may instead mean a decision has been made to go through with it. The patient may feel the conflict has been taken out of the situation. Be careful.

A helpful approach is to let the patient talk, and avoid judgmental or opinionated replies. "Oh, you don't want to do that!" is *not* what a suicidal person needs to hear. Instead, try to generate questions that elicit more than single-word answers, such as, "Can you explain what makes you interested in suicide?" In general, treat the person normally. Expect the person to continue doing a fair share of the group's tasks. But watch him or her carefully—and initiate an evacuation.

Assaultive Behavior

For many reasons people sometimes become menacing or assaultive. In a time of crisis people typically rely on old behavioral patterns for survival. Those who become aggressive may act on those feelings and become physically dangerous to others on the expedition. Sometimes a period of venting is helpful if the nature of the outburst is not threatening to the others. Wilderness life can be stressful, people live in close quarters, and eruptions of intense emotions are not unusual.

Resolving such a situation requires good rapport and tact. The goal is to calm the individual, not to trigger worse behavior. In some cases it may be better not to intervene until later. At other times someone who is out of control may need physical restraint. Evidence of homicidal intentions, for example, should be treated seriously and immediately. When enough people are available, use all the resources available and coordinate the effort through good communication. One person per extremity is best—hold the elbows and knees, not the feet and hands. Let restraint accomplish its intended purposes without seeming punitive or brutal.

Evacuation Guidelines

Wilderness care of psychiatric and behavioral emergencies relies on good judgment and common sense among the group in deciding which patients need professional help—and thus evacuation—and which can be tolerated until the trip is scheduled to end. These emergencies demand open, honest communication among group members.

Conclusion

Harold sits slumped against a tree in this seldom-seen section of Nevada when you join him. A few minutes of attempting honest but general conversation with him elicits little more than grunts and nods.

Then you ask, "Harold, are you really thinking about killing yourself?"

Harold nods a "yes."

"How do you plan on doing it?" you ask.

After several moments of silence Harold says quietly, "I don't know. Maybe I'll just walk off and die. It's cold. It shouldn't take too long."

You hold back a small sigh of relief, remembering that the more specific a plan of suicide the more likely the person is to attempt a suicide.

"I would really appreciate it if you'd talk to me about why, Harold."

Without looking up, Harold spends the next half hour relating his world of devastating emotional experiences while you listen. You avoid comments other than to let him know you're truly listening.

When he seems to wind down, you ask him to help you prepare tonight's dinner. He agrees.

You will sleep little this night, keeping an eye on Harold's tent after he has crawled into his bag. And tomorrow you'll hike the group out to a trailhead, beginning the process of accessing professional assistance for Harold.

Kate Dernocoeur, EMT-P, contributed her expertise to this chapter.

Section Seven

Special Concerns

CHAPTER THIRTY-FIVE

Emergency Procedures for Outdoor Groups

You Should Be Able To:

1. Describe the need for and demonstrate the writing of a trip plan.
2. Describe the factors affecting the organization of a wilderness evacuation.
3. Prepare a written evacuation plan and medical report that could be sent out to request a wilderness evacuation.
4. Describe the management of a lost person incident.

It Could Happen to You

This group, you decide, is ready and able to spend most of a day without you and your co-instructor. Dividing the 12 participants, aged 16 to 18 years, into 3 groups of 4, you go over the maps, explaining the routes you want them to take to tonight's campsite. Each group will hike a different route, following landmarks and a compass bearing cross-country. Then you and your co-instructor head out on a fourth route.

Having given yourself the shortest route, you reach camp in time to relax, catch up on some personal work, and await the arrival of the 3 small groups.

All goes well until the last team of participants arrive. Instead of 4 participants, only 3 backpackers hike into camp.

Johnny is missing!

Emergency and long-term care of the patient is only one of the responsibilities of the Wilderness First Responder. Leadership of the evacuation—from shelter construction, communication, and group organization to planning and patient transportation—may also demand attention.

Wilderness evacuations are mental as well as physical challenges. Outdoor leader and educator Paul Petzoldt gave wise advice for any emergency: Step aside for a moment and review the situation. He was well aware of the crucial role that planning plays in wilderness emergency procedures and of the dangers of haste. Forgotten maps have caused rescue teams to become lost. Misinformation has sent rescue teams hiking through the night for a patient with minor injuries. Verbal messages have been distorted from passing through several people or by misstatements from exhausted messengers. Errors in organization and technique may multiply over time, compounding the difficulty of the situation. The rescue team bivouacking without sufficient clothing, shelter, or a stove may themselves need rescue. The poorly constructed litter that falls apart a mile down the trail slows and stresses both the group and the patient. A misdiagnosed medical problem can initiate an unnecessary wilderness rescue. Rescuers have been injured, and some have died, responding to nonemergencies.

Wilderness evacuations involving rescuers in strenuous and potentially dangerous activities need the support of sound planning, organization, and leadership. This chapter presents a plan for organizing the management of a simple wilderness emergency and addresses some of the options the leader should consider.

Trip Plans

A trip plan is the foundation of an effective wilderness evacuation. An organized expedition or rescue group should have a written trip plan. In an ideal wilderness world, even a short personal trip with a few friends has a written plan. This document should be adaptable to a wide variety of contingencies including a lost person, an ambulatory patient, and a nonambulatory patient. It should cover situations requiring a technical rescue. It should be frequently reviewed and updated.

As your trip plan is written, consider possible scenarios and draft guidelines for their management. Investigate your options for getting help to respond to a medical emergency on your expedition or in your rescue area. Research and catalogue resources and evacuation options. Contact local rescue groups and check out their capabilities. Know if helicopters, technical rescue teams, and paramedical support are available. Know when self-rescue is the only option and when outside help must be utilized. Know who is responsible for rescue in your wilderness area.

On a personal level, prepare for an emergency by always carrying water, shelter, map, compass, matches, flashlight, knife, spare clothing, extra food, and a first-aid kit—as well as humility and competence.

Trip plans should include the following:

1. Guidelines for how to respond to emergency and nonemergency situations.
2. Lost person and technical rescue protocols.
3. Special instructions for serious injury, illness, or a fatality.
4. Resource lists—such as rescue services—with names, addresses, and telephone numbers.
5. Maps with roadheads and locations of nearest phones marked.

Evacuation Organization

After immediate medical and safety needs have been addressed, many things begin happening simultaneously. Evacuations always have more tasks than people, and every task is important. Someone needs to assume leadership and delegate responsibilities, and it is recommended that the leader of the evacuation refrain from becoming involved in the details of the emergency care of the patient. He or she needs to maintain a higher stance, a broader perspective. Your priorities are the safety of the rescuers, the care of the patient, and the organization of the rescue. Although they are discussed here separately and as a sequence, these tasks typically happen at the same time.

Safety

Review scene safety. This is as important as the initial scene survey. Look around, identify environmental hazards, and take steps to manage them. Inclement weather may make on-the-spot shelter a priority. Rockfall or avalanche danger may dictate an immediate move to a safer location. You may need to caution overzealous rescuers in steep terrain or on a slippery riverbank to be careful. Take care of immediate scene stabilization needs: Delegate someone to manage the patients, someone to write the SOAP note, and someone to gather equipment. Stop and make eye contact with everyone: Slow down, be thoughtful, and be careful.

Organizing the Evacuation

Develop a plan for the evacuation and delegate the tasks necessary to accomplish the plan. These may include briefing the group, selecting and briefing a messenger team who will go for help, writing an evacuation plan, building a fire, finding and organizing a campsite, feeding the group, preparing the evacuee's pack, building a litter, scouting a route, breaking trail through snow, finding and marking a landing zone for a helicopter, or packing soft snow for a landing pad (see Chapter 36: Wilderness Transportation of the Sick or Injured). To make the best plan you need an inventory of all available resources. Gather and inventory all your equipment: maps, first-aid kits, food, water, stoves, shelters, and technical outdoor gear.

When planning the evacuation consider (1) the severity (urgency) of the medical problem, (2) the distance to the roadhead, (3) the terrain difficulty, (4) the strength and stamina of the group, (5) the weather, (6) the time of day, (7) communication options, and (8) rescue possibilities (see Chapter 36: Wilderness Transportation of the Sick or Injured).

■ SEVERITY OF THE MEDICAL PROBLEM

It is fundamental to determine the severity of the medical situation. How soon does the patient need to be in the hospital? Minor medical problems may be well handled by a group walking out or carrying out a patient. Life- or limb-threatening problems may require the use of helicopters, fixed-wing aircraft, radios, and other means of outside support.

■ DISTANCE TO THE ROADHEAD

Consider how far it is to the roadhead and to the nearest telephone. Do *not* underestimate the time it will take to travel the distance, initiate your evacuation plan, and have help return. Messengers in good physical shape on a good trail can hike approximately 3 miles an hour. Paddlers might need to negotiate rapids and portages. Rescuers burdened with litters and medical gear hike slowly. Darkness, deep snow, boulders, weather, river crossings, and other technical obstacles must be factored into the time estimate. Speed on litter carriers is often less than 0.5 mile an hour.

■ TERRAIN DIFFICULTY

Think about the terrain difficulty. Essentially, the same factors that add to time add to difficulty. The time of day when you travel can be as important as its difficulty. Crossing rough or technically demanding terrain is best when you are fresh, rather than later in the trip when you're exhausted.

■ STRENGTH AND STAMINA OF THE GROUP

Consider the group's physical strength and stamina, as well as technical abilities and experience. Group members must be strong enough for the hard physical work and skilled and experienced enough to be safe when crossing rivers, glaciers, scree, boulders, or other technically demanding terrain. They must also be psychologically strong enough to manage the difficulty of an evacuation.

■ WEATHER

Consider the weather. Will you be able to deal with deteriorating weather? Will weather conditions slow, stop, or alter your timetable or your chosen evacuation method or route?

■ TIME OF DAY

Is it morning, afternoon, or evening? Do you have enough daylight to get to a vehicle or enough time to get to an acceptable campsite? Or perhaps there's not even enough light to move the patient. The time of day affects your evacuation plan.

Radios and Cell Phones

Radios and cell phones are more and more effective in remote areas, although they continue to have limitations. If you do have the chance to give a message with a cell phone, the message must be clear and concise. Write the message before you call. You need at minimum to communicate (1) where you are (the more exact the better), (2) a brief description of the patient and the urgency, (3) what you want or need from the rescuers, and (4) what you plan to do in the meantime. For example: "I'm Joe Backpacker. I'm at Red Bear Lake in the Lost Wild Mountains at 124, 34, 56 west, 40, 45, 32 north. I have a 24-year-old male with a broken left femur. The patient is splinted and is in mild shock. We need a helicopter evacuation. We plan to stay at this location until help arrives."

■ COMMUNICATION

In many cases radio or telephone communication is not available or not reliable from remote areas. Messages must be delivered on foot. If there is no chance to talk to a rescue group, the written message must be accurate, concise, and complete.

Determine if messengers should be sent. Designate a leader and a large enough messenger group to be safe and effective. Four is typically considered an appropriate number for safety and traveling efficiency, but assigning four messengers is often not practical.

When deciding on the composition of the messenger group consider physical stamina, first-aid skills, night travel and navigation skills, foul-weather experience, and any experience in the technical terrain that may be encountered.

The messenger group should carry written instructions, including copies of the evacuation plan (see sidebar here) and the SOAP note (see Chapter 3: Patient Assessment, page 27).

Group members should be prepared for extenuating circumstances with food, clothing, sleeping bags, or bivouac gear, as well as maps marked with the accident site, the destination roadhead, and the intended line of travel.

Written Reports

Written reports should include both a SOAP note detailing the patient's condition and the evacuation plan detailing considerations for getting the patient out.

The evacuation plan should include the following:

1. Type of evacuation (walking, helicopter, litter).
2. Marked maps showing the location of the accident, the present location of the group and patient, the anticipated route out of the wilderness, and the roadhead destination. If you don't have a map to send, describe the locations in as much detail as possible.
3. Any special requests (oxygen, medications, doctor).
4. Plans for the messenger group returning to the expedition.
5. Plans for the rest of the group remaining in the field.
6. An alternate plan that allows any rescue party to anticipate and support your actions if the initial plans go awry.

Sample Evacuation Plan

We have a patient with an obvious fracture of the lower right leg. Patient is stable but unable to walk (see SOAP note). We are a group of 5, and we are requesting a team of rescuers and a litter.

We are camped on the west side of Geneva Lake about 3 miles from the Geneva Lake Trailhead. The elevation gain is about 1,000 feet (see map).

The patient is in a great deal of pain, and pain killing medication would be appreciated.

Three members of our party will carry out this message. One will be remaining with the patient. At the Geneva Lake campsite we have a tent, sleeping bags, and food for 2 days. The 3 people carrying out the message will return to the trailhead in case assistance is needed carrying the litter.

Due to the disability of the patient, we do not anticipate an alternate plan.

Thank you for your help.

Managing a Lost Person Incident

When a member of your party is missing, you have an emergency on your hands. Lost persons not found within 24 hours tend to show up on their own or show up dead. Well-developed trip plans include the actions to be taken when someone is lost.

The Interview

The most important—and sometimes forgotten—first step in dealing with a missing person is to gather as much information as possible. This information primarily comes from members of the party during an interview, and the interviews should take place as soon as possible. The information gathered needs to answer several questions, including:

1. Who is missing, and for how long? And why?
2. Where was the missing person last seen?
3. What is the missing person's level of experience? What, if anything, was the person instructed to do if she or he got lost?
4. What gear, clothing, food, or other equipment did the person have? What was he or she wearing?
5. What was the emotional state of the missing person?
6. What was the missing person's physical condition? Are there any known medical problems?
7. What footwear is the person wearing, and what does the sole look like?

The Search

Studies indicate that a lost person walks at approximately 2 miles per hour. Based on that estimate, after 1 hour the search area covers approximately 12 square miles, after 2 hours the search area is 48 square miles, and after 3 hours it is 108 square miles. A search should be initiated as quickly as safely possible, using the following guidelines.

1. Start at the last seen point.
2. Insist that all searchers keep at least one other searcher in view at all times. Have an exact plan for each searcher.
3. Consider geographic containment, such as rivers and streams, high ridges, steep passes, and maintained trails. Lost persons tend to take the path of least resistance and stay within natural boundaries. They tend to follow trails and waterways.
4. Look for clues, such as tracks, broken branches, and dropped clothing or gear.
5. Make the search attractive by using fires, lights, bright cloth, whistles, and vocal calls. When making noise, take silent breaks to listen for responses. Leave written messages at obvious points, such as trail junctions.
6. Search natural attractions, such as old cabins, mines, or hot springs.

Going for Help

If you decide to send for help for any reason
1. Send a team, preferably four strong, well-equipped people.
2. Send a written SOAP note if you have a patient already at hand.
3. Send a written search plan if you are searching or an evacuation plan if you need to evacuate a patient.
4. Send in writing an exact location, preferably on a map, of your campsite and/or your search area.
5. Ask in writing for what you think you'll need.

Summary

Several things should be abundantly and perhaps painfully clear. There are typically no simple answers to the questions of who needs to be evacuated, how fast should the evacuation be, and how is the evacuation to be handled (for instance, can and/or should the patient walk). Many factors weigh heavily in your decision, including time, distance, terrain, and other such factors. The effort you put into planning what to do *before* the decisions need to be made is priceless.

The same is true for handling a lost person incident. Does a lost person in your charge know what to do if he or she is lost? Have you instructed the people you are responsible for to stay put, to "hug a tree," to carry and periodically blow a whistle three times? If a rapid search carried out by your group fails to find the lost person, and even while that search is in progress, do you have access to a search-and-rescue group? Trained teams have a much higher rate of success than a hastily organized group, even if that group has made prior plans in case of a lost person.

Your mantra might be to *plan ahead and prepare*.

Conclusion

Suppressing the knot of fear in your stomach, you begin systematically interviewing the 11 remaining participants, gathering as much information as possible about Johnny. As the story unfolds you learn Johnny was last seen walking away from the previous campsite with a trowel and a roll of toilet paper. No one remembers seeing him since then. When Johnny's group was ready to go, and Johnny was missing, they assumed that he, an eager hiker, had left with an earlier group.

Leaving 3 participants at the new campsite, you and the rest of the party return to the point where Johnny was last seen, stopping periodically to call his name and wait momentarily for a response. You follow the route Johnny's group followed.

Arriving just before dark you find Johnny waiting beside his pack. Realizing he had been left behind, and not knowing the route, Johnny wisely chose to remain at the old camp site, certain someone would eventually return for him.

All's well that ends well.

Tod Schimelpfenig, WEMT, contributed his expertise to this chapter.

CHAPTER THIRTY-SIX

Wilderness Transportation of the Sick or Injured

You Should Be Able To:

1. Describe the possibilities for transporting sick or injured patients from the wilderness, including walk-outs, carries, horse transports, and vehicle transports.
2. Demonstrate the improvisation of various means of transporting sick or injured patients from the wilderness, including improvised one-rescuer carries, two-rescuer carries, and litters.

It Could Happen to You

Scotland's Ben Nevis rises 4,000 feet from a peaty bog, and that's why you're there with 2 companions to climb in the hard cold of winter. You hiked in yesterday, approaching across the frozen marsh to the climber's hut at the base. Today you climbed light and fast, ascended a snow couloir without difficulty, got blasted by icy winds off the North Sea at the summit, and began your descent by midafternoon.

Shortly after beginning the descent, Alan slips on the steep upper slope. It could have been a fatal ride to the rocky bottom of the couloir, but it wasn't Alan's day to die. It was Alan's day, however, to shatter his left ankle when he slammed into an exposed rock after a wild 100-foot slide.

The cold is now bitter, and the wind is growing fierce. Alan might survive while you descend for help and climb back up with it. But he might not.

In the urban environment, patients are rushed in a speedy vehicle to a hospital, where expert medical advice and care are immediately available. In the wilderness transporting a sick or injured patient requires time to plan, organize, and prepare. Everyone benefits, especially the patient, when as much as possible of the preparation has been done prior to a trip (see Chapter 35: Emergency Procedures for Outdoor Groups).

Once the critical decision to evacuate a patient has been made, the next decision may be just as important: How? In general, the choices are (1) do it yourself, (2) go for help, or (3) a combination of the first two options when you have a larger group that can begin an evacuation while sending out messengers who request additional help to meet the evacuation en route.

There are many different modes of wilderness transportation, ranging from skis to helicopters. Here are some of the possibilities.

Walk-Out / Ski-Out

The ambulatory patient is the easiest evacuation. Patients with upper extremity injuries or stable athletic injuries to the knee or ankle can often carry their own weight. The best person to make the decision is often the patient, who knows how he or she feels compared with how he or she used to feel. The patient knows best how usable an injury is. Improvised crutches and canes can be of assistance. The patient's gear may need to be distributed to other members of the group.

An optimum group size for a walking patient is at least four, including the patient. If the patient becomes unable to continue walking, one person can remain with the patient while the others press on to the roadhead.

One-Rescuer Carries

For disabling yet still relatively minor injuries, such as seriously sprained ankles or knees or a fractured but stable lower leg, the group may evacuate the patient by sharing not only the weight of the patient's gear but also the weight of the patient. The simplest techniques involve one rescuer carrying one patient and require nothing more than physical strength and perhaps a few materials.

Patients can be moved short distances with the age-old piggyback method. Carrier and carried soon tire, and longer distances demand a better system of holding the patient in place than strength of arm and back.

Backpack Carry

Internal frame packs, the kind with a sleeping bag compartment in the bottom, can be slipped on the patient like a pair of crude shorts, allowing the patient to be carried out of the wilderness in relative comfort on the back of the rescuer.

INJURED PERSON

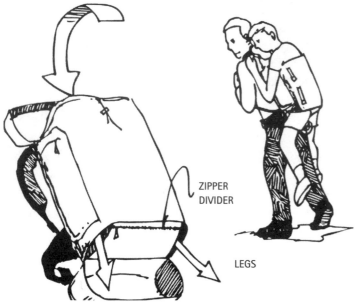

ZIPPER DIVIDER

LEGS

Figure 36-1: *Backpack carry.*

Webbing Carry

With 15 to 20 feet of nylon webbing, a carrying system can be improvised. The center of the length of webbing should be placed at the center of the patient's back, brought under his or her arms, and crossed over the chest. The webbing then passes over the rescuer's shoulders and back around the rescuer and between the patient's legs. When brought around the patient's legs and tied in front of the rescuer, a seat is formed. When slack develops in the system, the webbing can be untied, the patient hitched up, and the webbing tied again. Pressure points in the system, such as the armpits of the patient, the shoulders of the rescuer, and the legs of the patient, should be padded to increase the distance the system can be used before both participants collapse in pain.

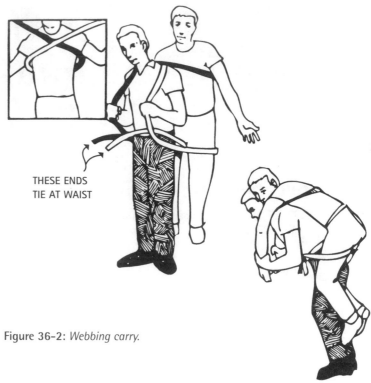

THESE ENDS
TIE AT WAIST

Figure 36-2: *Webbing carry.*

Split-Coil Carry

If a climbing rope is available, it can be tied in a mountaineer's coil and split into two approximately equal halves connected at the knot of the coil. When laid on the ground, a split-coil looks something like a ropey butterfly. The rescuer sticks one arm through each of the "wings" of the butterfly, wearing the coil like a rucksack. The patient's legs go through the lower part of each "wing," allowing the patient to be carried like an awkward backpack. Once again, padding at the pressure points adds comfort to the system.

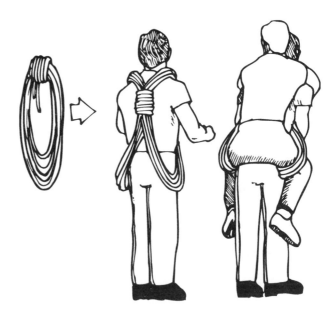

Figure 36-3: *Split-coil carry.*

Two-Rescuer Carries

When the strength of two people is required to move a heavier patient, the weight can be distributed evenly by the rescuers standing on either side of the patient. Then each rescuer reaches for the other rescuer's hand under the patient's arms and behind the back. With the other hand, each rescuer reaches under the patient's knees. Using their legs and not their backs, the rescuers stand with the patient seated in a temporary chair.

If one rescuer is obviously stronger than the other, the weight of the patient can be distributed unevenly by having one rescuer stand at the patient's head and the other rescuer at the patient's feet. The first rescuer reaches beneath the patient's arms and the second beneath the patient's knees. Although useful for short distances, neither of these carries is especially comfortable for the patient or rescuers over long distances.

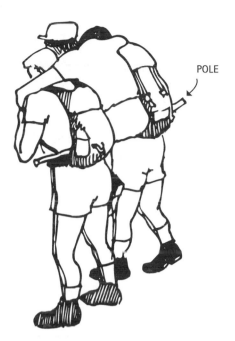

POLE

Figure 36-5: *Two-rescuer pole carry.*

Figure 36-4: *Two-rescuer carry.*

Note: None of the carrying techniques involving only one or two rescuers are recommended for seriously injured patients, and neither do they represent all the possible ways for one or two rescuers to move a patient. You are limited only by your imagination and the needs of patient and rescuers.

Two-Rescuer Pole Carry

Easier on both rescuers and patient is the two-rescuer pole carry (or ice ax carry). If two rescuers are wearing backpacks, a pole or long ice ax can be shoved into the strap system or tied to the bottom of each pack. When the pole is well padded, the patient can sit with relative comfort with his or her arms over the shoulders of the rescuers. The rescuers can carry in relative comfort, but this system is awkward when the carriers are of significantly different height.

Litters

Patients unable to walk are often best treated by camping and attending to their needs while a team goes for help, even though the patients may be seriously ill or injured. Some examples of such patients are those with myocardial infarction, injuries to the spine, multiple fractures that are difficult to stabilize, or any condition that is exacerbated by movement. These patients can be endangered by moving them without adequate means.

The most important situation in which to consider moving someone immediately, even when the only mode of transportation falls short of ideal, involves a patient whose condition is steadily deteriorating. You may also choose to move any patient under less than ideal conditions when the method for transporting the patient promises to do no further harm.

Litter evacuations are slow, safe, and effective. Litters may be commercially made or improvised in the field. In either case the litter must be well constructed, the patient must be packaged appropriately, and the carry must be well organized and properly managed.

Commercial Litters

Several manufacturers offer litters made from metals, plastics, fiberglass, or combinations of these materials. They are all the same general shape. Many of them can be attached to large wheels and pushed like a cart, converted into sleds and pulled over snow, attached to a rope system and lifted by a helicopter, or hauled up and lowered down a vertical slope. Commercial litters are much easier to handle and carry than improvised litters, and they eliminate the risk that the litter will fall apart during patient transport. Commercial litters offer a huge advantage over any litter you can make in the wilderness because they provide better patient care.

Pole Litters

As old as injuries, the pole litter, made from materials provided by the wilderness, probably represents the first litter ever improvised. It starts with two poles longer than the patient, preferably poles of dry dead wood. (Green wood is flexible, making the carry difficult for litter bearers and miserable for the patient who bounces up and down with each flex.) Shorter pieces of wood are lashed parallel to the longer poles until a "bed" is created. Materials for lashing include vines, grasses, rope, cord, string, boot laces, or anything else long and flexible enough. For patient comfort, the "bed" needs lots of padding.

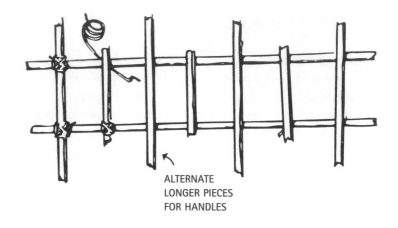

ALTERNATE LONGER PIECES FOR HANDLES

Figure 36-6: *Pole litter.*

Blanket Litters/Tarp Litters

With the advent of weaving, large pieces of cloth became available to improvise litters. Blankets seldom appear in wilderness areas these days, but tarps and tent flies do. A simple wrap around two long poles creates a workable litter when the patient's weight binds the tarps to the poles by way of friction. Cross-braces lashed to the ends of the poles add stability and ease for carrying.

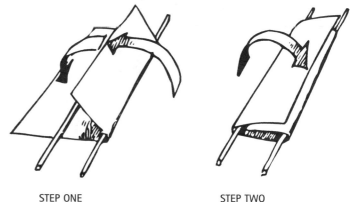

STEP ONE   STEP TWO

Figure 36-7: *Blanket litter.*

Rope-and-Pole Litters

Something like a hammock can be woven between two poles if there is enough rope on hand. Start by laying two long poles on the ground about shoulder width apart. Cross-braces are needed at each end to give the litter stability. The cross-braces can be sticks, ice axes, tent poles, or anything else available. With one end of the rope, lash one end of one cross-brace into place. Wrap the rope around and around one pole, pulling the loop formed by each wrap out to an imaginary center line between the two poles. Finish one side by tying one end of the second cross-brace into place. You can cut the rope and lash the second cross-brace into place or continue with the rope whole. As you work your way down and around the second pole, run the rope through the loops you've already made on the first pole. Tie off the second end of the first cross-brace. The number of loops is completely relative to the length of the rope, but the more loops, the more solid the patient's "bed." Take all the slack out of the system. Pad the "bed" well.

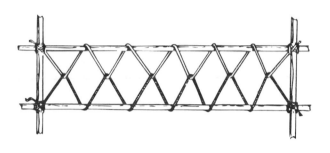

Figure 36-8: *Rope-and-pole litter.*

Rope Litters

When poles aren't available but a rope is, a nonrigid litter can be woven using only the rope. There are numerous ways to weave a rope into a litter, but they all have two things in common: They are the least easy to carry, and they are the most uncomfortable for the patient to ride in.

One method of constructing a rope litter requires a length of rope of approximately 150 feet (50 meters). Find the center of the rope, and stack the two halves 10 to 12 feet apart. From the center, make 14 to 18 bends in the rope, 7 to 9 of them on either side of the center point. The bends should be about as wide as the patient. If you have a full-length sleeping pad, lay it on the ground first, beneath the bends, to serve as a guide for the size of the litter. Bring the rope down past each bend, tying a clove hitch in the rope at the end of each bend. Pull the loop of the bend through the clove hitch until all the bend loops have been placed through a clove hitch. Weave the remainder of both ends of the rope around the litter and through the bend loops that extend through the clove hitches. Tighten the hitches and tie off the ends of the rope. Pad the "bed" of the litter.

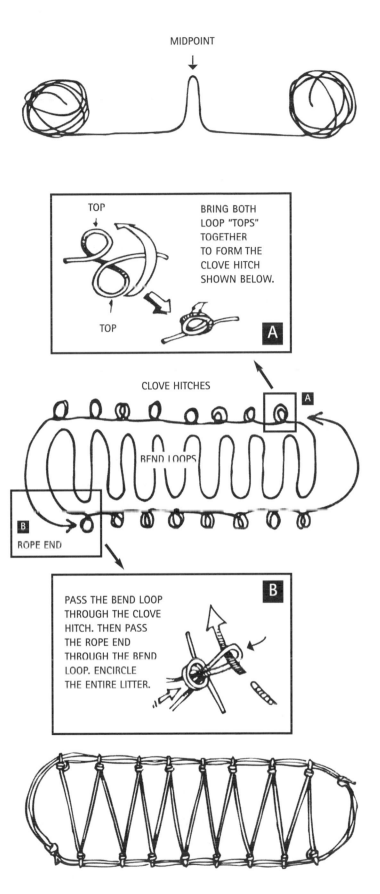

Figure 36-9: *Constructing a rope litter.*

Pack Litters

With three or four external frame packs, you can improvise a fairly substantial litter. Since the frames are molded to fit the shape of the wearer's back, they should be placed together in a way that best matches the shape of the person who will be carried. Take time to ensure the lashings are secure. Duct tape works well for lashing pack frames. Once again, adequate padding is required to create patient comfort.

Internal frame packs can be strapped to two long poles to form a litter, often by using the abundance of straps attached to the packs themselves. In an extreme situation, you can slit holes in the bottoms of the packs, slipping them over the poles and adding cross braces at the ends of the poles for stability.

maintaining an adequate airway, something a stable side position offers.

All patients need to be secured on a litter, and that means straps. Straps should be placed on the bony structures of the patient: lower leg, upper leg, pelvis, and chest. *Do not place straps across the abdomen or neck.* Straps across the chest must not impede breathing. The arms should be left free if the patient is conscious. Freeing the arms increases the patient's feeling of security while decreasing the claustrophobia inherent inside a litter. The tightness of the straps depends partially on patient comfort and partially on the ruggedness of the terrain to be crossed. Straps can be loosened or tightened as the evacuation proceeds, depending on need. Pad well beneath the straps where they press on the patient.

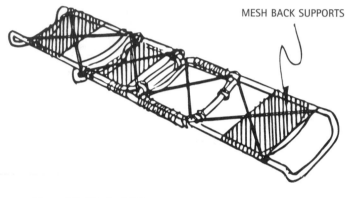

MESH BACK SUPPORTS

Figure 36-10: *Pack litter.*

Litter Packaging

Riding on a litter is at best an uncomfortable experience for the patient. All the attention you can pay to the details of packaging a patient for a carry is time well spent.

If you have a sheet of plastic, a tarp, or tent fly, place it spread out in the litter first. It can be rolled up against the side of the patient, providing extra padding, and it can quickly be rolled out and wrapped around the patient in case of rain, wind, and/or cold. In most situations—but certainly not all—you will package a patient on his or her back because (1) it is easier, in general, to provide care, and (2) it is easier to fit the patient on the litter. Padding the "bed" of the litter is essential. Soft pads behind the knees and the small of the back help ease discomfort. Create a pillow when a neck injury is not suspected, and place a slim pad beneath the head when a neck injury is suspected.

Sometimes a patient requires packaging on his or her side. This patient is usually unconscious, and you are concerned about

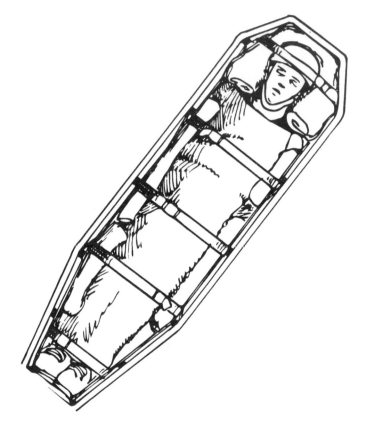

Figure 36-11: *A patient packaged in a litter.*

Litter Carrying

The carry of a litter provides an excellent opportunity to create a second patient. When it's time to move the loaded litter, remind the bearers to lift with their legs, not their backs. Space the bearers out evenly, preferably three to a side. The National Association for Search and Rescue (NASAR) recommends at least three teams of 6 to 8 people for litter carries. Experience suggests that a group of at least 10 to 12 is necessary

Figure 36-12: *Carrying a litter.*

Care of the Littered Patient

The person being carried by litter is under great stress. There is pain and discomfort, and sometimes darkness, and the inability to move around to find just the right spot. The muffled sounds that come from the people doing the carrying are often unintelligible. The patient feels terrible to cause this much trouble, but there's nothing he or she can do about it.

As a litter bearer, assume the responsibility for the mental as well as the physical comfort of the patient. Talk to the patient, especially when the litter is not being carried. Lean down near the patient's face and speak quietly and confidently. If someone else is talking to the patient, don't butt in—it's difficult to carry on two conversations even under normal circumstances. Ask about the injury. What can be done to make it more comfortable? Is it better, worse, the same? Ask about the amount of warmth inside the litter: Too little? Too much? Answer the patient's questions honestly and to the best of your ability.

If the patient is unconscious, talk to him or her anyway. Reach inside the covering of the litter to check for warmth. Listen to the breathing and check the quality of the pulse gently at the carotid artery in the neck. Unless it is blistering cold, you may choose to periodically uncover an unconscious patient to make sure any splints or bandages you've put on are not cutting off circulation.

In terms of bodily needs, the basic rule of the rescue is, If you need something, so does the patient. When you drink, offer the patient a drink, if the injury or illness allows. When you eat, try to get the patient to eat, again if the injury or illness allows. If you feel a chill coming on, there is even more of a chance that the patient is cold. If you are too hot, check the patient's comfort. If you need sunglasses, or sunscreen, or a toothbrushing, consider the needs of the patient, too.

When nature calls you to the bushes for a break, it's time to check in on your patient's need to urinate. He or she is often too embarrassed or too unwilling to cause a commotion to ask. You might have to unwrap your neat litter package to accommodate the patient, but that's part of the game. The job can usually be accomplished for a man by turning the litter on its side if the day is warm and he isn't too deep in coverings. He may be able to stay on his back and use a water bottle as a urinal. The problem is less easily solved for a woman, but she can sometimes catch most of her urine in a wide mouth water bottle. If something absorbent is placed between her legs first, any spill can immediately be caught. Perhaps you can improvise a bedpan from a frying pan. In the least easily managed scenario, you have to remove the patient from the litter for urination or defecation and repack him or her when the job is done.

for litter carries of more than a few miles. Ease of carrying can be increased if 8 to 10 feet of rope or webbing is attached to the litter for each bearer. This strap is thrown over the shoulder and across the back of the bearer and held in the hand away from the litter. Use of the carrying strap allows each bearer to shift the weight of the litter to a shoulder, easing the stress on the hand holding the litter. The carrying strap also allows bearers of different heights to adjust the attitude of the litter.

One or two people walking in front of the litter can find the route and clear obstacles for the litter party. People following behind the litter can carry the food, fuel, shelter, and clothing of the litter bearers.

In awkward situations, such as carrying in boulders or deadfall or across streams, stop the carry and move the free members of the party to the front of the litter. The litter can then be passed hand to hand over obstacles. In caves and other tight passages, the litter can be slid on the ground (pulled by a line attached to one end) or passed caterpillar style over rescuers lying in the passageway. In technical terrain, litters are lowered, belayed, and slung across ravines and rivers with rope systems.

Horse Transports

Evacuation by horse is an option in some areas. The patient must be conscious and able to sit on the horse without falling off. This method is injury-dependent. Lower leg, foot, and upper extremity injuries can be transported this way, but pelvic, hip, thigh, or back injuries cannot. The patient with a splinted leg may be unable to avoid striking the limb against trees, rocks, and other obstructions and may have difficulty controlling the horse. The chance of the patient falling off the horse must be considered. Horses have limits in steep and rocky terrain.

Vehicle Transports

You may be able to utilize a four-wheel-drive vehicle, snowmachine, fixed-wing airplane, boat, helicopter, or other vehicle to transport a patient if there is a suitable loading/landing site nearby or if the patient can be carried by other rescuers before transferring into a vehicle.

Snowmachines

In some winter situations a large snowmachine with an enclosed cabin may be able to access and transport a patient. Even though limited by unpacked snow and steep terrain, it is far more common to find a snowmobile used for patient transport. For the patient unable to sit behind the driver of a snowmobile, there are sleds designed specifically to be pulled by the machine to haul cargo and people. The patient has to be well protected from wind and cold, and the ride is not especially smooth.

Boats and Airplanes

Boat evacuations can range from paddling a sick person in a double kayak to flagging down a large vessel. Airplanes are options in many areas. The patient must first be moved, of course, to an adequate landing area.

Helicopters

Nothing has done more to change the face of wilderness rescue than helicopters. They land in remote areas that were inaccessible to aircraft only a few years ago. If the spot isn't flat enough, helicopters have been known to land on one skid while a patient is quickly loaded. When there is no spot to land, they have hovered with a rescuer hanging from a rope or cable, a rescuer equipped to attach the patient to the hauling system for evacuation.

Helicopters go where the pilot wants because of the rapid spinning of two sets of blades. The large overhead blades create lift by forcing air down. The pilot can vary the angle at which the blades attack the air and the speed at which they rotate to vary the amount of lift. The entire rotor can be tilted forward, backward, or sideways to determine the direction of travel. Without a second set of blades spinning in an opposite direction, the helicopter would turn circles helplessly in the air. Some large helicopters have two large sets of blades spinning in opposite directions, one fore and one aft, but most helicopters used in the wilderness maintain stability with one small tail rotor.

When they are close to the ground, the spinning blades build a cushion of air that helps support the helicopter. This cushion of air varies in its ability to work, depending on its density. Rising air temperatures and increasing altitude reduce air density. So trying to land a helicopter on a mountaintop on a hot day is dangerous, and the weight of one person may prevent liftoff.

Air density also is altered by the nearness of a mountainside. The downward shove of air by the blades can recirculate off the mountainside and reduce lift.

One of the greatest fears of mountain flying is a sudden downdraft of air that can slam a helicopter toward the ground. Downdrafts are not only dangerous but also unpredictable.

Add to air density and downdrafts the possibility of darkness and fog and wind, and you can understand that even if a helicopter is available it may not be able to come to your rescue.

When you're in need of a rescue the approaching thump-thump-thump of rapidly rotating blades is a joyous sound. To give the helicopter rescue the greatest chance of success, a suitable landing zone will have to be found. The ideal landing zone should not require a completely vertical landing or takeoff, both of which reduce the pilot's control. The ground should slope away on all sides, allowing the helicopter to immediately drop into forward flight when it's time to take off. Landings and liftoffs work best when the aircraft is pointed into the wind because that gives the machine the greatest lift. The area should be as large as possible, at least 60 feet across for most small rescue helicopters, and as clear as possible of obstructions such as trees and boulders. Clear away debris (pine needles, dust, leaves) that can be blown up by the wash of air, with the possibility of producing mechanical failure. Light snow can be especially dangerous if it fluffs up dramatically to blind the pilot. Wet snow sticks to the ground but also sticks to the runners of the helicopter and adds dangerous weight. If you have the opportunity, pack snow flat well before the helicopter arrives—the night before would be ideal—to harden the surface of the landing zone. Tall grass can be a hazard because it disturbs the helicopter's cushion of supporting air and hides obstacles such as rocks and tree stumps.

To prepare a landing zone, clear out the area as much as possible, including removing your equipment and all the people except the one who is going to be signaling the pilot. Mark the landing zone with weighted bright clothing or gear during the day or with bright lights at night. In case of a night rescue, turn off the bright lights before the helicopter starts to land—they

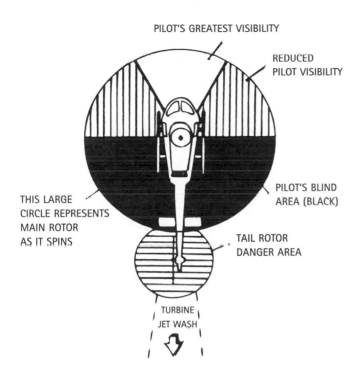

PILOT'S GREATEST VISIBILITY

REDUCED PILOT VISIBILITY

THIS LARGE CIRCLE REPRESENTS MAIN ROTOR AS IT SPINS

PILOT'S BLIND AREA (BLACK)

TAIL ROTOR DANGER AREA

TURBINE JET WASH

Figure 36-13: *Helicopter safety.*

can blind the pilot. Use instead a low-intensity light to mark the perimeter of the landing area, such as chemical light sticks, or at least turn the light away from the helicopter's direction. Indicate the wind's direction by building a very small smoky fire, hanging brightly colored streamers, throwing up handfuls of light debris, or signaling with your arms pointed in the direction of the wind.

The greatest danger to you occurs while you're moving toward or away from the helicopter on the ground. *Never approach from the rear and never walk around the rear of a helicopter.* The pilot can't see you, and the rapidly spinning tail rotor is virtually invisible and soundless. In a sudden shift of the aircraft, you can be sliced to death. Don't approach by walking downhill toward the helicopter, where the large overhead blade is closest to the ground.

It is safest to come toward the helicopter from directly in front, where the pilot has a clear field of view, and only after the pilot or another of the aircraft's personnel has signaled you to approach. Remove your hat or anything that can be sucked up into the rotors. Stay low because blades can sink closer to the ground as their speed diminishes. Make sure nothing is sticking up above your pack, such as an ice ax or ski pole. In most cases someone from the helicopter will come out to remind you of the important safety measures.

One-skid landings or hovering while a rescue is attempted are solely at the discretion of the pilot. They are a high risk at best, and finding a landing zone and preparing it should always be given priority.

Summary

Wilderness transportation of the sick or injured is at best a risky business for everyone involved. Litter bearers may stumble and end up being patients themselves. Improvised carrying systems can fall apart, causing injury to bearers and/or further injury to the patient. Helicopters may crash. When an evacuation is necessary, choose a method that provides the least possibility of harm to all participants—or, better yet, carry a message or call out with your needs and allow someone with skill and experience to make the decision.

Conclusion

Alan waits, anchored to the side of Ben Nevis, wearing what little extra clothing is available, grimacing in pain while discussing the possibilities with you and your other climbing partner. The decision is made. The safest alternative, you decide, is to lower Alan to the bottom of the couloir.

It is a laborious job that runs well into the frigid darkness, setting anchor after anchor from which you can belay Alan down a rope's length. Alan slides on his back, aided by the other climber.

Finally reaching the bottom, you begin to weave the rope into something vaguely resembling a litter while the third member of your party hikes the hour-and-a-half round-trip to the hut to bring back help in the form of other climbers.

By midnight Alan is warming in the hut, and you make plans to cross the bog to enlist a rescue party with a commercial litter.

CHAPTER THIRTY-SEVEN

Wilderness Medical Kits

You Should Be Able To:

1. Describe the importance of and the general guidelines for packing a wilderness medical kit.
2. List the possible contents of a general wilderness medical kit.

It Could Happen to You

By the time Cedar Rapids on the upper Flambeau River of Wisconsin spits out your paddling companion and his overturned canoe, you are ready to do an assessment—not of his skills but of his medical condition. It doesn't take much experience to recognize a broken lower left arm: a splinter of radius is sticking out through the skin.

Taking the medical kit from your dry bag, you open it and are at once reminded of your forethought. It looks like everything you'll need is right here. You have an irrigation syringe to clean the bone end, and, once traction in line has reduced the fracture, you'll use it to clean the wound. Supplies aplenty to dress and bandage the wound. Even an analgesic to start your patient on the trail to feeling better.

But wait! You used your SAM Splint and elastic wrap a couple of trips ago, and you forgot to replace those items. With the arm asking for immobilization, the things you want most for treatment are missing.

Carpenters and mechanics may only be as good as their tools, but that truism need not apply to the Wilderness First Responder. The able WFR may indeed be best exemplified by someone who can make do with what is available. That said, there is still plenty of room in a backpack or drybag for an adequate medical kit, and the well prepared WFR certainly has one at her or his disposal. Being able to improvise everything is an excellent skill—but sometimes not having to improvise is even better. The choice of contents for a wilderness medical kit can vary from the minimalist kit—a roll of duct tape and a roll of sterile gauze to the all inclusive kit that could burden the back of a strong porter. Every kit carried by a WFR should at least be the result of much forethought.

General Guidelines for Wilderness Medical Kits

1. *Accept the fact that there is no such thing as the perfect wilderness medical kit.* Many factors should determine your choices of specific contents. No matter how much you plan and prepare, someday you will want something that is not there and/or discover you've carried an item for years and never used it. When considering the contents of a kit, take into account (1) the environmental extremes you will face (altitude, cold, heat, endemic diseases); (2) the number of people that may require care; (3) the number of days the kit will be in use; (4) the distance from definitive medical care; (5) the availability of rescue services; (6) your medical expertise and/or the expertise of other group members; and (7) preexisting problems of group members, such as individuals with diabetes.

2. *Evaluate and repack your wilderness medical kit before every trip.* Renew medications that have reached expiration dates. Replace items that have been damaged by heat, cold, or moisture. Remove items that are unnecessary for the proposed trip, such as insect repellent on winter trips, and add items that may be useful on the upcoming adventure.

3. *Do not fill your kit with items you do not know how to use.* Maintain a high level of familiarity with the proper uses of all the items in your wilderness medical kit.

4. *Choose specific items for the wilderness medical kit, whenever possible, that are versatile rather than particular.* For example, a wide variety of sizes and shapes of Band-Aids is nice, but wound coverings can be created from pads of gauze and strips of tape. Triangular bandages are useful, but safety pins and T-shirts can be used to make slings. Medical adhesive tape has limited usefulness compared with duct tape.

5. *Encourage each group member to pack and carry a personal first-aid kit* to reduce the size and weight of the general wilderness medical kit.

Specific Considerations for Wilderness Medical Kits

Specific considerations for a wilderness medical kit can be divided into four categories: injury management supplies, tools, miscellaneous supplies, and medications. Keep in mind that these are suggestions and not requirements (see Chapter 12: Fractures; Chapter 14: Athletic Injuries; Chapter 15: Wound Management; and Chapter 31: Common Simple Medical Problems).

Injury Management Supplies to Consider

1. Adhesive strips, such as Band-Aids.
2. Sterile gauze pads and/or sterile gauze rolls.
3. Athletic tape, 1 inch by 10 yards, and/or duct tape.
4. Tincture of benzoin compound.
5. Wound closure strips.
6. Microthin film dressings, such as Tegaderm or Opsite.
7. Large trauma dressings and/or an individually wrapped sanitary napkins.
8. Moleskin and/or molefoam.
9. Gel wound coverings, such as 2nd Skin.
10. Soap-impregnated cleaning sponges, such as Green Soap Sponges.
11. Antimicrobial towelettes and/or alcohol wipes.
12. Lightweight splint, such as SAM Splint.
13. Elastic wraps and/or Coban.
14. Rubber gloves.

Tools to Consider

1. Trauma shears.
2. Forceps (tweezers).
3. Irrigation syringe.
4. Disposable scalpels.
5. Safety pins.

Miscellaneous Supplies to Consider

1. Pad and pencil.
2. Sawyer Extractor (see Chapter 21: North American Bites and Stings).
3. Stethoscope.
4. Blood pressure cuff.
5. Pocket rescue mask.
6. Thermometer.
7. Sunscreen and lip protection.
8. Insect repellent.
9. Insect bite treatment, such as StingEze.
10. Water disinfection system (see Chapter 30: Communicable Diseases and Camp Hygiene).
11. Small flashlight or penlight.

Medications to Consider

Note: The administration of any drug falls outside the scope of practice for Wilderness First Responders unless the WFR is acting under specific protocols established by and managed by a medical adviser.

1. Analgesics (painkillers), such as acetaminophen, aspirin, ibuprofen, ketoprofen, and naproxen, are available without a prescription and work for mild to low-moderate pain. All reduce fever, and all except acetaminophen have anti-inflammatory properties. Percocet and other narcotics available by prescription only should be considered for high-moderate to severe pain.
2. Anti-inflammatory drugs, such as aspirin, ibuprofen, ketoprofen, and naproxen, are available without a prescription (see Chapter 14: Athletic Injuries).
3. Antipyretics (anti-fever), such as aspirin, acetaminophen, and ibuprofen.
4. Antihistamines, such as Benadryl (see Chapter 28: Allergic Reactions and Anaphylaxis and Chapter 31: Common Simple Medical Problems).
5. Antibiotics, such as cephalexin (Keflex), erythromycin, ciprofloxacin (Cipro), are all available by prescription only. Also consider antibiotic ointments.
6. Anti-diarrheals, such as Imodium, are available over-the-counter (see Chapter 31: Common Simple Medical Problems).
7. Anti-emetics (anti-nausea and anti-vomit), such as prochlorperazine (Compazine), are available by prescription only.
8. Antifungals, such as Tinactin Antifungal and Monistat, are available over-the-counter (see Chapter 31: Common Simple Medical Problems and Chapter 32: Gender-Specific Emergencies).

9. Anti-vertigo drugs, such as meclizine (Antivert), are available over-the-counter (see Chapter 31: Common Simple Medical Problems).
10. Decongestants, such as Afrin and Sudafed, are available over-the-counter (see Chapter 22: Diving Emergencies and Chapter 31: Common Simple Medical Problems).
11. Antacids are available over-the-counter.
12. Throat lozenges and/or hard candy.
13. Anti-altitude medications, such as acetazolamide (Diamox), dexamethasone (Decadron), and nifedipine (Procardia), are available by prescription only (see Chapter 18: Altitude Illnesses).
14. Anti-anaphylaxis medications, such as injectable epinephrine (EpiPen), are available by prescription only (see Chapter 28: Allergic Reactions and Anaphylaxis).

Suggested Drug Information Card

Make sure to carry information on every medication in your kit. You can do this by writing out the following information on 3-by-5-inch index cards:
1. Brand name of medication.
2. Its generic name.
3. Description (color, shape, etc.).
4. Dose (how much/how often).
5. Indications (reasons to dispense).
6. Contraindications (reasons not to dispense)
7. Side effects (possible reactions, adverse and otherwise).
8. Interactions (what to not mix with it).

Sample Drug Information Card

Brand name: Bayer

Generic name: Aspirin

Description: White, round, flattened tablet.

Indications: Pain, inflammation, fever. At the s/s of a heart attack.

Contraindications: Aspirin allergy.

Side effects: Gastric upset/bleeding—take with food and water.

Interactions: Do not take within 2 hours of another NSAID such as ibuprofen.

Final Thoughts on Kits

Unless you are able to acquire medical supplies in bulk and/or free of charge, you probably will save time and money by purchasing one of many excellent, commercially prepared wilderness medical kits. In addition, well-made kits offer durability, organization of supplies, easy access to supplies, and space to add on and take away from the included supplies.

Remember that the wilderness medical kit that saves lives rarely comes from a bag but, instead, from a brain packed with medical expertise. Aside from a few emergencies, such as anaphylaxis, your use of a medical kit is primarily to ease pain, speed healing, and prevent further injury, but it is knowledge and the ability to use that knowledge that makes the difference between life and death in a critical situation. Learn what you can do for the seriously hurt or sick person and carry that information with you all times.

Conclusion

On the Flambeau River you cut about 10 inches off the end from your paddling companion's foamlite sleeping pad. Folding it once, the foamlite becomes the basis of your splint. Extra clothing, neatly folded, is added to the splint above and below the broken arm. With lots of padding to protect the arm, the duct tape you apply to hold the arm to the folded foamlite doesn't threaten circulation to the fracture site and beyond.

To create a sling and swathe to hold the injured arm in place against the patient's body, you fold the bottom of the dry T-shirt you have put on your friend up and over his arm. Two safety pins, dug from the bottom of your kit, hold the shirt in place, and ultimately the arm in place.

Things could be better, but with a combination of the supplies you brought, the skills you learned, and a touch of creativity, you're ready to paddle your patient safely to the next take-out.

Appendixes

Oxygen and Mechanical Aids to Breathing

There are times when the patient treated by the Wilderness First Responder will benefit greatly from the administration of supplemental oxygen. An understanding of oxygen and its associated delivery equipment is fundamental, therefore, to providing optimum care to many patients.

Oxygen and Respiration

Oxygen is an odorless, colorless gas that makes up approximately 21 percent of the atmosphere. You inhale this atmosphere, and use approximately 25 percent of the available oxygen. The air you exhale then is approximately 16 percent oxygen, 5 percent carbon dioxide, and the rest composed of gases, mostly nitrogen, that you do not use.

At rest, most adults have a total inspired volume of about 500 milliliters per breath. This amount is referred to as *tidal volume*. The respiratory system can use about 350 milliliters of this at the alveolar level, where gas exchange takes place. The remaining 150 milliliters is contained within the bronchi and bronchioles, an area known as dead air space since none of the air is used for gas exchange. One person may take great gasps of air much larger than 500 milliliters, and another may take very shallow breaths of 200 milliliters. In either case the "dead air space" remains approximately 150 milliliters.

Supplemental Oxygen Safety

Because oxygen promotes combustion, it should never be used near an open flame, and petroleum products should not be used with oxygen equipment. Petroleum products include any types of greases, oils, or tapes with petroleum-based adhesives in them. The rapid escape of oxygen from a tank, combined with the petroleum products, could create friction heat and start a dangerous fire.

Oxygen cylinders are potentially dangerous because the oxygen is stored in them under great pressure. They should always be kept laying flat, both when in use and in storage, unless they are supported by some sort of rack or by your hands. If the neck of the cylinder should be accidentally cracked and the pressure rapidly escapes, the tank could become a deadly missile.

The Tank

Oxygen is stored for use in special tanks (cylinders) under a pressure, when the tank is full, of 2,000 pounds per square inch (psi). These cylinders are green, or shiny aluminum with a green label, for ease of identification. Some portable tanks are available for use in the field, conveniently sized to fit in backpack-like carrying cases. These are known as C, D, or E tanks, depending on size. Even larger tanks are found on board ambulances and aeromedical units.

Oxygen cylinders must be inspected every 5 years for the presence of corrosion or damage that can affect safe delivery of oxygen to patients. To prevent corrosion resulting from moisture, oxygen cylinders need to be maintained with a small amount of pressure within them at all times. This amount is referred to as *safe residual* and is usually 200 psi. Oxygen is very drying, and a small amount left in the tank prevents the surrounding atmosphere from entering the tank, thus preventing moisture from corroding the interior of the tank.

The Regulator

Oxygen tanks are fitted with special regulators that reduce the pressure of the gas as it leaves the cylinder, normally down to 40 to 75 psi. An important safety feature of these regulators is a system of pin fittings. All cylinders used for gases have specific pin settings, and regulators for one gas cannot be attached to a tank used for another gas. When fitting a regulator onto an oxygen tank, it is important to check for the correct pin setting. Another feature of regulators is the plastic gasket, or washer, that is used to provide a seal onto the tank. It is important to check that the washer is present and in good condition before fitting the regulator onto the tank.

Before securing a regulator onto a tank, open the tank slightly to allow a small amount of oxygen to escape. This is done to blow any dust or particulate matter away from the opening, keeping the regulator clear of obstructions. The regulator is then secured in place—finger-tight only—and the tank can be opened one full turn to allow for oxygen delivery. All regulators have a gauge that shows the current pressure within the tank. When the regulator is opened, the pressure gauge should be facing *away from* the WFR and/or patients. This prevents any danger should the gauge come apart due to the sudden change in internal pressure.

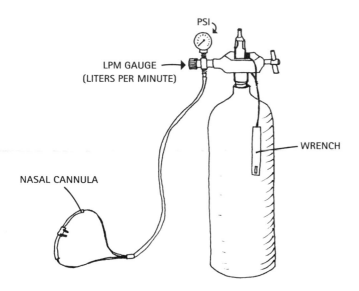

Figure A-1: *Oxygen delivery system.*

In addition to a pressure gauge, regulators have a flow meter that allows the WFR to dial in the amount of oxygen to be delivered to the patient. This flow meter measures the amount of delivered oxygen in liters per minute (lpm). Regulators will deliver up to 15 lpm and sometimes 25 lpm.

Once you have finished using a tank and regulator, the tank should be shut off. Any remaining pressure in the regulator and lines should be bled. This is done by opening the liter flow gauge on the regulator (or by pressing the delivery button on the demand valve). Once the needle on the pressure gauge returns to zero, the regulator and lines are clear.

Oxygen Therapy

Assessment of the patient determines the type of oxygen therapy that is needed in each situation. During the initial assessment, the WFR discovers if the patient has an open airway and if measures are needed to maintain the airway in an open position. Further assessment determines the patient's level of consciousness, the best indicator of oxygen perfusion to the brain. In addition to level of consciousness, knowledge of the mechanism of injury can aid you in determining the most appropriate method of oxygen therapy. For example, a rock climber who has fallen a short distance and fractured a rib may need only a low level of supplemental oxygen. But a climber with several severely fractured ribs and giving evidence of respiratory distress needs a much greater level of oxygen to maintain adequate perfusion of the brain.

Airway Maintenance

In extended care situations, body position may be your only option for maintaining an open airway. Patients in a stable side position—the recovery position—have the best chance of keeping their airway open.

The Oropharyngeal Airway

Many wilderness medical kits contain an *oropharyngeal airway* (OPA). The OPA, short and curved, and also known as an oral airway, is used to keep the tongue away from the back of the throat, and thus keep the airway open. They come in sizes from infant to large adult. The OPA is constructed so that it is possible to suction either through it or alongside it. Correct sizing is measured either from the center of the mouth to the angle of the jaw, or from the corner of the mouth to the earlobe. Either way is acceptable.

Oral airways should be used only with patients who are unconscious and have no gag reflex. When used with patients who have a gag reflex, there is danger of vomiting and aspiration and of vocal cord spasm. If the patient gags during insertion of the OPA, it should be immediately removed, and reinsertion should not be attempted.

Insertion of the OPA is performed first by opening the mouth with a jaw thrust or the *cross-finger technique* (placing your thumb and index finger between the patient's front teeth and pushing in opposite directions to open the mouth). The OPA may be inserted initially with the curved tip facing the roof of the patient's mouth. Once it is halfway in, the OPA is rotated 180 degrees and placed so that the flange rests on the patient's lips. Should the patient have dentures, they may be left in place. Dentures provide for better maintenance of a patent airway because the OPA will stay in place more easily.

The Nasopharyngeal Airway

Soft and flexible, a *nasopharyngeal airway* (NPA) is for use with conscious patients and those who may be unconscious but have an intact gag reflex. They are of value when the openness of an airway is threatened or compromised. They do not, however, ensure that the tongue will stay out of the back of the throat, so careful patient monitoring is of the utmost importance. Correct sizing is measured from the tip of the nose to the earlobe.

The airway should be lubricated with a sterile, water-based material, preferably containing a topical anesthetic. Do not use petroleum-based lubricants that may damage nasopharyngeal mucosa. Check to see if one nostril is noticeably larger, and utilize the larger one. If both nostrils appear the same size, it doesn't matter which one you use. The NPA is inserted, with the beveled side of the NPA toward the patient's septum, by gently pushing along the normal curvature of the nose. After insertion, the NPA can be rotated gently if the bevel is on the "wrong" side. If resistance is met, do not force the airway further. Try the other nostril. The airway is inserted until the flange is flush with the outside of the nostril. It is normal for a small amount of bleeding to occur during insertion of the airway. If cerebrospinal fluid is leaking from the nose or other signs of skull fracture are present (see Chapter 9: Head Injuries), it is considered unsafe to insert a nasal airway because it may introduce infection into the brain cavity.

Passive Delivery

Passive delivery describes the process in which oxygen is delivered to patients who must be breathing adequately for the gas to be of benefit. Keep in mind your patient is likely to be hurting and afraid and finding it somewhat difficult to breathe. And you might be strapping a plastic mask over his or her face. You need to increase your level of reassurance to the patient.

Pocket Mask

Most pocket rescue masks, usually used for artificial respirations,

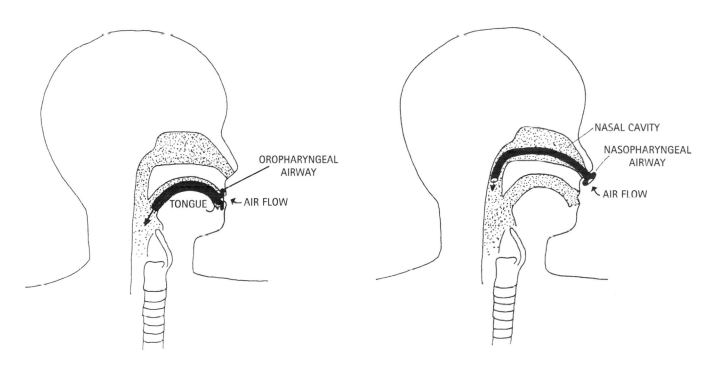

Figure A-2: *Oropharyngeal airway.*

Figure A-3: *Nasopharyngeal airway.*

can be used as simple oxygen delivery masks. After starting a flow of oxygen, place a tube from the tank under the edge of the mask. Some rescue masks have a port specifically for attaching supplemental oxygen, and you should, of course, use the port if it exists.

Nasal Cannula

The *nasal cannula* is soft tubing with two prongs that curve into the patient's nostrils. It is fastened around the ears and cinched up under the chin in a fashion similar to a bolo tie. It is used at a flow rate of 2 to 6 lpm and can deliver up to 40 percent oxygen at 6 lpm. A patient is free to talk while using a cannula, and it is relatively comfortable. The oxygen can be drying to nasal mucosa, a fact that may cause some itching. The cannula is used most often with patients in mild to moderate respiratory distress. It is especially useful with a patient who may vomit because it doesn't block the patient's mouth.

Non-Rebreather Mask

A *non-rebreather mask* is a simple mask with an oxygen reservoir bag attached. Thin rubber gaskets at the exhalation ports prevent the intake of any ambient air. The reservoir is filled with 100 percent oxygen through the oxygen flow connection at the tank. With a good seal around the patient's mouth, the non-rebreather mask delivers up to 95 percent oxygen. The liter flow setting must be high enough to ensure that the reservoir bag does not fully deflate when the patient inhales. A setting of 10 to 12 lpm is usually sufficient for this. This mask is excellent for patients with lowered tidal volume or decreased respiratory effort.

Positive Pressure Delivery

When a patient cannot breathe adequately, or at all, oxygen must be delivered under pressure for it to get into the lungs.

Pocket Mask

There are several types of lightweight barriers available to the WFR for use in ventilating patients who are not breathing on their own. These devices fit directly over the patient's mouth and nose. You breathe for the patient mouth to mask instead of mouth to mouth. If the pocket mask has a one-way valve, secretions, even exhaled air from the patient, cannot enter your mouth. OPAs or NPAs are appropriate in many cases in which a pocket mask is used. Proper head position of the patient, one that keeps the airway open, must be maintained by the rescuer while using the mask alone. Pocket masks are available with oxygen inlet ports that allow for delivery of up to 50 percent oxygen to the patient at a rate of 10 lpm. One advantage of a pocket mask is that the rescuer is able to tell when the patient's lungs are fully inflated by the back-pressure felt with mouth-to-mask ventilation.

Bag Valve Mask

The *bag valve mask* uses a semirigid bag, filled with either ambient air or supplemental oxygen, and a transparent mask that fits tightly onto the patient's mouth and nose. These devices come in adult, child, and infant sizes. There are reservoir tubes or bags that fit onto most bag valve masks that increase the potential for delivery of up to almost 100 percent oxygen to the patient. They may be used on patients who need respiratory assistance or on nonbreathing patients.

When a bag valve mask is used as a respiratory-assist device, the WFR synchronizes squeezing the bag with the beginning of the patient's inspirations. The use of a bag valve mask allows you to feel lung resistance as the lungs reach full, preventing overflow of air into the stomach.

Proper technique of the bag valve mask allows you to maintain an open airway for the patient and ventilate adequately at the same time. While holding the transparent mask in the fleshy area between the thumb and forefinger, you can maintain a proper mask seal by putting slight pressure onto the patient's face. At the same time, the other fingers of the hand can hold along the patient's jaw line, maintaining proper head position. This technique is called the *anesthesiologist grip.* With the bag firmly attached to the mask, you can ventilate the patient with the other hand by squeezing the bag. If your hands are small, the bag can be swiveled so that it is near the forearm and pressed against your body or leg to fully deflate the bag for each desired ventilation.

Demand Valve

The *demand valve* is a fitting that attaches to the oxygen tank via a length of sturdy tubing. It has within it a special valve that allows oxygen to be released with a slight amount of negative pressure. A patient, therefore, can self-administer oxygen simply by inhaling through the mask, the oxygen being delivered "on demand." Oxygen can also be delivered via a demand valve by depressing a button on the device. Thus it is frequently used with nonbreathing patients, as in cardiopulmonary resuscitation. An OPA is typically inserted to maintain an open airway, and the patient is ventilated as is appropriate to the situation.

Since the demand valve works directly off pressure through the regulator (at 40 to 75 psi, depending on the regulator), there is danger of too great a pressure entering the patient's respiratory system. For this reason a demand valve should *never* be used with patients under 12 years of age. There is less ability to feel when the lungs are fully inflated with a demand valve, so there is more danger of overflow into the stomach. This overflow promotes vomiting, a dangerous situation with an unconscious patient.

Oxygen Delivery

Device	Flow Rates (Liters per Minute)	Percent Oxygen
Pocket mask	10–15 lpm	16–50%
Nasal cannula	2–6 lpm	25–40%
Non-rebreather	10–15 lpm	90–95%
Bag valve mask	10–15 lpm	100%
Demand valve	10–15 lpm	100%

Suction

An open airway is of paramount importance in patient care. Whether from bleeding, vomiting, broken teeth, food particles, or buildup of saliva, potential problems may exist in maintaining the patient's airway. At times a patient can be positioned so that potential complicating fluids can flow out—the stable side position. When this is not possible, and even when it is, the use of a suction device to remove matter from the mouth may be extremely important.

There are numerous portable suction devices available for use in wilderness rescue. Some are battery powered, some work from the oxygen in a portable tank, and others are hand-operated. As with anything taken into the wilderness, weight and durability must be taken into account when choosing equipment. You may even choose a simple turkey baster for use as a suction device.

Suction devices use either a rigid or soft *catheter* (a tube inserted into the body) to collect materials threatening the airway. The rigid or "tonsil tip" catheter attaches onto suction tubing. The soft catheter is sometimes part of suction tubing or may need to be attached onto the tubing. In either case the catheter is inserted only to the back of the throat. Once the catheter has been inserted through the mouth and to the back of the throat, suction is then initiated.

You should not suction a patient for more than 15 seconds, unless greater time is needed, to clear an airway for ventilation. Suction devices have their limits, and large particulate matter or thick secretions may require rolling a patient and/or the use of finger sweeps in order to clear the airway. As you may remember from your last visit to a dentist, a conscious patient who must remain supine may even help to suction himself or herself.

Daniel DeKay, RN, WEMT, and Joel Buettner, WEMT, contributed their expertise to this chapter.

Automated External Defibrillation

You Should Be Able To:

1. Describe the advantages of automated external defibrillation.
2. Describe the use of an automated external defibrillator.

Ventricular fibrillation (v fib) is rapid, quivering, incomplete contractions of the muscle fibers of the ventricles of the heart. V-fib often results from blockage in coronary blood vessels, but it may also be the result of other causes, such as a substantial electric shock or certain drugs. A patient in v-fib is in cardiac arrest: pulseless and breathless. V-fib typically converts to *asystole* (the absence of all heart activity) within a few minutes. Asystole cannot be converted back to a normal rhythm with a defibrillator, but v-fib can be converted back to a normal rhythm.

Electrical defibrillation stops fibrillation by stopping the heart via a shock through electrodes placed on the chest wall. If the shock is applied in time, the heart sometimes starts again under its own electrical impulse and in a normal rhythm. V-fib is the most common rhythm encountered by bystanders initiating cardiopulmonary resuscitation (CPR), and electrical defibrillation gives a patient in v fib the single most significant chance of survival.

Although wilderness medicine situations today seldom offer patients in cardiac arrest the opportunity to be defibrillated, that fact is changing. Shipboard clinics, rural clinics, aeromedical units, and even some remote camps are acquiring devices to defibrillate and the skills to defibrillate. It is recommended that the Wilderness First Responder carry at least a basic understanding of the easiest form of defibrillation, the *automated external defibrillator* (AED). AEDs are sophisticated, computerized devices, an example of remarkable new technology.

The AED

Conventional defibrillators require a relatively high degree of skill in heart rhythm recognition and device operation. An AED only requires the operator to recognize cardiac arrest in the patient, properly attach the device to the patient, and either memorize a short treatment sequence or respond appropriately to the directions "spoken" by the recorded voice within the device itself. An AED interprets the cardiac rhythm for the rescuer, recommending a shock when appropriate. Learning to use an AED is much easier than learning CPR, but CPR skills remain critical. After a sudden cardiac arrest, says the American Heart Association, the chance of surviving decreases by 7 to 10 percent for each minute that passes without defibrillation. But AEDs are even less successful when no CPR is provided.

Rates of Patient Survival with AED Use

Time from Collapse of Patient	Survival Rate
1 minute	70 to 90%
5 minutes	50%
7 minutes	30%
9 to 11 minutes	10%
12 minutes or more	2 to 5%

American Heart Association, Basic Life Support for Healthcare Providers, 2001.

AEDs use a cardiac rhythm analysis system. All AEDs attach to the patient with cables connected to two adhesive pads that are stuck to the patient's chest. When the device is turned on, the patient's cardiac rhythm is analyzed. If the device reads ventricular fibrillation—or ventricular tachycardia above a certain rate—it charges its capacitors and delivers, or recommends that the rescuer deliver, an electric shock.

Operational Procedures

Four simple steps are followed in using all AEDs: (1) turn the device on, (2) attach the device to the patient, (3) "stand clear" and allow the device to analyze the patient's heart rhythm, and (4) deliver the shock if a shock is indicated.

CPR should be initiated immediately on pulseless, breathless patients. If possible, it is recommended that CPR be performed from the patient's right side by one rescuer while a second rescuer sets up the AED on the patient's left side. The left side gives the AED operator better access to the patient. When the device is in place beside the patient's head, the AED operator activates the device. The cables are attached to the device and to the adhesive pads. The two pads are stuck to the patient, one on the chest wall beneath the right shoulder (the upper right sternal border), the other on the chest wall beneath the left nipple (the lower left ribs over the apex of the heart). The pads themselves have illustrations showing where each should be placed. When the pads are attached, CPR is halted, and the analysis control button is pressed. All rescuers must avoid contact with the patient during the 5 to 15 seconds it takes the machine to do the analysis. If a shock is indicated, the device will so indicate in one of several ways, including a printed display message or a synthesized voice command, depending on the AED model being used.

If the device indicates a shock is needed, the AED operator must loudly remind everyone to remain clear of contact with the patient, usually by saying, "Clear!" or something similar, and then check to make sure everyone is clear before pressing the shock delivery button. Contact with the patient during the shock delivers a shock to rescuers as well as the patient. For that reason the patient should not be in water or in contact with a conducting surface or a conducting material such as a wet rope.

After the first shock, rescuers do *not* resume CPR. The analysis control button should be pressed a second time for an immediate check on the patient's heart rhythm; in more and more models, the device automatically rechecks the heart rhythm. If the patient remains in v-fib, the device recharges and indicates the need for a second shock. The same sequence may be repeated, and a third shock delivered if necessary. A sequence of 3 shocks takes about 90 seconds. AEDs deliver the shocks in increasing numbers of joules, typically starting at 200. (A few older devices require the operator to increase the joules manually.)

If the third shock does not resuscitate the patient, CPR should be resumed for 60 seconds before the AED is activated or automatically activates for a second round of analyses and shocks. The three shock sequences should be repeated until the machine gives a "No Shock Recommended" message. Local protocols, however, may require immediate transport of the patient, if available, after the first round of shocks. In most cases CPR should be continued during transport if the AED was unsuccessful. The use of the AED during transportation is not acceptable since movement interferes with heart rhythm analysis.

If the rescuer is alone, CPR should be performed for 1 minute first. After a minute of CPR, assistance should be summoned, if it hasn't already been summoned. Then the AED should be utilized.

Basic Life Support for Healthcare Providers

You Should Be Able To:

1. Demonstrate two-rescuer CPR.
2. Demonstrate CPR on children and infants.
3. Demonstrate clearing the obstructed airway of adults who go unconscious.
4. Demonstrate clearing the airway of children and infants.

Most educational institutions that certify Wilderness First Responders are satisfied with the student who demonstrates competence in American Heart Association Heartsaver CPR (see Chapter 1: Airway and Breathing and Chapter 5: Cardiopulmonary Resuscitation). The guidelines for Basic Life Support (BLS) for Healthcare Providers are presented here for those who might want or need to know more.

Adult Two-Rescuer CPR

1. Check for responsiveness.
2. With no response, Rescuer One opens the airway and checks for breathing for about 10 seconds.
3. With no breathing, Rescuer One delivers 2 slow, full breaths. Watch chest rise. Allow chest to deflate.
4. Rescuer One checks for a carotid pulse or other signs of circulation—breathing, coughing, or movement—for about 10 seconds.
5. With no signs of circulation, Rescuer Two, on the opposite side of the patient from Rescuer One, starts cycles of 15 compressions to 2 breaths. The breaths come from Rescuer One.
6. After 4 cycles of 15:2 (approximately 1 minute), recheck for signs of circulation. With no signs of circulation, continue cycles of 15:2 beginning with chest compressions.

If Rescuer Two arrives while CPR is in progress, Rescuer Two should state that he or she knows CPR. If Rescuer One is performing chest compressions, Rescuer Two should check for and find a weak carotid pulse, an indication that Rescuer One is delivering adequate compressions. Then Rescuer Two should take over the ventilations. If Rescuer One is doing ventilations, Rescuer Two should assume the position to do chest com-

pressions. Rescuer One should finish his or her ventilations and check for signs of circulation before Rescuer Two begins chest compressions.

When a switch is needed, no exact sequence determines how the switch should be made. Let it suffice to be said *you should reduce the amount of time the patient goes without CPR to a minimum.*

Child and Infant CPR

Cardiac arrest in children (1 to 8 years of age) and in infants (up to 1 year of age) rarely results from heart disease. The usual cause is an insufficiency of oxygen from suffocation, injuries, or illnesses. CPR in children and infants differs slightly from adult one-rescuer CPR (see Chapter 5: Cardiopulmonary Resuscitation).

Child CPR

1. Check for responsiveness.
2. With no response, open the airway and check for breathing for about 10 seconds.
3. With no breathing, deliver 2 slow breaths. Watch chest rise. Allow chest to deflate.
4. Check for a carotid pulse or other signs of circulation for about 10 seconds.
5. With no signs of circulation, start cycles of 5 compressions to 1 breath. Compressions should depress the chest 1 to 1.5 inches at a rate of 100 compressions per minute. Most adults should be able to deliver the compressions with one hand while the other hand maintains the child's head and airway position.
6. After approximately 1 minute of CPR, recheck for signs of circulation. With no signs of circulation, continue cycles of 5:1 beginning with chest compressions.

Infant CPR

1. Check for responsiveness.
2. With no response, open the airway and check for breathing for about 10 seconds.
3. With no breathing, deliver 2 slow breaths. Watch chest rise. Allow chest to deflate.
4. Check for a brachial pulse (inside the upper arm) or other signs of circulation for about 10 seconds.

5. With no signs of circulation, start cycles of 5 compressions to 1 breath. The compression site is approximately a finger width below an imaginary line between the infant's nipples. Use two of your fingers to perform the compressions. Compressions depress the chest 0.5 to 1 inch at a rate of at least 100 compressions per minute. The infant's head should not be higher than the rest of the body.

6. After approximately 1 minute, recheck for signs of circulation. With no signs of circulation, continue cycles of 5:1 beginning with chest compressions.

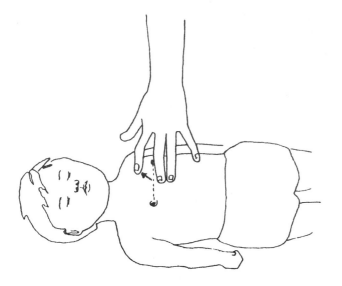

Figure C-1: *Finger placement for infant CPR.*

Foreign–Body Airway Obstructions

Adult Patient Becomes Unconscious

If you are performing abdominal thrusts on a conscious adult (see Chapter 4: Airway and Breathing) and the patient becomes unconscious, do the following:

1. Place the patient in the supine position and, in the urban environment, send someone to call 911.
2. Open the airway. Perform a finger sweep by hooking your index finger down on side of the cheek, across the back of the airway, and up the side of the other cheek.
3. Attempt to ventilate. If air fails to go in, reposition the airway and attempt to ventilate again. If the airway remains blocked . . .
4. Kneel astride the patient's thighs.
5. Place the heel of one your hands on the midline of the patient's abdomen, just above the navel, well below the xiphoid at the bottom of the sternum.

6. Place your second hand on top the first hand.
7. Press quickly in and upward 5 times with a powerful motion.
8. Repeat Steps 4 to 7 until you clear the airway.

Unconscious Adult

If the patient is found unconscious, immediately open the airway and check for visible obstructions. Send someone to call 911. Then look, listen, and feel for breathing. If the patient is not breathing, then do the following:

1. Attempt to ventilate the patient. If air does not go in, reposition the head and make a second attempt to ventilate.
2. If the airway remains blocked, straddle the patient's thighs and give 5 abdominal thrusts. Be sure your hands are positioned well below the xiphoid.
3. Open the airway and perform a finger sweep by running your index finger down the inside of one of the patient's cheeks, hooking your finger along the base of the tongue and back out in hopes of removing the obstructing object.
4. Repeat Steps 1 to 3 until the airway has been successfully opened.
5. If the patient does not resume normal breathing, give 2 full breaths and check for signs of circulation. With signs of circulation, begin rescue breathing. If the patient does resume normal breathing and circulation, place the patient in the recovery position.

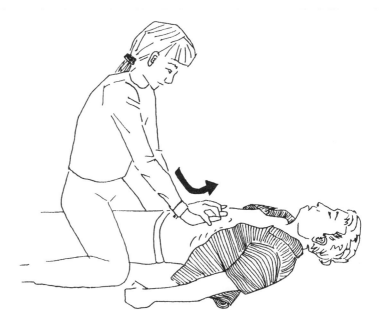

Figure C-2: *Abdominal thrusts.*

Child

An airway obstruction in a child is managed the same as in an adult with one exception: If you *see* an object, you should remove it from the airway, but blind finger sweeps to remove unseen objects are *not* performed. A blind finger sweep could lodge the object deeper in a child's airway.

Infant

Infants are not little adults, and management of an infant with a foreign-body airway obstruction differs from an adult's management. If the infant is conscious:

1. Determine if there is a serious breathing difficulty: lack of breathing, ineffective cough, weak cry.
2. Hold the infant face down on your forearm, supporting the head firmly with your hand. You may sit with your forearm on your thigh to increase support. The infant's head should be lower than his or her trunk.
3. With the heel of your other hand, deliver up to 5 forceful back blows between the infant's shoulder blades.
4. After the back blows, place your free hand on the infant's head, making a baby "sandwich" between your forearms and hands. Turn the infant carefully into a supine position with the head still lower than the trunk.
5. Deliver up to 5 quick chest thrusts with two fingers of your free hand on the lower half of the sternum, approximately one finger width below the infant's nipple line.
6. Repeat the back blows and chest thrusts until the airway is cleared or the infant goes unconscious.

If the infant goes unconscious:

1. Open the airway with a tongue-jaw lift (see Chapter 4: Airway and Breathing), look for the obstruction, and, if you see it, remove it with your little finger. Do *not* perform a blind finger sweep.
2. Open the airway and attempt to ventilate by sealing your mouth over the infant's mouth and nose and blowing in with only enough force to cause the infant's chest to rise. If the chest does not rise, reposition the infant's head and attempt a second ventilation.

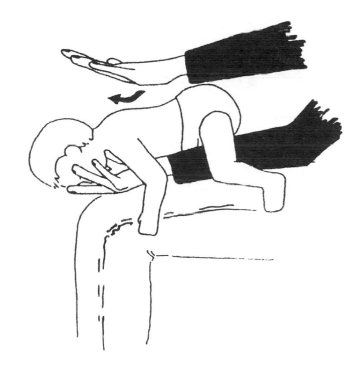

Figure C-3: *Back blows for choking infants.*

3. If the airway remains blocked, give up to 5 back blows followed by 5 chest thrusts as described for a conscious infant.
4. Repeat Steps 1 to 3 until the airway is cleared.

If the infant is found unconscious:

1. Assess responsiveness, open the airway, and look, listen, and feel for breathing.
2. If the infant is not breathing, attempt to ventilate. If your air will not go in, reposition the infant's head and attempt a second ventilation.
3. If the airway remains blocked, give up to 5 back blows followed by 5 chest thrusts as described for a conscious infant.
4. Open the airway with a tongue-jaw lift, look for the obstruction, and, if you see it, remove it with your little finger. Do *not* perform a blind finger sweep.
5. Repeat Steps 2 to 4 until the airway is cleared.

Figure C-4: *Rescue breathing for infants.*

Rescue Breathing for Children and Infants

Few differences distinguish rescue breathing for adults, children, and infants (see Chapter 4: Airway and Breathing). These differences may be summarized as follows:

1. Small children and infants should *not* have their heads fully extended to open an airway. A neutral or slightly extended position is enough. Too much extension will close off a small airway.

2. For infants seal off the mouth and nose with your mouth to perform mouth–to–mouth-and-nose rescue breathing.

3. Small lungs take less time to fill with your breaths. Slow breaths on your part will help you control ventilation.

4. You may see the chest *and* abdomen of a small child and infant rise when you ventilate.

5. For an infant, "puffs" of air from your mouth may adequately fill the lungs. Remember, the rise of the chest and abdomen are the indicators of a full breath.

6. The rate of breathing for small children and infants is approximately 1 breath every 3 seconds, or approximately 20 breaths per minute.

Glossary

abandonment: Leaving a patient in need of medical care alone or with someone who is less capable of providing care.

abdominal breathing: Breathing using only the diaphragm; "belly breathing."

abduction: Movement away from the midline of the body.

abrasion: A wound in which one or more layers of skin are scraped away.

acclimatization: The process of physiologically adjusting to a new environment, such as a higher altitude.

acetabulum: Hip socket.

Achilles tendon: The tendon connecting the heel bone to the muscles of the lower leg.

acidosis: A condition produced by the accumulation of acid or the reduction of base in the body.

ACL: See **anterior cruciate ligament.**

acute: An immediate problem with a short duration; not chronic.

acute mountain sickness (AMS): Nonspecific problems caused by a failure to acclimatize to higher altitudes; the first stage in a progression that can lead to **high-altitude cerebral edema.**

adduction: Movement toward the midline of the body.

AED: Shorthand for automated external defibrillator.

AGE: See **arterial gas embolism.**

alkalosis: A condition produced by the accumulation of base or the reduction of acid in the body.

allergen: An allergy-causing substance.

ALS: Shorthand for advanced life support.

alveoli: Microscopic sacs of the lungs where gas exchange takes place.

ambulatory: Able to walk.

amenorrhea: Absence of a menstrual cycle.

AMI: Acute myocardial infarction; a sudden heart attack.

amnesia: Loss of memory.

amniotic sack: The thin membrane covering the fetus and placenta and containing amniotic fluid.

amputation: A separation of a body part from the rest of the body.

AMS: See **acute mountain sickness.**

analgesic: A medication that relieves pain.

anaphylaxis: A severe allergic reaction to a foreign protein characterized by bronchoconstriction and vasodilation.

anesthetic: An agent that produces a partial or complete loss of sensation.

aneurysm: Abnormal widening of a blood vessel, usually in an artery.

angina pectoris: Pain in the chest caused by the heart demanding more blood than is available via the coronary circulation.

angioedema: Swelling of the mucous membranes of the lips, mouth, or other parts of the respiratory system.

ankle hitch: An anchor at the sole of the foot from which traction can be pulled for a traction splint.

anomia: The inability to remember names.

anorexia: Loss of appetite.

anovulatory cycle: An abnormal menstrual cycle during which no ovulation occurs.

anterior: Front surface.

anterior cruciate ligament (ACL): One of two ligaments that cross within the knee.

antibiotic: A substance that inhibits growth of or destroys microorganisms.

anticholinergic: An agent that blocks parasympathetic nerve impulses.

anti-diarrheal: A substance used to prevent or treat diarrhea.

anti-emetic: A substance used to prevent or treat vomiting.

antihistamine: A drug that counteracts the effects of histamines.

anti-inflammatory: A substance used to counteract inflammation.

antipyretic: An agent that reduces or relieves fever.

antivenin: Serum containing antitoxin for a venom.

aorta: The main artery that leaves the heart and travels to the body.

aortic valve: The valve between the left ventricle and the aorta.

apnea: Absence of breathing.

appendicitis: Inflammation of the appendix.

aqueous humor: Salty fluid that fills the cornea.

arachnoid: Weblike membrane that surrounds the brain and spinal cord; the middle of the three meninges.

arterial gas embolism (AGE): An air bubble in the arterial blood stream.

arteriole: Small artery.

arteriosclerosis: Thickening, hardening, and loss of elasticity of the arteries.

artery: A vessel carrying blood from the heart.

asphyxia: A condition caused by an insufficient intake of oxygen.

aspirate: To draw in or out by suction; to inhale particulate matter.

asthma: A condition resulting in shortness of breath and wheezing due to swelling of bronchi and their mucous membranes.

asystole: Absence of all activity in the heart.

ataxia: Loss of muscle coordination leading to difficulty in maintaining balance.

atherosclerosis: Clogging of the arteries from fatty deposits and other debris.

atrium: The upper chamber of each half of the heart.

auscultate: Listen.

AVPU scale: For *Alert, Verbal, Pain, Unresponsive*—a scale for assessing a patient's level of consciousness.

avulsion: The forcible tearing away of a part or structure; a piece of skin left hanging as a flap.

axial: The imaginary line through the middle of the body or through the middle of a part of the body.

bacteria: Single-cell microorganisms.

baroreceptors: Nerve endings sensitive to changes in pressure.

barosinusitis: Pain or inflammation of nasal sinuses due to pressure changes; also called sinus squeeze.

barotitis: Inflammation of the ear due to changes in pressure.

barotitis media: Inflammation of the middle ear due to changes in pressure; also called middle ear squeeze.

barotrauma: Injury caused by a change in pressure, as in scuba diving.

basal metabolic rate (BMR): the constant rate at which a human body consumes energy to drive chemical reactions and produce heat to maintain an adequate core temperature at rest.

Battle's sign: Bruising behind and below the ears indicating a fracture to the base of the skull.

BEAM: For *Body Elevation And Movement;* a technique for lifting spinally injured patients as a unit.

biliary duct: The duct that carries bile to the small intestine.

bipolar disease: Cyclical pattern of depression and mania.

bleb: Fluid-filled blister.

Blood pressure: The pressure of circulating blood against the walls of the arteries.

BLS: Basic life support.

BMR: See **basal metabolic rate.**

bony crepitus: Sound or feel of broken bone ends grating against each other.

BP: See **blood pressure.**

brachial: The arm from the shoulder to the elbow.

bradycardia: Slow heart rate.

brain stem: Part of the brain connecting the cerebral hemispheres to the spinal cord.

breech presentation: A birth in which the baby presents feet or buttocks first.

bronchiole: A small airway of the lung.

bronchitis: Inflammation of the mucous membranes of the bronchi.

bronchospasm: Spasms in the muscles of the bronchi.

bronchus: One of the two large airways branching off from the trachea.

BSI: For *Body Substance Isolation;* universal precautions against communicable disease.

bubo: Inflamed, enlarged lymph node.

bursa: Fluid-filled sac or cavity commonly found in joints.

BVM: Bag valve mask, a patient ventilation device.

calcaneofibular ligament: The ligament connecting the heel and the fibula.

calcaneus: Heel bone.

capillary: A tiny blood vessel where gas exchange occurs between the bloodstream and the tissues.

cardiac arrest: The cessation of heart muscle activity.

cardiogenic: Originating in the heart.

cardiopulmonary resuscitation (CPR): Artificial respirations and manual chest compressions that stimulate lung and heart activity.

cardiovascular system: The heart, the blood vessels, and the blood.

carotid artery: Primary blood vessel supplying the head and neck; the body has two carotid arteries.

carpopedal spasms: Muscular spasms in the hands and feet.

cartilage: Tough, elastic tissue that forms protective pads where bone meets bone.

catheter: A tube inserted into the body for the removal or insertion of fluids.

CC: For *chief complaint.*

Centruroides: Arizona bark scorpion, a particularly potent species of scorpion.

cerebellum: The part of the brain that helps control movement and balance.

cerebral: Relating to the cerebrum.

cerebrospinal fluid (CSF): The fluid surrounding the brain and spinal cord.

cerebrovascular accident (CVA): Interruption of normal blood flow to a part of the brain; stroke.

cerebrum: The largest part of the brain; consists of two hemispheres.

cervical vertebrae: The first seven bones of the spinal column; the neck bones.

cervix: The narrow opening at the low end of the uterus.

CHF: See **congestive heart failure.**

chondromalacia: Disintegration of cartilage under the kneecap.

chronic obstructive pulmonary disease (COPD): A collection of diseases sharing the common symptoms of airway obstruction in the small to medium airways, excessive secretions, and/or constriction of the bronchial tubes.

chronic: Slow progression or of long duration; not acute.

cilia: Hairlike processes.

CISD: For *Critical-Incident Stress Debriefing.*

CISM: For *Critical-Incident Stress Management.*

clavicle: Collarbone.

closed pneumothorax: A tear in the lining of the lung that results in air accumulating in the pleural space with no open wounds.

cnidoblast: A capsule containing the stinging nematocyst of **coelenterates.**

cnidocil: The trigger on a **cnidoblast.**

CNS: For *Central Nervous System;* the brain and spinal cord.

CO: Carbon monoxide.

coccyx: The last bone of the spinal column, consisting of four fused vertebra; the tail bone.

cochlea: Cone-shaped tube in the inner ear.

coelenterates: Phylum of sea creatures including jellyfish, fire coral, stinging medusa, Portuguese man-of-war, sea wasp, hairy stinger, and stinging anemone.

collateral ligament: A ligament on the inside of the knee (medial collateral ligament) or on the outside of the knee (lateral collateral ligament).

coma: An unconsciousness level from which the patient cannot be aroused.

comminuted fracture: A fracture in which the bone is splintered or crushed.

conduction: Heat lost from a warmer object when it comes in contact with a colder object.

congestive heart failure (CHF): A condition in which blood and tissue fluids congest due to the heart's inadequacy.

conjunctiva: The membrane lining the eyes.

constipation: Infrequent or difficult bowel movements.

contusion: A bruise.

convection: Heat lost directly into air or water by movement.

COPD: See **chronic obstructive pulmonary disease.**

coronary artery: The vessel that supplies the heart muscle with blood.

CPR: See **cardiopulmonary resuscitation.**

crackle: Abnormal breath sound heard by listening to the chest produced by air passing through small airways filled with secretions.

cranium: The skull.

critical incident: An incident so distressing it surpasses an individual's normal coping mechanisms.

crown: The portion of a tooth above the gum.

CRT: For *Capillary Refill Time.*

cruciate ligament: One of the crossed ligaments that hold the knee joint together. The anterior cruciate ligament attaches to the rear of the femur and the front of the tibia; the posterior cruciate ligament attaches to the front of the femur and the rear of the tibia.

Cryptosporidium: Protozoa found in surface waters.

CSF: See **cerebrospinal fluid.**

CSM: For *Circulation, Sensation, Motion.*

cuboid: Bone of the instep of the foot.

cuneiforms: Small bones of the foot.

CVA: See **cerebrovascular accident.**

cyanosis: Blue hue to the skin indicating a lack of oxygen in the blood.

DAN: For *Divers Alert Network.*

DCS: See **decompression sickness.**

decompression sickness (DCS): an accumulation of nitrogen bubbles in tissues after breathing compressed air and ascending too quickly.

DEET: N,N-diethyl-meta-toluamide; an insect repellent.

defibrillation: The act of stopping fibrillation of the heart by using electricity or drugs.

defusing: An informal discussion of a critical incident.

deltoid ligament: The ligament attaching the tibia to bones on the inside of the ankle.

dentin: The calcified area surrounding the **pulp** of a tooth.

dependent lividity: Discoloration of the skin, as from a bruise, where noncirculating blood has settled via gravity.

depression: Altered mood characterized by lack of interest in pleasurable things.

dermatitis: Inflammation of the skin.

dermatophytes: Fungi that grow on skin.

dermis: True skin.

diabetes mellitus: A disease in which the pancreas secretes an insufficient or ineffective amount of insulin.

diaphoresis: Profuse sweating.

diaphragm: The large muscle separating the abdominal and thoracic cavities; the primary muscle used in breathing.

diarrhea: Frequent passage of unformed, watery bowel movements.

diastolic pressure/diastole: Pressure exerted by blood on the walls of arteries during the relaxation phase of the heart activity.

DICC: For *Disoriented, Irritable, Combative, Comatose.*

dislocation: The complete or partial disruption of the normal relationship of a joint.

distal: Away from the center.

distal interphalageal joint: The distal joint of the finger; also called the DIP joint.

diuretic: A substance that causes increased urination.

DNR: For *Do Not Resuscitate.*

dorsalis pedis pulse: The pulse located on the top of the foot.

ductus deferens: Excretory duct of the testicle.

dura mater: The outer membrane of the **meninges,** covering the brain and spinal cord.

dysentery: Bacterially caused diarrhea that may produce blood and mucus in the stool.

dysmenorrhea: Pain associated with menstruation.

dyspnea: Breathing difficulty.

ecchymosis: Bruising.

ectopic pregnancy: A pregnancy in which the fetus implants outside of the uterus, commonly in a fallopian tube.

edema: Swelling caused by a buildup of fluid.

efface: Thin, as in the cervix during labor.

embolism: Obstruction of a blood vessel by a clot of blood, a gas bubble, or a foreign substance.

emetics: Vomit inducers.

emesis: Vomit.

emphysema: Chronic pulmonary disease characterized by destruction of the alveoli.

enamel: Hard covering of the crown of a tooth.

endometrium: The lining of the uterus.

envenomation: Introduction of poison into the body by a bite or sting.

epidermis: The thin outer layer of the skin.

epididymis: The organ lying behind the testicle that stores sperm.

epididymitis: Inflammation of the epididymis.

epidural: Above the dura mater.

epigastric: The upper abdomen.

epiglottis: The structure overlying the larynx that prevents food or liquid from entering the airway during swallowing.

epistaxis: Nosebleed.

eschar: Scab.

esophagus: The tube carrying food and liquid from the mouth to the stomach.

ETOH: Alcohol.

eugenol: Oil of cloves.

eustachian tube: The tube extending from the middle ear to the back of the throat.

evaporation: The process in which a liquid changes into vapor.

eversion: Turning outward.

evisceration: A condition in which abdominal contents are exposed and protruding from an open wound.

exhalation: Breathing out.

fallopian tube: The tube extending from the ovary to the uterus.

fascia: Membrane that covers muscle.

febrile: Feverish.

fecal impaction: A condition in which hardened feces form a blockage in the descending colon, preventing passage of fecal material.

femoral: Relating to the femur.

femur: Thigh bone.

fibrin: A protein that forms a matrix for clotting and scabbing.

fibula: Small lower leg bone.

fimbria: Small, fingerlike structures at the end of the **fallopian tube.**

flail chest: A condition in which two or more ribs are fractured in two or more places, creating a floating section of rib.

FOAM: For *Free Of Any Movement.*

foramen magnum: The large opening at the base of the skull through which the spinal cord passes.

fracture: A break in the normal continuity of a bone.

frostbite: Localized tissue damage caused by freezing.

frostnip: Superficial frostbite.

fungus: A primitive life form that feeds on living plants, decaying organic matter, and animal tissue.

fx: Shorthand notation for *fracture.*

gallstone: A concretion formed in the gall-bladder or bile duct.

gastric distension: A condition that occurs when air overinflates the stomach.

gastritis: Inflammation of the stomach.

gastrocnemius: Calf muscle.

gastrocolic: Relating to the stomach and colon.

gastroenteritis: Inflammation of the gastrointestinal tract.

germ: A microscopic organism that might infect a human and cause disease.

Giardia lamblia: Protozoa with flagella found in surface water.

gingiva: The gum of the mouth.

glottis: The vocal cords and space between them at the opening of the trachea.

glucagon: A hormone that increases the concentration of glucose in the blood.

gluteals: Muscles of the buttocks.

greenstick fracture: A fracture that does not extend all the way through the bone.

grief: Emotional response to loss.

gross negligence: Extreme deviation from the accepted standard of care.

guarding: Protecting an area of injury through muscle tension and/or physical positioning.

HACE: See **high-altitude cerebral edema.**

hamstring: A muscle in the back of the upper leg.

hantavirus: A virus transmitted by inhaling aerosolized microscopic particles of dried rodent saliva, urine, or feces, causing severe respiratory distress.

HAPE: See **high-altitude pulmonary edema.**

heat cramps: Painful spasm of major muscles being exercised caused by dehydration in association with electrolyte depletion.

heat exhaustion: Weakness produced by fluid loss from excessive sweating in a hot environment that causes compensatory shock.

heatstroke: A life-threatening condition produced by exposure to hot environments and/or excessive heat production characterized by an elevated core temperature that causes the brain to "cook."

hematemesis: Blood in the vomit.

hematoma: Pooling of blood in the form of a tumor.

hematuria: Blood in the urine.

hemiparesis: Weakness affecting one side of the body.

hemiplegia: Paralysis affecting one side of the body.

hemoptysis: Coughing up blood.

hemorrhage: Bleeding.

hemostasis: Control of bleeding.

hemothorax: Blood in the pleural space.

high-altitude cerebral edema (HACE): A condition in which fluid collects in the patient's brain as a result of extremes of elevation gain.

high-altitude pulmonary edema (HAPE): A condition in which fluid shifts from the pulmonary capillaries and fills the alveolar spaces as result of extremes of elevation gain.

HIRICE: For *Hydration, Ibuprofen, Rest, Ice, Compression, Elevation.*

histamine: A natural substance in the body released by the immune system in response to an injury or foreign protein.

HR: For *Heart Rate.*

humerus: Upper arm bone.

hx: Shorthand for *history.*

hydrophobia: Fear of water; once a common name for rabies.

Hymenoptera: Bees, wasps, and fire ants.

hyperbaric: High pressure.

hyperglycemia: High blood sugar.

hypertension: High blood pressure.

hyperventilation: Breathing unusually fast and/or deeply.

hyperventilation syndrome: Breathing fast and deeply without an apparent cause, such as chest trauma or respiratory illness.

hypoglycemia: Low blood sugar.

hypostome: The feeding apparatus of a tick.

hypotension: Lowered blood pressure.

hypothalamus: The part of the brain responsible for controlling certain metabolic activities, including temperature regulation.

hypothermia: Lowered core temperature.

hypovolemic: Low blood volume.

hypoxia: Low oxygen level.

ICP: For *Intracranial Pressure.*

ICS: For *Incident Command System.*

iliotibial band: The long tendon that runs from the region of the buttock, down the lateral aspect of the thigh, across the knee, attaching to the outside of the tibia.

IM: For *Intramuscular.*

immersion foot: A nonfreezing cold injury from prolonged contact with cold, and typically, moisture that causes inadequate circulation and tissue damage.

immersion syndrome: Sudden death following immersion in cold water.

impacted fracture: A fracture in which broken bone ends are wedged together.

implied consent: Legal assumption that an unconscious and/or unreliable patient would desire treatment if he or she were conscious or reliable enough to make a decision; also applies to minors when a parent or guardian is unavailable to offer consent.

incision: A smooth-edged cut through the skin made by a sharp edge.

incontinence: Loss of bowel and/or bladder control.

inferior: Below.

inferior vena cava: The large vein carrying blood to the heart from the pelvis, abdomen, and lower limbs.

inflammation: Successive changes in tissue in response to injury.

informed consent: Consent to treat obtained from a patient who has been informed of the potential risks and benefits of a proposed treatment.

inguinal hernia: A hernia in the region of the groin.

inhalation: Breathing in.

insomnia: Inability to sleep.

insulin: The hormone manufactured in the pancreas required to move glucose out of the bloodstream and through cell walls.

integumentary system: Skin.

intercostal muscles: Muscles between the ribs.

interstitial space: Space between cells.

inversion: Turning inward.

ischemia: Lack of blood supply.

IV: For *Intravenous.*

joint: The point at which two bones come in contact with each other.

JVD: For *Jugular Vein Distension.*

keratin: Hard protein found in skin, nails, and hair.

kidney stone: Concretion formed in the kidney and excreted through the urinary tract.

laceration: An irregular cut or tear through the skin.

lacrimal: Relating to tears.

LAF: For *Look, Ask, Feel.*

laryngospasm: A constrictive spasm of the muscles of the upper airway.

larynx: The structure of cartilage that holds open the upper end of the trachea; the "voice box."

lassitude: Weariness, exhaustion.

lateral: Away from the midline of the body; the outer aspect.

lateral collateral ligament: The outside ligament supporting the knee.

lesion: An area of pathologically altered tissue; a wound or injury.

LCL: See **lateral collateral ligament.**

ligament: Connective tissue holding bone to bone.

LLQ: For *Left Lower Quadrant* of the abdomen.

LNT: For *Leave No Trace.*

LOC: For *Level Of Consciousness.*

lumbar vertebrae: Five bones of the lower spine found between the **thoracic vertebrae** and the sacrum.

LUQ: For *Left Upper Quadrant* of the abdomen.

lymphadenitis: Inflammation of the lymph nodes.

lymphangitis: Inflammation of the lymph channels or vessels.

malaise: A general feeling of discomfort or indisposition.

malleolus: Rounded distal end of the tibia or fibula; known as the ankle bone.

mania: A mood disorder characterized by excessive elation, agitation, accelerated speaking, and hyperactivity.

MCL: See **medial collateral ligament.**

medial: Near the midline; the inner aspect.

medial collateral ligament: The inside/inner ligament supporting the knee.

medical adviser: A licensed physician who advises an unlicensed medical practitioner, such as a **Wilderness First Responder.**

medulla: The lower portion of the brain stem.

menarche: The initial menstrual period.

meninges: The three membranes surrounding the brain and spinal cord—the **dura mater,** the **arachnoid,** and the **pia mater.**

meniscus: Crescent-shaped fibrocartilage in the knee joint.

metacarpophalangeal joint: The joint where the finger or thumb joins the hand.

MI: See **myocardial infarction.**

miscarriage: Spontaneous abortion.

mitochondria: Intracellular "furnaces" where metabolism takes place.

mitral valve: The valve between the left atrium and the left ventricle of the heart.

mittelschmerz: Pain associated with ovulation.

MOI: For *Mechanism Of Injury.*

mucous membranes: Membranes that lines passages and cavities communicating with the air.

mucus: Viscous fluid secreted by the **mucous membranes.**

myocardial infarction: Death of a portion of the heart muscle resulting from a blockage in the coronary circulation; a heart attack.

myocardium: Heart muscle.

NASAR: National Association of Search and Rescue.

nasopharynx: Posterior part of the nose; the part of the pharynx above the soft palate.

naviculars: Bones at the base of the wrist and ankle.

necrosis: Tissue death.

negligence: A careless, unintentional act that harms another person to whom you owe a duty of care.

nematocyst: The stinging cell of a **coelenterate.**

neurogenic: Originating in the nervous tissue.

nitrogen narcosis: Increased concentration of nitrogen gas in body tissues from scuba diving that produces a sensation of euphoria and impaired judgment.

NOLS: National Outdoor Leadership School.

NPA: Nasopharyngeal airway. An airway adjunct placed in the nose and upper throat.

NRB: Non-rebreather mask. An oxygen delivery device that supplies oxygen from a reservoir bag.

NSAID: Non-steroidal anti-inflammatory drug. A drug administered for pain, fever, and inflammation.

N/V/D: For *Nausea/Vomiting/Diarrhea.*

oblique fracture: A diagonal break in a bone.

odontoid process: The bony projection of the second cervical vertebra that the first cervical vertebra rotates upon.

olecranon: The bony process of the proximal end of the ulna.

OPA: Oropharyngeal airway. An airway adjunct placed in the mouth and throat.

open fracture: Any fracture over which the skin is broken.

open pneumothorax: An open wound causing a tear in the lining of the lung that results in air accumulating the pleural space.

operculum: "Trap door" on a **cnidoblast.**

ophthalmic: Relating to the eye.

oropharynx: The posterior part of the mouth; the part of the pharynx between the soft palate and the epiglottis.

orthopnea: Difficulty breathing while lying down.

orthostatic changes: Dizziness, increase in heart rate, and/or increase in blood pressure that result from sitting or standing from a supine position.

otitis externa: Inflammation of the outer ear; "swimmer's ear."

otitis media: Inflammation of the middle ear.

ovary: Gland that produces eggs in the female reproductive system.

ovulation: Release of an egg from an ovary.

ovum: Egg.

P: Shorthand for *pupils*.

palliate: To reduce pain or make feel better.

palpate: To feel by touching.

paradoxical respiration: Asymmetrical chest wall movement associated with a **flail chest.**

paraparesis: Partial paralysis of the lower portion of the body.

paraplegia: Paralysis of the lower portion of the body.

parasite: An organism that lives on or within another organism.

paresthesia: Loss of or unusual sensations; "pins and needles."

parietal pleura: The portion of the **pleura** that lines the chest cavity.

patella: Kneecap.

patellar compression syndrome: A painful condition caused by too much pressure on the back of the kneecap, typically resulting from too much walking, especially downhill.

patent: Clear and open, as in an airway.

pathogen: A microorganism capable of producing disease.

pathological: Concerning disease.

PCL: See **posterior cruciate ligament.**

PE: See **pulmonary embolism.**

pedal: Relating to the foot.

pelvic inflammatory disease (PID): Inflammation of the fallopian tubes, ovaries, and/or uterus.

perfusion: The passing of fluid, typically well-oxygenated blood, through an organ or tissue.

pericardial sac: The sac surrounding the heart.

pericardial tamponade: Filling of the pericardial sac with fluid.

perineal: Region of the vagina and anus.

periodontal abscess: An area of infection and pus formation found on the gum.

peripheral: Away from the center; near the edge.

peripheral nervous system (PNS): The network of nerves not including the brain and spinal cord.

peritonitis: Inflammation of the abdominal lining (peritoneum).

permethrin: Insect repellent spray applied to clothing; an insecticide.

PERRL: For *Pupils Equal Round and Reactive to Light.*

PFD: Personal flotation device.

phalanges: Bones of the fingers or toes.

pia mater: The inner membrane of the **meninges** that surrounds the brain and spinal cord.

PID: See **pelvic inflammatory disease.**

placenta: The organ that provides nourishment to the fetus.

placenta previa: A placenta that implants over the cervix.

placental abruption: Separation of the placenta from the uterine wall.

plague: Disease caused by the bacteria *Yersinia pestis.*

pleura: The membrane that surrounds both lungs and extends to line the chest cavity.

pleural space: The potential space between the parietal and visceral **pleura.**

PMS: See **premenstrual syndrome.**

pneumonia: Infection or inflammation in the lungs.

pneumonic: Concerning the lungs.

pneumothorax: A tear in the lining of the lung, resulting in air accumulating the pleural space.

PNS: See **peripheral nervous system.**

polyuria: Increased volume of urine output.

posterior: Back surface.

posterior cruciate ligament (PCL): One of two ligaments that cross within the knee.

postictal: The period of recovery following a seizure.

premenstrual syndrome (PMS): A cluster of symptoms that occur prior to menstruation.

priapism: Painful, constant, emotionally unprovoked erection of the penis due to damage to nerves that control the genitals.

prolapsed cord: A presentation of the umbilical cord in the vaginal canal prior to the baby.

prone: The face-down position.

prostatitis: Inflammation of the prostate.

proximal: Nearest the midline of the body; nearest the center.

proximal interphalangeal (PIP) joint: The proximal joint of the finger.

psychogenic: Originating in the mind.

pulmonary edema: Fluid in the lungs.

pulmonary embolism: Blockage in a pulmonary artery or arteriole.

pulmonary overinflation syndrome: A dangerously rapid increase in pressure within the lungs caused by too rapid an ascent when scuba diving.

pulmonary valve: The valve between the right ventricle and the pulmonary artery.

pulp: The soft portion of the center of a tooth containing nerves and blood vessels.

pulpitis: Inflammation of pulp.

pulse pressure: The difference between **systolic pressure** and **diastolic pressure.**

puncture: A wound made by a pointed object such as knife, bullet, or ice ax.

px: Shorthand for *prevention.*

quadriceps: Thigh muscles.

quadriplegia: Paralysis of all four extremities.

rabies: An acute viral infection of the central nervous system transmitted by the bite of an infected mammal.

raccoon eyes: Bruising around the eyes that is indicative of a skull fracture.

radial: Pertaining to the **radius.**

radiation: Heat given off by a warm object.

radius: The shorter lower arm bone.

reduction: A return to normal anatomical relationship.

respiratory arrest: Cessation of breathing.

rhinoviruses: A group of viruses that cause the common cold.

rigor mortis: Stiffness that occurs in a body after death.

RLQ: For *Right Lower Quadrant* of the abdomen.

root: The portion of a tooth below the gum.

rotator cuff: The muscles (supraspinatus, infraspinatus, subscapularis, and teres minor) that hold the head of the humerus in the shoulder socket.

RR: For *Respiratory Rate.*

RUQ: For *Right Upper Quadrant* of the abdomen.

sacrum: The five fused bones of vertebrae between the lumbar spine and the coccyx; one of three bones of the pelvic ring.

SAMPLE: For *Symptoms, Allergies, Medications, Pertinent medical history, Last intake/output, Events.*

scapula: Shoulder blade.

sclera: The fibrous tissue covering the "white of the eye."

scrotum: The sac containing the testicles and some other parts of the reproductive system in males.

SCTM: For *Skin Color, Temperature, Moisture.*

scuba: Self-contained underwater breathing apparatus.

seizure: A sudden, abnormal electrical discharge in the brain.

separation: Enlargement of the spaces between bones.

sepsis: An illness resulting from a buildup of microorganisms or their toxins in the blood.

septicemia: "Blood poisoning" from buildup of bacteria or their toxins in the blood.

serum: The watery portion of blood.

shin: The front of the lower leg.

shin splints: Persistent pain in the shin area.

sinusitis: Infection of the sinuses.

SOAP: For *Subjective, Objective, Assessment, Plan.*

SOB: Shorthand for *short of breath.*

sphygmomanometer: A blood pressure cuff.

spirochete: A spiral-shaped bacterial microorganism.

spontaneous pneumothorax: A tear in the lining of the lung that occurs without trauma and that results in air accumulating the pleural space.

sprain: The stretching or tearing of ligaments.

sputum: Substance coughed up from the airway.

S/S: Shorthand for *signs and symptoms.*

status asthmaticus: A prolonged asthma attack unrelieved by conventional treatment.

status epilepticus: A persistent seizure or series of seizures with no time for adequate breathing.

sternum: The breastbone.

stoma: Artificial opening surgically placed in the trachea to assist breathing.

strain: Stretching or tearing of muscle fibers or tendons.

stridor: High-pitched sound created by constrictions in the airway.

subarachnoid: The space between the **arachnoid** and the **pia mater.**

subcutaneous: Below the skin.

subcutaneous emphysema: Air bubbles underneath the skin.

subdural: Below the **dura mater.**

subungual: Under a fingernail or toenail.

sucking chest wound: See **open pneumothorax.**

superior: Above.

superior vena cava: The large vein carrying blood to the heart from the head neck, upper limbs, and thorax.

supine: The face-up position.

syncope: Fainting.

synovial fluid: Clear lubricating fluid, as in a joint.

systolic pressure/systole: Pressure exerted by blood on the walls of arteries during the contraction phase of heart activity.

T: Shorthand for *temperature* of the core of the body.

tachycardia: Rapid heart rate.

talofibular ligament: The ligament connecting the talus to the fibula.

talus: The true ankle bone.

TBSA: For *Total Body Surface Area.*

tendinitis: Inflammation of a tendon.

tendon: Connective tissue that holds muscle to bone.

tension pneumothorax: Buildup of air in the pleural space that collapses the injured lung and begins to exert pressure on the uninjured lung and the heart.

testis: The gland that produces sperm in the male reproductive system.

tetanic contraction: Continuous uterine contraction.

thermoregulation: Body core temperature regulation.

thoracic vertebrae: The 12 vertebrae between the cervical vertebrae and the lumbar vertebrae with ribs attached on both sides.

thorax: Chest cavity.

TIA: See **transient ischemic attack.**

tibia: The large lower leg bone.

tibiofibular ligament: The ligament holding the tibia and fibula together.

tidal volume: Volume of air inhaled during normal inhalation.

TIL: Shorthand for *traction in line.*

tinnitus: Ringing in the ear.

torsion of the testis: Twisting of the testis on the spermatic cord.

toxic shock syndrome (TSS): Infection caused by the bacterium *Staphylococcus aureus.*

trachea: Windpipe.

transient ischemic attack (TIA): A temporary stroke, with signs and symptoms lasting less than 24 hours.

transverse fracture: A horizontal break in a bone.

triage: A sorting, as in a sorting of patients to determine the order of treatment and transport.

tricuspid valve: The valve between the right atrium and the right ventricle of the heart.

TSS: See **toxic shock syndrome.**

tularemia: A plague-like disease caused by a bacteria transmitted by tick bites.

tumor: A swelling or enlargement.

tx: Shorthand for *treatment.*

tympanic membrane: The eardrum.

ulcer: An open sore or lesion on skin or mucous membranes.

ulna: The smaller lower arm bone.

umbilical cord: A cord with two arteries and one vein connecting the fetus to the placenta.

ureter: The tube connecting the kidney to the bladder.

urethra: The tube from the bladder to the opening where urine leaves the body.

urethritis: Inflammation of the urethra.

URI: For *Upper Respiratory Infection.*

urinary tract infection (UTI): Infection of the urethra, bladder, and/or ureters.

urticaria: Hives.

urushiol: A resinous oil found in poison ivy, poison oak, and poison sumac that causes an allergic reaction.

uterus: The organ of the female reproductive system that supports a fetus.

UTI: See **urinary tract infection.**

UV: Ultraviolet light.

UVA: Ultraviolet A.

UVB: Ultraviolet B.

UVC: Ultraviolet C.

UVR: Ultraviolet radiation.

vagina: The birth canal.

Valsalva maneuver: Forced exhalation with a closed mouth, nose, and glottis that causes the pulse to slow down and pressure in the ears to equalize.

vasoconstriction: A narrowing of blood vessels.

vasodilation: A widening of blood vessels.

vasogenic: Originating in the vessels.

vein: A vessel carrying blood toward the heart.

ventricle: The lower chamber of each half of the heart.

ventricular fibrillation (v-fib): Rapid, uncoordinated, quivering of muscle fibers in the ventricles of the heart.

ventricular tachycardia: A heartbeat faster than the normal range.

venule: A small vein.

vertigo: Dizziness caused by a disturbance in equilibrium.

vestibular system: The middle part of the inner ear that helps determine balance.

V-Fib: See **ventricular fibrillation.**

virus: A microorganism that depends on another organism's cells as a host for replication.

visceral pleura: The portion of the pleura that surrounds the lungs.

vitreous humor: Clear fluid that fills the eyeball.

VT: See **ventricular tachycardia.**

WFR: Wilderness First Responder.

xiphoid process: The flexible, cartilaginous protuberance at the lower end of the sternum.

Bibliography

American Heart Association. *Basic Life Support for Healthcare Providers.* Dallas: American Heart Association, 2001.

Benenson, Abram S., ed. *Control of Communicable Diseases in Man.* Washington, D.C.: American Public Health Association, 1990.

Buttaravoli, Philip, and Thomas Stair. *Minor Emergencies: Splinters to Fractures.* St. Louis, Mo.: Mosby, 2000.

Cauchy, Emmanuel, Eric Chetaille, Vincent Marchand, and Bernard Marsigny. "Retrospective study of 70 cases of severe frostbite: a proposed new classification scheme." *Wilderness & Environmental Medicine* 12, no. 4 (2001), p. 248.

Forgey, William W. *Basic Essentials of Hypothermia, 2nd ed.* Guilford, Conn.: The Globe Pequot Press, 1999.

_____. *Wilderness Medicine, 5th ed.* Guilford, Conn.: The Globe Pequot Press, 2000.

Forgey, William W., ed. *Wilderness Medical Society Practice Guidelines for Wilderness Emergency Care, 2nd ed.* Guilford, Conn.: The Globe Pequot Press, 2001.

Fradin, Mark, and John Day. "Comparative efficacy of insect repellents against mosquito bites." *New England Journal of Medicine* 347 (July 4, 2002), 13–18.

Giesbrecht, Gordon. "Prehospital treatment of hypothermia." *Wilderness & Environmental Medicine* 12, no. 1 (2001), p. 24.

Gill, Paul G., Jr. *Pocket Guide to Wilderness Medicine.* New York: Simon & Schuster, 1991.

_____. *Waterlover's Guide to Marine Medicine: How to Identify and Treat Aquatic Ailments and Injuries.* New York: Simon & Schuster, 1993.

Gold, Barry, Richard Dart, and Robert Barish. "Bites of venomous snakes." *New England Journal of Medicine* 347, no. 5 (August 1, 2002), p. 347.

Gookin, John. *NOLS Backcountry Lightning Safety Guidelines.* Lander, Wyo: National Outdoor Leadership School, 2000.

Hackett, Peter, and Robert Roach. "High-altitude Illness." *New England Journal of Medicine* 345, no. 2 (July, 12, 2001), p. 107.

Hampton, Bruce, and David Cole. *Soft Paths: How to Enjoy the Wilderness without Harming It.* Mechanicsburg, Pa.: Stackpole Books, 1995.

Houston, Charles. *Going Higher: Oxygen, Man, and Mountains, 4th ed.* Seattle: The Mountaineers Books, 1998.

Hultgren, Herb. *High Altitude Medicine.* Stanford, Calif.: Hultgren Publications, 1997.

Jong, Elaine, and Russell McMullen. *The Travel and Tropical Medicine Manual.* Philadelphia: W. B. Saunders Co., 1995.

McGivney, Annette. *Leave No Trace: A Guide to the New Wilderness Etiquette.* Seattle: The Mountaineers Books, 1998.

Roberts, James R. *Roberts' Practical Guide to Common Medical Emergencies.* Philadelphia: Lippencott-Raven, 1996.

Renner, Jeff. *Lightning Strikes.* Seattle: The Mountaineers Books, 2001.

Schmutz, Ervin, and Lucretia Breazeale Hamilton. *Plants That Poison.* Flagstaff, Ariz.: Northland Publishing, 1979.

Setnicka, Tim J. *Wilderness Search and Rescue.* Boston: Appalachian Mountain Club, 1980.

Shimanski, Charley. *Helicopters in Search and Rescue Operations: Basic and Intermediate Levels.* Golden, Colo.: Mountain Rescue Association, 2002.

_____. *Search and Rescue for Outdoor Leaders.* Golden, Colo.: Mountain Rescue Association, 2002.

Smith, David. *Backcountry Bear Basics: The Definitive Guide to Avoiding Unpleasant Encounters.* Seattle: The Mountaineers Books, 1997.

Steele, Peter. *Backcountry Medical Guide, 2nd ed.* Seattle: The Mountaineers Books, 1999.

Stewart, Charles E. *Environmental Emergencies.* Baltimore: Williams & Wilkins, 1990.

Sutherland, Stuan, and Guy Nolch. *Dangerous Australian Animals.* Flemington, Victoria, Australia: Hyland House, 2000.

Tilton, Buck. *Backcountry First Aid and Extended Care, 4th ed.* Guilford, Conn.: The Globe Pequot Press, 2002.

_____. *Basic Essentials of Avalanche Safety.* Guilford, Conn.: The Globe Pequot Press, 1992.

_____. *Basic Essentials of Rescue from the Backcountry.* Guilford, Conn.: The Globe Pequot Press, 1990.

_____. *Don't Get Bitten: The Dangers of Bites and Stings.* Seattle: The Mountaineers Books, 2003.

Tilton, Buck, and Rick Bennett. *Don't Get Sick: The Hidden Dangers of Camping and Hiking.* Seattle: The Mountaineers Books, 2002.

Tilton, Buck, and Frank Hubbell. *Medicine for the Backcountry: A Practical Guide to Wilderness First Aid, 3rd ed.* Guilford, Conn.: The Globe Pequot Press, 1999.

Trott, Alexander. *Wounds and Lacerations: Emergency Care and Closure.* St. Louis, Mo.: Mosby, 1991.

Van Tilburg, Christopher. *Emergency Survival: A Pocket Guide.* Seattle: The Mountaineers Books, 2001.

Weiss, Eric. Wilderness 911: *A Step-by-Step Guide for Medical Emergencies and Improvised Care in the Backcountry.* Seattle: The Mountaineers Books/*Backpacker* magazine, 1998.

Wilderness Medicine Institute of NOLS. *Wilderness Medicine Handbook, 8th ed.* Lander, Wyo.: National Outdoor Leadership School, 2003.

Wilkerson, James A. "Rabies Update." *Wilderness & Environmental Medicine* 11, no. 1 (2000), p. 31.

Wilkerson, James A., ed. *Medicine for Mountaineering and Other Wilderness Activities, 5th ed.* Seattle: The Mountaineers Books, 2001.

Index

B

generalized, 116
immersion, 116, 135, 136
mild, 117, 118–19
moderate, 117, 118–19
preventing, 120
severe, 117, 119–20
signs/symptoms, 116, 117
supplemental oxygen for, 39, 119
treating, 118–20
urban, 116
wraps, 119
hypovolemic (low-volume) shock, 46–47
hypoxia, 45, 129, 188

I

ibuprofen, 94, 168
icing injuries, 94
ICP. See intracranial pressure (ICP)
iliotibial band syndrome, 99
immersion foot, 122
immersion incidents, 135–36. See also submersion
    incidents
  defined, 135
  hypothermia, 116, 135, 136
  immersion syndrome, 136
  reach, throw, row, tow, go and, 135–36
  rescuing patients, 135–36
impaled objects, 71, 104, 107–8
impingement, 100
implied consent, 9
infant life support, 267–68, 269–70
infection, abdominal, 189
infection management, 104, 105, 106, 111
influenza, 210
informed consent, 8–9
inguinal hernia, 216
inhaled poisons, 183–84
inhalers, 169–70
initial assessment, 15–16
injuries. See also specific injuries
  comparative negligence (fault), 9
  from negligence, 8, 9
injury management supplies, 254
inner ear barotrauma, 158
inquire (ask), 18
inspect (look), 18
insulating pad, 19
insulin emergencies. See diabetic emergencies
insurance protection, 9

internal asthma, 169
internal bleeding, 44
interstitial spaces, 47
intracranial pressure (ICP), 63–64, 174
irreversible shock, 48–49
irrigating wounds, 104–5
ischemia, 46

J

jaundice, 22
jaw. See mandible (jaw)
jaw-thrust maneuver, 16, 31
jimsonweed poisoning, 182
joint death, 91
jugular vein distension (JVD), 70

K

Kendrick Traction Device (KTD), 82
kidney stones, 192
knee. See also patella (kneecap)
  ACL, 98, 99
  anatomy of, 98
  athletic injuries, 98–100
  dislocations, 91
  iliotibial band syndrome, 99
  patellar compression syndrome, 99
  PCL, 98, 99
  walking splint, 99
kneecap. See patella (kneecap)
KTD. See Kendrick Traction Device (KTD)

L

labor. See obstetrical emergencies
lacerations, 104, 106–7
LAF (look for, ask about, feel for), 18, 78, 88, 93–94
laryngospasm, 136
lassitude, 130
legal issues, 7–10
  abandonment, 9
  assault/battery, 9
  breach-caused injuries, 8
  breach of duty, 8
  civil law, 7–8
  confidentiality, 9
  consent, 8–9
  contract law, 7–8
  documenting events, 9
  duty to act, 8
  insurance protection, 9

PMS. *See* premenstrual syndrome (PMS)

About the Author

Buck Tilton is co-founder of the Wilderness Medicine Institute of the National Outdoor Leadership School in Lander, Wyoming and current chairperson of the school's Medical Advisory Panel. He has extensive hands-on experience in pre-hospital medicine and wilderness search and rescue. He has also written numerous books, including *Medicine for the Backcountry* and *Backcountry First Aid and Extended Care*. The first edition of his book, *Wilderness First Responder*, garnered an award for Excellence in Medical Publications from the American Medical Writers Association. In addition, Buck is the recipient of the Paul K. Petzoldt Award for Excellence in Wilderness Education, bestowed by the Wilderness Education Association. He has written more than 1,000 magazine articles on outdoor subjects and has been a contributing editor to *Backpacker Magazine* since 1989. Buck lives in Lander, Wyoming.

About WMI and NOLS

Wilderness Medicine Institute

For over a decade the Wilderness Medicine Institute (WMI) has been the most recognized and respected teacher of wilderness medicine, training more than 30,000 students around the world to respond to medical emergencies in remote settings. At WMI students learn treatment principles and decision-making skills for times when there are few resources, no help on the way, and they have to make their own medical decisions. Dynamic lessons and realistic scenarios simulate the type of medical situations people face in wilderness settings, when calling 911 is not an option.

National Outdoor Leadership School

WMI is an institute of the National Outdoor Leadership School (NOLS), the premier teacher of outdoor skills and leadership. NOLS offers courses that last from ten days to a full semester in the world's most spectacular wilderness classrooms. Since 1965 more than 75,000 people have turned to NOLS to learn sea kayaking, backpacking, sailing, mountaineering, horsepacking, and canoeing. NOLS students come from all over the world and have diverse backgrounds; they are students, entrepreneurs, educators, astronauts. College credit and scholarships are available for many NOLS and WMI courses, transferrable to more than 400 U.S. colleges. For more information on NOLS and WMI, contact 1-800-710-NOLS or visit www.nols.edu.